The Ecology of Building Materials

The Ecology of Building Materials

Bjørn Berge

Translated from Norwegian by Filip Henley

With Howard Liddell

To my three girls,
Marianne, Sofia Leiresol and Anna Fara

Architectural Press
OXFORD AUCKLAND BOSTON JOHANNESBURG MELBOURNE NEW DELHI

Architectural Press
An imprint of Butterworth-Heinemann
Linacre House, Jordan Hill, Oxford OX2 8DP
225 Wildwood Avenue, Woburn, MA 01801-2041
A division of Reed Educational and Professional Publishing Ltd

 A member of the Reed Elsevier plc group

British Library Cataloguing in Publication Data
A catalogue record for this book is available from the British Library

ISBN 0 7506 3394 8

Library of Congress Cataloging in Publication Data
A catalogue record for this book is available from the Library of Congress

Composition by Scribe Design, Gillingham, Kent
Printed and bound in Great Britain by The Bath Press, Bath

contents

author's foreword

The Ecology of Building Materials came out originally in 1992 in Scandinavia. It has now been revised and adapted for the English-speaking world.

The book is far-reaching in its subject matter: too far, maybe, for some readers. There may well be the inevitable m istake or certain inaccuracies, if one dissects the information. On discovery of any such mistakes, I would greatly appreciate the corrected information being sent to me via the publishers, so any new editions will not repeat the same mistake. Any other comments, additions or ideas are also very welcome. Many have helped me in preparing this new edition, first and foremost my colleagues in our two Norwegian offices, Gaia Lista and Gaia Oslo. Howard Liddell in Gaia Scotland has given a great deal of worthwhile and necessary help in the preparation of the English edition.

I would also like to thank those who have read through the whole or part of the manuscript and given me useful comments and corrections, among them: Dag Roalkvam, Varis Bokalders, Jørn Siljeholm, Hans Granum, Arne Nass, Karl Georg Hoyer, Geir Flatabø, Peer Richard Neeb, Ødd Overeng and Tom Heldal.

And I would like to give an extra special thank you to the Translator, Filip Henley. He has achieved a use of language that surpasses the Norwegian original.

Bjørn Berge
Lista, 1999

foreword

The Ecology of Building Materials is a seminal contribution to the built environment survival kit. This important reference source has been confined to the Nordic countries for too long and I am delighted to be involved in its introduction to the English-speaking readership. It is one of a select but growing group of "Tools for Action" towards a sustainable construction industry.

There is a long tradition of books that have been influential catalysts towards a change in attitudes to our human habitat. I believe, for example, that the 20th century environmental movement was catapulted into centre stage by Rachel Carson's *Silent Spring* in 1966. It was, however, side-tracked into an obsession with energy issues during the 70's and 80's. It is only since the Rio Summit in '92 that the epidemic scale losses of natural bio-diversity, and the realisation of the criticality of toxicity, in all its forms (including inappropriate and polluting forms or fuel), have led to the re-discovery of our inappropriate relationship with our planet.

I would like to think that this book will have an impact on the building industry as effective as that which Carson had on agriculture. We have all become aware of the benefits of healthy eating even if we do not practice it as well as we should, but how far has even the awareness of toxicity in buildings penetrated the public's conscious perception of the places in which they spend 90% of their lives? Sick Building Syndrome is, however, a generalised catch-all in the mind of the public at large – but it is already the case that they are expecting their environment to be free of risk and they are asking for the industry to sign on the dotted line to that effect. In such circumstances the precautionary principle appears to be inevitable and specifying benign a pre-requisite. Therefore we need the tools to do the job.

Understanding the life cycle of the materials we use every day has never been more complicated, and therefore its ready interpretation was never more essential. As a major consumer of both primary and secondary resources and a major producer of waste, the construction industry has been made well aware of its responsibilities with regard to its enormous potential contribution to sustainable development, and its part in the threat to all human existence if it fails to meet the challenge. It is important therefore that it acquires the expertise now and not at some unidentified time in the future to lessen its impact. This book is a significant source in the wide range needed for immediate and effective action.

The clear fusion of well-researched fact with experienced opinion in this book is certainly timely and indeed probably overdue, since it scores in much more than the strictly numerical sense. A practising architect as well as a researcher and author, Bjørn Berge presents a carefully considered view of a whole range of key building materials – from the basis of his own underpinning, technical expertise. The Life Cycle Analysis research industry is replete with academic and impenetrable LCA scoring systems, which run the gauntlet of seeking to establish mechanisms that will give equal valency to the infinitely measurable and the essentially subjective and almost unmeasurable – usually ending in a three point scale (good/neutral/bad or plus/zero/minus) that leaves specifiers as confused as if they had not been given the information in the first place; this is especially so when they see products scoring well, which instinctively they consider to be very questionable. Selective or, worse, misinformation is now a significant problem as companies realise the sales pitch benefits of having an environmental profile – whilst the more cynical amongst them regard green issues merely as a marketing opportunity rather than what is becoming more and more clearly, at the very least, a health and safety issue.

The great strength of this book is that it is written in a style which is neither stodgy nor pulling its punches. Bjørn Berge simply states his view on building materials and processes in a way which leaves the reader in no doubt as to what their environment impact is.

I am reminded of the quotation by Richard Feynman: In technology it is not enough to have good Public Relations because Nature will not be fooled.

It is particularly refreshing to have a reference source which sifts and evaluates key components and is not then afraid to seek to influence our thinking and give both opinion and guidance.

It has taken a while to convert this book into the English-speaking public domain. Its Norwegian language precursor was published in 1992 and translation has been much more than a straightforward language exercise. Firstly Bjørn Berge himself has updated and amended much of the original text, then Filip Henlcy has done a tremendous piece of work in the primary translation from the original Norwegian and I have then sought to contribute a bit of cultural translation, albeit Norway's building industry – with its long timber tradition – is subject to all the same influences and trends as the rest of Europe, and hence the need for the book in that context in the first instance.

Howard Liddell
Edinburgh 1999

preface

The building industry has not only become a major consumer of materials and energy; it has also become a source of pollution, through the production of building materials and the use of pollutant substances. This book demonstrates that alternatives to modern building materials are available and that today it is possible to produce building materials and select raw materials from an ecological perspective.

At a time when environmental labelling is becoming increasingly popular and the producers of building materials are urged to be more environmentally aware, it is obviously important to be acquainted with these alternatives.

Important issues discussed in this book include:

- Can raw materials from non-renewable sources be replaced with raw materials from widely available or non-depletable sources?
- Can environmentally-friendly chemicals replace environmentally-damaging ones?
- Can the make-up of building materials be altered so that their individual components can be re-used?

The following aspects will be illuminated in this book:

- *Work*: production methods of today and tomorrow
- *Raw materials*: deposits and their potential for reuse
- *Energy*: energy consumption in production and transportation
- *Pollution*: pollution in production, use and demolition.

With the aid of tables, each of the most important building materials in use in Scandinavia will be given a characteristic *environmental profile*.

This book will be of special interest to environmentally-minded producers and suppliers of building materials and to engineers, architects and building workers, but it may also be of use to readers who are interested in housing but who lack specialist technical knowledge.

introduction

'We cannot cure illnesses, but we can help Nature cure herself' Hippocrates

'I object! I do not agree that the Earth and everything that exists on her shall be defined by the law as man's living environment. The Earth and all that is hers, is a special being which is older, larger and stronger than us. Let us therefore give her equal rights and write that down in the constitution and in all other laws that will come . . .A new legal and moral status is needed where Nature herself can veto us through her own delegates . . .One must constitute the right of all things to be themselves; to be an equal with Nature, that is totally unarmed; do well out of it in a human way and only in accordance with their own nature. This means that one must never use a tree as a gallows, even if both its form and material fit the purpose excellently. . .What practical consequences should a law like this have? Before all economic considerations, this law would decide that nothing will be destroyed or severely damaged, all outstanding natural forms, landscape characteristics and naturally linked areas shall remain untouched. No economic or leisure concern shall be developed at the cost of nature, or worsen the living conditions of man and other beings. Everything that man wants to do in the future, he must do at his own cost and with his own strength. As a result of this law we may return to old methods of production or discover new ones which do not violate the law. The manufacturing society will crumble and multiply, the meaningless superfluity of similar products on the world market will give way to the local market, independent of transcontinental connections.'

Ludvìk Vaculìk, Czech author, in his essay *An alternative constitution*

The Greek terms economy, ecology and ecosophy belong together:

Oikos House
Nomos Management
Logos Understanding
Sofos Wisdom

If we consider the world to be our common house, we can say that we have managed too much and understood too little. In Nature – the existential base of humanity – the consequences of this are becoming clearer: forest death, desertification, marine pollution. These are things of which we are all aware. The growing incidence of mental problems among the populations of industrialized nations would indicate that we have not even understood the nature of ourselves – that we, too, have become the victim of too much management.

Ecosophy expands the Kantian imperative 'to see every person as a goal, rather than a means' to include other living beings. In this way, it defends the value of Nature in itself, but is fully aware that it is impossible to escape the third law of ecology: 'All things are connected' (Commoner, 1972).

The problem consists of establishing a perspective on Nature that has a genuine influence or, alternatively, establishing a general morality which is acceptable to all. The ecologist Aldo Leopold maintains: 'A thing is right when it tends to preserve the integrity, stability and beauty of the biotic community. It is wrong when it tends otherwise.'

This represents an ethic for which, in ancient times, there was no need. Trond Berg Eriksen (1990) describes the situation in antiquity:

> 'In antiquity, commanding the forces of Nature and bringing discipline to human nature were two sides of the same coin. In neither area did the interveners need to fear that they would succeed completely. The power of Nature was overwhelming. It took care of itself. Humans had to battle to acquire the bare necessities. Nature's order and equilibrium was unshakeable. Man was, and considered himself, a parasite on an eternal life system. The metropolis was a hard won corner, a fortified camp under threat from earthquakes, storms, drought and wild animals. The metropolis did not pose a threat to Nature, but was itself an exposed form of life. . . In such a perspective, technology was ethically neutral. Morality comes into play only when one can cause damage, in relation to someone or something that is weaker or equally strong. Therefore, the consequences of human actions for non-human objects lie beyond the horizon of moral issues.'

Our ancestors' morality was based on the axiom that man himself was the only living being that could be harmed by human actions. Ethics focused on this; ethics dealt with interpersonal relationships. At the same time this morality was limited to the moment – only the immediate consequences of an action were of significance. Long-term effects were of no interest and beyond all regulation. Today, man's position and influence is drastically changed. The way in which we manage natural resources may have irremediable consequences for future generations of all life forms. Paradoxically, we still cling to antiquity's anthropocentric moral philosophy, often mingled with some of the Enlightenment's mottos of man's sovereign supremacy.

'Four conditions to achieve a sustainable society', according to L.P. Hedeberg from the movement 'The Natural Step', are:

1. *Do not take more out of the crust of the Earth than can be replaced.* This means that we must almost totally stop all mining and use of fossil fuels. Materials that we have extracted from beneath the Earth's surface, for example metals, coal and oil, are difficult for Nature to renew, except in a very small part. And that takes time. On the surface the rubbish pile gets higher because we have not followed this condition. And matter does not disappear – even if we reduce it to very fine particles, by burning for example, it is only transformed into molecular waste. Every single atom of a completely rusted car continues to exist, and has to find a new home somewhere else. Everything just spreads, nothing disappears.
2. *Do not use man-made materials which take a long time to decompose.* Materials that Nature can break down and change into nutrients belong to the natural lifecycle. Man-made materials, which have never been a part of Nature, are very difficult for Nature to break down. Certain synthetic materials such as PCB, dioxines, DDT, freones and chloroparaffins will never be broken down by Nature.
3. *Maintain the conditions for Nature to keep its production and its diversity.* We must stop impoverishing Nature through forest clearing, intensive fishing and the expansion of cities and road systems. A great diversity of animals and plants are a necessity for all life cycles and ecosystems, and even for our own lives.
4. *Use resources efficiently and correctly – stop being wasteful.* The resources that are available must be divided efficiently and fairly.

The ecology of building materials

Is it realistic to imagine a technology that functions in line with holistic thoughts while also providing humanity with an acceptable material standard of living? This book is an attempt to suggest the possible role and potential of building materials in such a perspective. And, in the same context, to illuminate the following aspects:

- *Work.* The methods used to produce each building component. How production takes place and can take place.
- *Raw materials.* Occurrence of material resources, their nature, distribution and potential for re-use.
- *Energy.* The energy consumed when producing and transporting the materials, and their durability.

- *Pollution.* Pollution during production, use and demolition, the chemical fingerprint of each different material.

How to use the book

This book is an attempt to present the possibilities for existing materials as well as evaluating new materials. A number of partly abandoned material alternatives have also been evaluated. In particular, we will look at vegetable products, with traditional methods of preparation marked by former technological development. In their present state, these methods are often of little relevance, and the reviews must therefore be regarded as experimental platforms on which to build.

Many factors relating to the materials discussed depend upon local conditions, so the book is mainly based on the climatic and topographical conditions in northern and central Europe. When considering the Earth as a whole, it will, however, become quite clear how little the use of materials varies.

The materials dealt with are those that are generally used by bricklayers, masons, carpenters and locksmiths. Under this category, all fixed components and elements that form a building are included, with the exception of heating, ventilation and sanitary installations. Materials proving high environmental standards are supplied with thorough presentations in the book while less attractive and often conventional alternatives are given less attention.

It is my hope that *The Ecology of Building Materials* can function as a supplement to other works on building. For this reason, only brief mention has been made of some factors of a more professional nature. These include such matters as fire protection and sound insulation, and other aspects which have no direct link with ecological criteria.

The book is divided into three sections:

Section 1: **Eddies and water-level markers**. *Environmental profiles and criteria for assessment* covers the tools which we will use to evaluate and select material on the basis of production methods, the raw material situation and energy and pollution aspects. Tables show the different material alternatives available and information relating to their environmental profile. The information contained in them derives from many different reliable European sources. They show quantifiable environmental effects and should be read in conjunction with the environment profiles in Sections 2 and 3. The final chapter gives an introduction to the chemical and physical properties of building materials.

Section 2: **The flower, the iron and the sea**. *Raw materials and basic materials* presents the materials at our disposition. The term 'raw materials' denotes the materials as they are found in Nature, as one chemical compound or as a combination of several such compounds. They form the basis for the production of 'basic materials' such as iron, cement, linseed oil and timber. These materials will form building blocks in complete products. The section is divided into chapters which present the different organic and mineral materials and discuss the ecological consequences of the various ways of utilizing them.

Section 3: **The construction of a sea-iron-flower**. *Building materials* discusses usage, such as roofing and insulation, and assesses the usability of the various alternatives from an ecological perspective. Descriptions are given of the practical uses of the best alternatives. This section is divided into seven chapters:

- *Structural materials* which support and brace
- *Climatic materials* which regulate warmth, humidity and air movement
- *Surface materials* which protect and shield structures and climatic materials from external and internal environments
- *Other building elements:* windows, doors and stairways
- *Fixing and connections* which join the different components
- *Surface treatment* which improves appearances and provides protection
- *Impregnating agents and how to avoid them:* the different impregnating substances and the alternatives.

The structural, climatic and surface materials covered in the first three chapters represent 97–99 per cent of the materials used in building, and environmental evaluations are given for each. The tables are based on available life span analyses and evaluations of building materials carried out in European research institutes (Fossdal, 1995; Kohler, 1994; Suter, 1993; Hansen, 1996; Weibel, 1995). In addition to many conventional environmental evaluations, this book also discusses the human ecological aspects through questions such as the feasibility of local production of building materials.

The evaluation tables are ordered so that each function group has a best and a worst alternative for each particular aspect of the environment, then a summary. The summarized evaluation means that priority is given to specific environmental aspects, which in turn relate to each particular situation. In such processes, political, cultural and ethical aspects come into play in a strong way. In Africa, the raw material question is usually given a high priority; in New Zealand and Argentina, all of the factors that affect the ozone layer are strongly considered; in Western Europe, the highest priority is likely to be acid rain. This book contains the author's own subjective views and the summarizing column should be taken as a suggestion. The main aim of the book is to give the reader the opportunity to quite objectively come to his or her own conclusions.

It is also necessary to realize that all information is *of the present moment*. The sciences that consider the different relationships in the natural environment are relatively young, and in many cases just beginning. There are new aspects coming into the picture continuously, all of which affect the whole situation. One example is chlorofluorocarbons (CFCs), which were not considered to be a problem in the 1970s before their effect on the ozone layer became known. The evaluations in the book are based on the before–after principle, the consequences of using a material should be understood before it is used. Any uncertainty over what a material actually is should not be to the material's advantage.

It must be emphasized that the evaluation tables account for isolated materials and not constructions consisting of several elements as they occur in the building. This may give a slightly distorted picture in certain cases, for example, in the case of ceramic tiles and mortar or joint mastic which cannot be considered independently, or of plasterboard and fillers. In most cases, however, the tables represent a thorough basis for comparisons between products at a fundamental level. It is also recommended to do further research into the sources of this book. A comprehensive list of further reading is to be found at the end of each of the three sections.

Life span evaluations of building materials

Many attempts have been made to establish evaluating methods to objectivise the environmental profile of building materials. These are based on a numbering and evaluating system for the different environmental effects of a material during its life span. These evaluations take into account national and international limits for polluting substances in air, earth and water, which are then added together. Methods include the EPS-Enviro-Accounting Method (IVL, 1992), the Environmental Preference Method (Anink, 1996) and the Ecoscarcity Method (Abbe, 1990).

In 1994 all three methods were tried in Swedish investigations on the floor materials linoleum, vinyl and pine flooring (Tillman, 1994). One concentrated on the materials impact on the external environment on the materials, and the different methods gave very different results. In all three methods the pine floor achieved the best result, while the linoleum floor proved better than the vinyl in the EPS method but worse in the Ecoscarcity method. In the Environmental Preference Method, the results for both floors were about the same.

Other guidelines for reading this book

Due to the arrangement of the groups of materials in this book, compound materials with components belonging to different substance groups will often be encountered, such as woodwool-cement boards, made up of wood shavings and cement. In such cases, the volume of each component will determine where that material will be listed.

There will also be instances where a material has, for example, both structural and climatic characteristics. The material will be included in both the main summaries and in the tables, but the main presentation will be found where it is felt that this material best belongs.

A number of approaches and recipes for alternative solutions are described. If no other sources are mentioned, these are the author's own statements, and have no judicial or economic bearing. In some cases, recipes with less well-documented characteristics are presented in order to give historical and factual depth.

Terms such as 'artificial/synthetic' and 'natural' materials are used. These are in no way an assessment of quality. In both cases, the raw materials used were originally natural. In artificial/synthetic materials, however, the whole material or part of it has undergone a controlled chemical treatment, usually involving high levels of heat. The extraction of iron from the ore is a chemical process, while the oxidization or corrosion of iron by air is a natural process.

References

ABBE S *et al*, *Methodik für Oekobilanzen auf de Basis Ökologishen Optimirung*, BUWAL Schriftenrephe Umwelt Nr 133, Bern 1990

ANINK D *et al*, *Handbook of sustainable building*, James & James, London 1996

COMMONER B, *The Closing Circle*, Jonathan Cape, London 1972

ERIKSEN T B *Briste eller bare*, Universitetsforlaget, Oslo 1990

FOSSDAL S *Energi-og miljøregnskap for bygg*, NBI, Oslo 1995

HANSEN K *et al*, *Miljonktig prospjektenhg*, Miljöstyrelsen, Københaum 1996

IVL, *The EPS Enviro-accounting method*, IVL Report B 1980:92

KOHLER N *et al*, *Energi- und Stoffflussbilanzen von Gebäuden während ihrer Lebensdauer*, EPFL-LESO/ifib Universität Karlsruhe, Bern 1994

LINDFORS *et al*, *Nordic Manual on Product Life Cycle Assessment – PLCA*, Nordic Ministry, Copenhagen 1994

SUTER P *et al*, *Ökoinventare für Energisysteme*, ETH, Zürich 1993

TILLMAN A *et al*, *Livscycelanalys au golvmaterial*, Byggforskiningsradet R:30, Stockholm 1994

WEOBEL T *et al*, *Okoinventare und Wirkungsbilanzen von Baumaterialen*, ETH, Zührich 1995

section 1

Eddies and water-level markers

Environmental profiles and criteria for assessment

1 Resources

The earth's resources are usually defined as being 'renewable' or 'non-renewable'. The renewable resources are those that can be renewed or harvested regularly, such as timber for construction or linseed for linseed oil. These resources are renewable as long as the right conditions for production are maintained. Thinning out of the ozone layer is an example of how conditions for the majority of renewable resources can be drastically changed. All renewable resources have photosynthesis in common. It has been estimated that man uses 40 per cent of the earth's photosynthetic activity (Brown, 1990).

Non-renewable resources are those that cannot be renewed through harvesting, e.g. iron ore, or that renew themselves very slowly, e.g. crude oil. Many of these are seriously limited – metals and oil are the most exploited, but in certain regions materials such as sand and aggregates are also becoming rare. The approximate sizes of different reserves of raw materials are given in Table 1.1, though there are many different estimates. Everyone, however, is quite clear about the fact that many of the most important resources will be exhausted in the near future.

Fresh water is a resource that cannot be described either as a renewable or non-renewable resource. The total amount of water is constant if we see the globe as a whole, but that does not present a drastic lack of water in many regions. This is especially the case for pure water, which is not only necessary in food production but also essential in most industries. Water is often used in industry in secondary processes, e.g. as a cooling liquid, and thereafter is returned to nature, polluted and with a lower oxygen content.

Usable and less usable resources

It is also normal to divide resources into 'usable' and 'less-usable'. The crust of the earth contains an infinite amount of ore. The problem of extracting ore is a question of economy, available technology, consequential effects on the landscape and environment and energy consumption. Around 1900 it was estimated that to make extraction of copper a viable process, there should be at least 3 per cent copper in the ore; by 1970 the level had

Table 1.1 Existing reserves of raw materials

Raw material		*Statistical reserve (years)*
Mineral		
1.	Aggregate (sand, gravel)	Very large
2.	Arsenic	21
3.	Bauxite	220
4.	Boric salts	295
5.	Cadmium	27
6.	Chrome	105
7.	Clay, for fired products	Very large
8.	Copper	36
9.	Earth, stamped	Very large
10.	Gold	22
11.	Gypsum	Very large
12.	Iron	119
13.	Lead	20
14.	Lime	Very large
15.	Mineral salts	Very large
16.	Nickel	55
17.	Perlite	Very large
18.	Quartz	Very large
19.	Silica	Very large
20.	Stone	Very large
21.	Sulphur	24
22.	Tin	28
23.	Titanium	70
24.	Zinc	21
Fossil		
25.	Carbon	390
26.	Natural gas	60
27.	Oil	40

(Source: Crawson 1992; World Resource Institute, 1992)

fallen to 0.6 per cent. Resources that have been uneconomical to extract in the past can become a viable proposition; e.g. a more highly developed technology of stone extraction would give this material a fresh start for use in construction. The sum of usable and less usable resources are also called 'raw material resources', while the usable resources are called 'reserves of raw material'.

There are also cases where developed technology has a negative impact on the extraction of raw materials; e.g. technological development in the timber industry has made hilly forests inaccessible. It is only by using a horse that one can get timber out of such a forest, but it is rarely the way of the modern timber industry, despite the fact that it causes the least damage to the forest. In the same way, modern technology cannot cope with small deposits of metallic ores – modern mining needs large amounts of ore to make it economical.

Political situations can also affect the availability of raw materials. The civil war in Zaire increased the price of cobalt by 700 per cent, as Zaire has the world's largest deposits of cobalt. Likewise the price of oil was affected by the war in the Persian Gulf. The United States Department of Domestic Affairs has made a list of 'critical minerals'. As well as cobalt, it also includes bauxite for aluminium production, copper, nickel, lead, zinc, manganese and iron; in other words, most metals (Altenpohl, 1980).

Used and unused resources

Resources can also be categorized as 'used' or 'unused'. Along a typical forest path, between 30 and 40 different species of plants, from moss and heather to trees and bushes, can be identified. The total number of different species for all of Norway is about 1500. Two to three of these are well used for building, 10 species are used occasionally while 60 further species have potential for use.

A further example is flint, which was once amongst the most important resources available, but today is virtually unused. At the same time it can be said that in 1840, oil was a totally unexploited non-resource.

The geographer Zimmermann stated in 1933: 'Resources are not anything static, but something as dynamic as civilization itself'. This conclusion gives no reason for optimism. With the accelerating rate of exploitation we are on the verge of bankruptcy in raw materials. Those at high risk of exhaustion are ores and oil, but prospects are not good for sustainable renewal of other resources. Problems related to tropical timbers are well known and discussions centre around the effect of different forms of management, tax rates, replanting, etc. Conditions for biological resources will change quickly as a result of the environmental effects of acid rain and a thinner ozone layer. In Europe the death of innumerable forests is already occurring. An estimate in 1990 stated that over 30 per cent of the existing forest population has been seriously damaged.

It is quite absurd that raw materials should be stripped and disappear in a fraction of the time span of human existence; important ores, minerals and fossil fuels are just used up! From this perspective, it is irrelevant whether these latent resources last two or ten generations. Even a traditional 'anthropocentric' morality with a limited time perspective demands that use of such raw materials be allowed only in very special circumstances, or that recycling is a mandatory requirement.

A differentiation is also made between 'material resources' – the actual constituents of a resource and 'energy resources' – the amount of energy needed to produce the material.

Material resources

The building industry is the largest consumer of raw materials in the world today after food production. A major guiding principle for the future should be a drastic

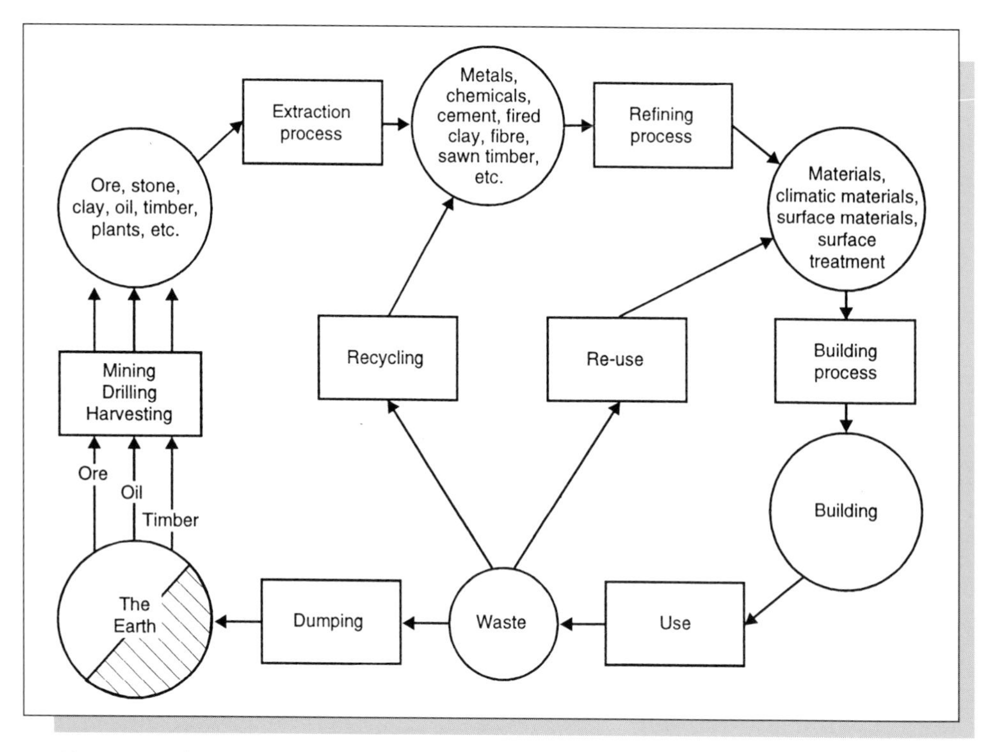

Figure 1.1: The cycle of materials.

reduction in the use of raw materials. This is best applied to the less common non-renewable resources, but is also necessary for others. Another important aspect to address is to reduce the loss of resources during production, the construction process and throughout the life of the completed building. The re-use of materials following demolition should also be taken into account. Recycling processes should be developed so that materials can be taken care of at their original level of quality, rather than downcycled.

Reduction of the use of raw materials in the production process

Increased exploitation of smaller sources of raw materials

This is mainly a question of technology. Even if modern technology is primarily geared up for large scale exploitation, there are certain areas of exploitation that have developed small scale technology, such as in mineral extraction.

Greater attention to unused resources and waste products

Resources that have been earlier classified as 'uneconomical', or never used, can be re-evaluated. Examples of such resources are:

- compressed earth as a construction material
- fibres from the seaweed eelgrass as an insulating material
- increased use of timber from deciduous trees.

A series of different sorts of waste from industry, agriculture and dwellings, e.g. straw, industrial sulphur and waste glass, can also be evaluated.

Increased exploitation of rich fields of resources
Not all resources are being totally exhausted. An example is stone, which is still a plentiful resource over the whole earth; another is blue clay, which has great potential and is in no way exhausted by the comparatively low production of bricks. The side-effects that the excavation of minerals exerts on their immediate environment, e.g. lowering the water table, damaging local ecological systems, must be taken into account.

Increased use of renewable resources
Many building components made from mineral raw materials have organic alternatives, e.g. timber can be used as an alternative to steel. This usually has an overall positive environmental impact.

Increased recycling of waste products during production
A series of good examples already show that this method can save valuable resources, such as the manufacture of plasterboard. Re-use of water in the production processes of certain industries also occurs, e.g. production of ceramic tiles.

Reduction of the use of resources in the building process and during building use

In these two phases there are the following possibilities for reducing the use of resources:

- to build with an economic use of materials
- to minimize loss and wastage of materials on site
- to use the materials in such a way as to ensure their durability
- to maximize re-use and recycling of materials from demolition.

Economical construction
Every structural system has its specific use of materials. The difference between systems can be quite significant. A lattice beam uses much less material than a solid beam, whether it is timber or steel.

There is also a tendency to build too large. There can be no doubt about the fact that smaller buildings use fewer resources! The same applies to energy consumption in a building which is of optimal size. There is a greater efficiency co-efficient in such a building compared with the use of heat pumps, solar panels and thick, insulated walls in a less optimized building. This is one of the greatest challenges for architects of the future – to make small buildings as comfortable as possible.

Reduced loss of building materials

Every material has a 'loss factor' which describes how much of a particular material is lost during storage, transport and installation of the final product. As well as indicating the amount of wastage the material undergoes, the loss factor gives an idea of the amount of resources lost. For many materials, increased prefabrication would decrease this loss, which would be further strengthened through an increased standardization of products.

Loss of materials on site is approximately 10 per cent of the total waste in the building industry. In Scandinavia in the last few years there have been a number of large projects where the amount of material loss has been reduced by more than 50 per cent through, amongst other things, having usefully planned site management. Sawn off timber lengths and waste products have been separated out and kept within the building process (Thonvald, 1994).

Within the building industry a great deal of packaging material is also used during transport and for storage on site. Some packaging serves no greater purpose than to hold the name of the firm. An important aspect of packaging is that it should be easy to recycle, and therefore should not comprise different materials such as aluminium or plastic emblems printed on cardboard.

Loss of material caused by wear and tear in the completed building will also occur. In Sweden in 1995, the Department of the Environment estimated that the loss of copper from roofs and pipes etc. through weathering amounts to more than 1000 tons per year. Apart from the pollution risk, there is also a huge loss of resources that could be recycled. Materials based on rare, non-renewable resources should not be used in exposed parts of the building.

High durability

By producing more durable products the use of raw materials is reduced by ensuring that materials of the same durability are used during the construction process, therefore not sacrificing better quality components in a building when there is decay elsewhere. If there are any materials of a lesser quality, then it is important that they are easily replaceable while the more durable materials can be dismantled for re-use or recycling in the case of demolition. As far as resources are concerned, there is a clear advantage in using robust materials and allowing buildings to last as long as possible.

Simply put, twice as much damage to the environment can be tolerated for a product that lasts 60 years compared with one that lasts 30 years. The lifespan of a material is governed mainly by four factors:

- the material itself, its physical structure and chemical composition
- construction and its execution; where and how the material is fitted into the building
- the local environment; the climatic and other chemical or physical conditions
- maintenance and management.

The life span of a roof tile, for example, is not only dependent on the type of clay used, but also on the immediate environment of the building in which it is used. A high moisture content during winter can cause frost damage even in the highest quality tiles.

The best way to find the anticipated life span of a material is through experience and tabulated results from real situations. The real situation must have a comparable local climate.

It is difficult to anticipate the life span of most new materials, e.g. plastics. It is possible to create accelerated deterioration in laboratories, but these generally give a simplified picture of the deterioration process than the more complex actual situation. Results from these tests can only be taken as a prognosis. It is necessary to evaluate the role of the material in construction very carefully for such a prognosis.

We should also remember that durability is not only a quantifiable technical property. Durability also has an aesthetic and fashionable side to it. It is quite a challenge to design a product that can outlast the swings of fashion. Especially with technical equipment, it is also important to consider an optimal durability rather than a maximum durability. Changes to new products can often show a net environmental gain in terms of energy-saving criteria.

Effects of the climate and durability

Even if we do not know all the durability factors, it is still certain that climate is a factor that regulates the life span of a material:

Solar radiation. Ultraviolet radiation from the sun deteriorates organic materials by setting off chemical reactions within the material and producing oxidation. This effect is stronger in mountainous areas, where the intensity of ultraviolet radiation is higher, and it also increases as you move further south.

Temperature. An old rule of thumb tells us that the speed of a chemical reaction doubles for every 10°C increase in temperature. Higher temperatures should therefore increase the deterioration of organic materials. Emissions of formaldehyde from chipboard with urea-based glue is doubled with every 7°C increase of temperature. Warmth also stimulates deterioration processes in combination with solar radiation, oxygen and moisture.

At low temperatures, materials such as plastic and rubber freeze and crumble. An exterior porous low-fired brick only lasts a couple of winters in northern Europe – in Forum Romanum in Rome the same brick has lasted 2000 years! The cycle of freezing and thawing is a deciding factor for this material. The coastal climate of the north is also very deleterious. Wide changes in temperature strain the material, even without frost, and will cause it to deteriorate.

Air pressure. Air pressure affects the volume of and tension within materials which have a closed pore structure, such as foam glass and different plastic insulation materials. Sealed windows will also react. Changes in size which occur have the same effect as temperature changes.

Humidity. Change of humidity effects deterioration by causing changes in volume and stress within the material. Increased humidity increases deterioration. This is why the manufacture of musical instruments such as pianos and violins can only take place in premises with a stable air moisture content. The same conditions should also be applied to other interiors to reduce the deterioration of cladding materials and improve cleaning.

Urea-based chipboard, mentioned above, doubles its emissions with an increase of 30–70 per cent in relative humidity.

Wind and rainfall. Are at their worst when both wind and rain come simultaneously. In this case damp can force its way into the material and start off the deterioration process. Strong winds cause pressure on materials which may even lead to fracture or collapse. Combined with sand, wind can have a devastating effect on certain materials. The weight of snow can also break down structures.

Chemicals. Along the coast the salt content of air can corrode plastics, metals and certain minerals. In industrial and built-up areas and along roads, aggressive gases such as sulphur dioxide can break down a variety of different materials. Concrete suffers from so-called 'concrete sickness' because the calcium content is broken down in such an environment. This even occurs with certain types of natural stone.

Recycling

Every material accumulates a resource effect and a pollution effect, particularly during production. Through recycling products, rather than manufacturing from new raw materials, a good deal of environmental damage can be prevented. A product that can be easily recycled has an advantage over a product that is initially 'green' but cannot be recycled.

In the building industry a great many products or materials have both low durability and low recycling potential. There are also products that can be recycled several times, but this potential is seldom used nowadays.

The level of recyclable products in Sweden in 1992 was 5 per cent. In Germany in 1990 as much as 29 per cent was recycled. Both these countries have a target of 60 per cent for the year 2000. In Holland, demolition contractors at tender stage have to state how much of the material will be sold for recycling, together with a presentation of how they will advertise this.

There are already a few examples of successful selective demolition projects. All the different materials and products have been separated out, and a level of recycling of 90 per cent has been made possible (Thormark, 1995). The buildings demolished have been older types with a simple use of materials. For modern buildings, it is doubtful whether the level of recycling will get as high as 70 per cent. There are also examples of successful projects in which buildings consist mainly of recycled materials and products (Bitsch Olsen, 1992).

Recycling levels

There is a hierarchical model of recycling levels; the goal is to achieve the highest possible degree of recycling:

A: Re-use

B: Recycling

C: Energy recovery

Re-use depends upon the component's life span and refers to the use of the whole component again, with the same function.

Development of re-usable structures or component design has not come very far. There are few quality control routines for re-usable products. Efficient re-use of materials or components demands simple or even standardized products. Very few products on the market today meet these requirements. In Germany there are as many as 300 000 products within the building industry, all with different design and composition.

The re-use of materials always used to be a part of building practice. In many coastal areas older buildings have been constructed using a great deal of driftwood and parts of wrecked ships. Log construction is a good example of a building method with high re-use potential. The basic principle of lying logs on top of each other makes them easy to take down and re-use, totally or in part. This building method uses a large amount of material, but the advantages of re-use balance this out.

Figure 1.2: A traditional summer village on the south coast of Turkey. The huts are made of driftwood, packing cases, packaging and other available free material, and are used as summerhouses by the local population.

Recycling is mainly dependent upon the purity of the material. Composites or multiple materials are no good for recycling. Recycling is done by smelting or crushing the

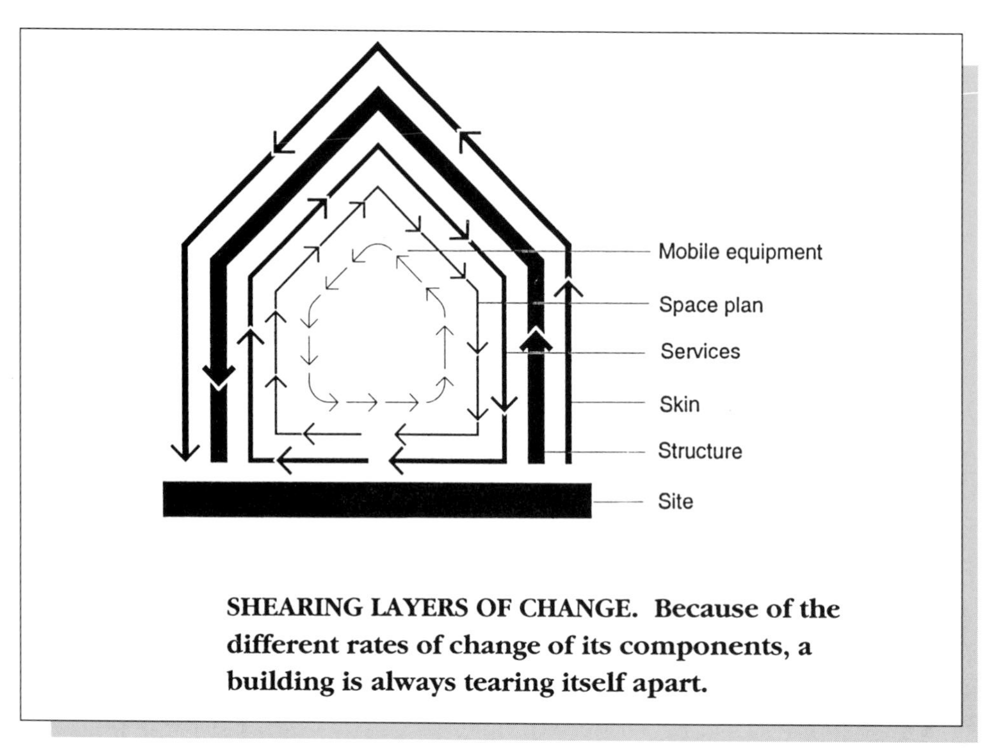

Figure 1.3: The main layers of a building. *Source: Brand 1994*

component, which then enters a new manufacturing process. This is a very efficient method for metals. For other materials different methods of down-cycling makes less valuable products, e.g. reducing high quality PVC articles to flower pots, or crushing light-weight concrete blocks into aggregate.

Where products claim to have a potential recycling, the statement is often based on theoretical figures. In practice there are often complications: thin aluminium fibre or containers burn up totally or evaporate when being melted down, while small amounts of impurities in the worst cases can lead to extra refining processes and a higher use of energy.

Energy recovery means burning the product to produce energy. It is an advantage if the material can be burned at a local plant and if the fire gases do not need special treatment, so that simple furnaces can be used.

Assembly for disassembly

In recycling, 'assembly for disassembly' (ADISA) is a very important factor. With this type of assembly, different components are separated into piles of the same material during demolition. Not only do ADISA assemblies separate the individual components for re-use, they also simplify recycling and energy recovery.

Designing for the direct re-use of building materials gives the best opportunity for slowing down the trip to the rubbish dump. Assembly for DISAssembly (ADISA) principles give some fundamental guidelines for optimizing the re-use possibilities of single components:

First principle: Separate layers

A building consists of several parallel layers (systems): interior, space plan, services, structure, skin (cladding) and site (see Figure 1.3). The main structure lasts the lifetime of the building – 50 years in Norway and Britain and closer to 35 in the USA (Duffy, 1990) – while the space plan, services etc. are renewed at considerably shorter intervals. In modern buildings the different layers are often incorporated in a single structure. Initially this may seem efficient, but the flow in the long-term cycles will then block the short-term cycles, and short-term cycles will demolish slower cycles via constant change. It is, for example, normal to tear down buildings where installations are integrated in the structure and difficult to maintain.

Space plans can be so specialized and inflexible that, for example, in central Tokyo modern office buildings have an average life span of only 17 years, (Brand, 1994).

We are therefore looking for a smooth transition between layers (systems), which should be technically separated. They should also be available independently at any given time. This is a fundamental principle for efficient re-use of both whole buildings and single components.

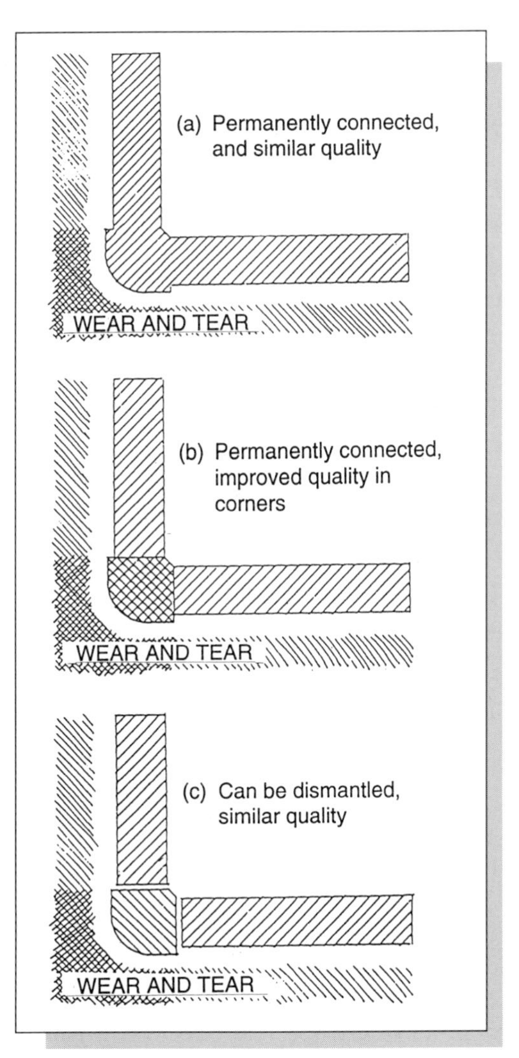

Figure 1.4: Three principles for connecting walls.

Second principle: Possibilities for disassembly within each layer

Single components within each layer should be easy to disassemble. Figure 1.4 shows three different principles for assembling a wall cladding at a corner. The shading shows where the mechanical wear and tear is greatest, from people, furniture, wind and weather. The normal choice today is the first solution, (a), where all parts are the same quality and permanently connected. When the corner is torn down the whole structure follows with it. In many expensive public buildings, solution (b) is chosen. By increasing the quality of the most worn area, the whole structure will have a longer lifetime. This is usually an expensive solution and makes changes in the space plan difficult, unless the whole structure is demolished. In solution (c), worn components can easily be replaced separately. The used component can then be re-used in another corner where the aesthetics are less important, or it can be sent directly to material- or energy-recycling.

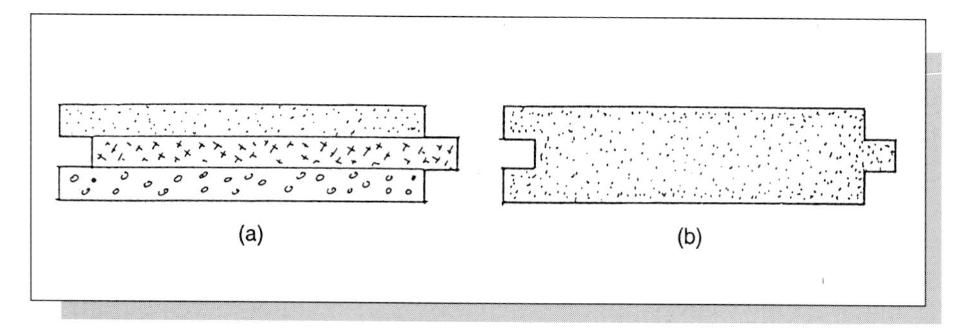

Figure 1.5: (a) Multimaterial component; (b) monomaterial component.

Third principle: Use of standardized monomaterial components
Before re-use of the components on the open market it is necessary to check their quality. This often presents problems. Many building components are composed of different materials laminated together (see Figure 1.5). Re-use of such products is difficult. Different rates of decay within the same component may result in one of the materials being partially broken down while the others are still in good condition. This problem is especially acute in large, prefabricated building elements where cladding, insulation and structure are integrated in a single component.

For re-usable structures only so-called primary and secondary monomaterials are used. A primary monomaterial is a single homogeneous material used in its natural state, e.g. untreated wood. A secondary monomaterial is a mixed material of homogeneous nature, e.g. concrete, glass or cellulose fibre.

By only using monomaterials it is usually easy to check the quality for re-use. Even if re-use products are thoroughly quality controlled, there still may not be a market for them. The shape of the components may be so unusual that they would need to be transported some distance to find a buyer. So this whole strategy can quickly become an energy problem. Re-usability is therefore determined by the generality of the component, i.e. its re-usability in a local market. This means that it has to comply with local standards, making it easy to use in new structures.

Most components of a building can be adapted for re-use in this way, though some, e.g. electrical installations, may be less suitable for re-use. In this case, new technology may meanwhile have taken over, for example in energy-saving, making re-use quite ecologically unsound.

In all levels of recycling there will be waste. And even when all the recycling is done, there are still materials left over which need to be taken care of. This can be a very large amount if the material quality is poor from the beginning, as in the case of waste paper pulp, which has already gone through several rounds of recycling. The alternatives for their use are dumping or global recycling. Global recycling means making compost of the materials, or in some other way reuniting them with nature, making them a potentially new resource. When cellulose, for example, is composted, it is first covered by earth. A series of complex biological processes follow in which mould deteriorates the cellulose structure.

Special enzymes in the mould release carbohydrates which enter the earth, stimulating bacterial growth, which in its turn attacks the molecular structure of the cellulose and releases soluble constituents of nitrogen. The end product is humus, which forms a foundation for different plant organisms, providing nutrients for the growth of new cellulose fibres.

In this way global recycling is based almost entirely on closed cycles, which means that there is hardly any waste in nature. These methods can also be considered a more sensible way of depositing a material compared with ordinary recycling or energy recovery.

Raw materials in a world context

The term 'under-developed country' is a totally misleading description when considered in an ecological light. In many cases the ecological cycles work much better in the so-called under-developed countries. Here we characterize countries by their degree of industrialization: high industrialization, medium industrialization, and low industrialization.

Most of today's global consumption of materials takes place in the northern temperate zone. But that does not mean that most of the raw materials are found in this part of the globe – it seems that the consequence of increased industrialization is an increased dependence on imported raw basic materials. Western Europe imports about 80 per cent of its minerals and 60 per cent of its energy. The suppliers are usually countries with low industrialization.

Looking at the accessibility of raw materials, it is quite clear that increased consumption and industry in countries with low or medium industrialization must lead to a de-industrializing of the northern part of the globe. Many western European concerns have exported all or part of their work operations to guarantee future development. Initially it looks as though they often choose a manufacturing process that has difficulty achieving Western environmental standards.

Energy resources

On current projections, there are sufficient gas and oil resources for another 40–60 years. Coal reserves will last for another 1000 years, but with the problem of related acid rain and carbon dioxide emission. Environmentalists predict a quick and violent ecological crisis if we use coal as an alternative energy resource. This means we have to keep to nuclear power in breeder-reactors, using uranium and thorium, or renewable energy resources such as the sun, wind, heat exchangers and water power. The conclusions are quite clear: nuclear power has a great many risks and waste problems, while the renewable natural resources are safe but difficult to harness. During recent years the threat of the increased 'greenhouse effect' has received a lot of attention. This problem relates directly to energy, which in turn is mainly produced by the fossil fuels. This theme is discussed more thoroughly in the following chapter.

The building industry is the giant amongst energy consumers. Use of energy is divided between the production, distribution and use of building materials.

Stages of energy consumption in building materials

The manufacture, maintenance and renewal of the materials in a standard timber-framed dwelling for three people over a period of 50 years requires a total energy supply of about 2000 MJ/m^2 (Fossdal, 1995). A house in lightweight concrete block construction needs over 3000 MJ/m^2. For larger buildings in steel or concrete the energy required is around 2500 MJ/m^2. The amount of energy that actually goes into the production of the building materials is between 6 per cent and 20 per cent of the total energy consumption during these 50 years of use, depending on the building method, climate, etc.

Energy consumption during the manufacture of building materials

The primary energy consumption (PEC) is the energy needed to manufacture the building product. An important factor in calculating PEC is the product's combustion value. This is based on the amount of energy the raw material would have produced if burnt as a fuel. The combustion value is usually included in the PEC, because the raw material would have had a high value as an energy resource, and if this combustion value is removed or heavily reduced in the product one gets a false picture of the energy equation.

PEC is usually about 80 per cent of the total energy input in a material and is divided up in the following way:

- *The direct energy consumption in extraction of raw materials and the production processes.* This can vary according to the different types of machinery for the manufacturing process.
- *Secondary consumption in the manufacturing process.* This refers to energy consumption that is part of the machinery, heating and lighting in the factory and the maintenance of the working environment.
- *Energy in transport of the necessary raw and processed materials.* The method of transport also plays an important role in the use of energy. The following table shows energy consumption per ton of material transported in Norway in 1990:

Type of transport	MJ/ton/km
Diesel: road transport	1.6
Diesel: sea transport	0.6
Diesel: rail transport	0.6
Electric: rail transport	0.2

Energy consumption during building, use and demolition

Transport and the use of completed products is usually about 20 per cent of the total energy input.

- *Energy consumption for the transport of manufactured products.* This can have a very decisive role in the total energy picture. One example is the export of lightweight concrete elements from Norway to Korea, which uses over 10 000 MJ/m^3, while the actual manufacture of the elements require a primary energy input of 3500 MJ/m^3. This confirms the principle that heavy materials ought to be used locally.

- *Energy consumption on the building site.* This includes consumption which is already included within the tools used, heating and lighting, plant, electricity and machines. It also includes the energy needed to dry the building construction such as in-situ concrete. The use of human energy varies from material to material just as it varies between the manufacture and use of a material. This will not have much of an impact on the overall picture. Assuming one person uses 0.36 MJ energy per hour, a single house would consume 270–540 MJ.

  The amount of energy used on the building site has grown considerably in recent years as a result of increased mechanization. Drying out of the building with industrial fans is relatively new. Traditionally the main structure of the building, with the roof, is completed during spring, so it could dry during the summer break. The moisture content of the different building materials also affects the picture. For example, it takes twice as long to dry out a concrete wall as it does a solid timber wall.

- *Energy consumption during maintenance.* Sun, frost, wind, damp, human use etc. wear away the different materials, so that the building needs to be maintained and renovated. Initially one treats the surfaces by painting or impregnation, materials that have an energy content themselves. The next stage is replacement of dilapidated or defective components.

- *Energy consumption of dismantling or removal of materials during demolition.* This is approximately 10 per cent of the energy input which is integral within the different materials.

Reduction of energy consumption in the building industry

It is quite possible to reduce drastically the amount of energy consumed in building. The following steps could achieve a great deal:

Energy saving during the manufacturing process

Decentralized production

This requires less transport and is especially appropriate when local materials are being processed (see Figure 1.6).

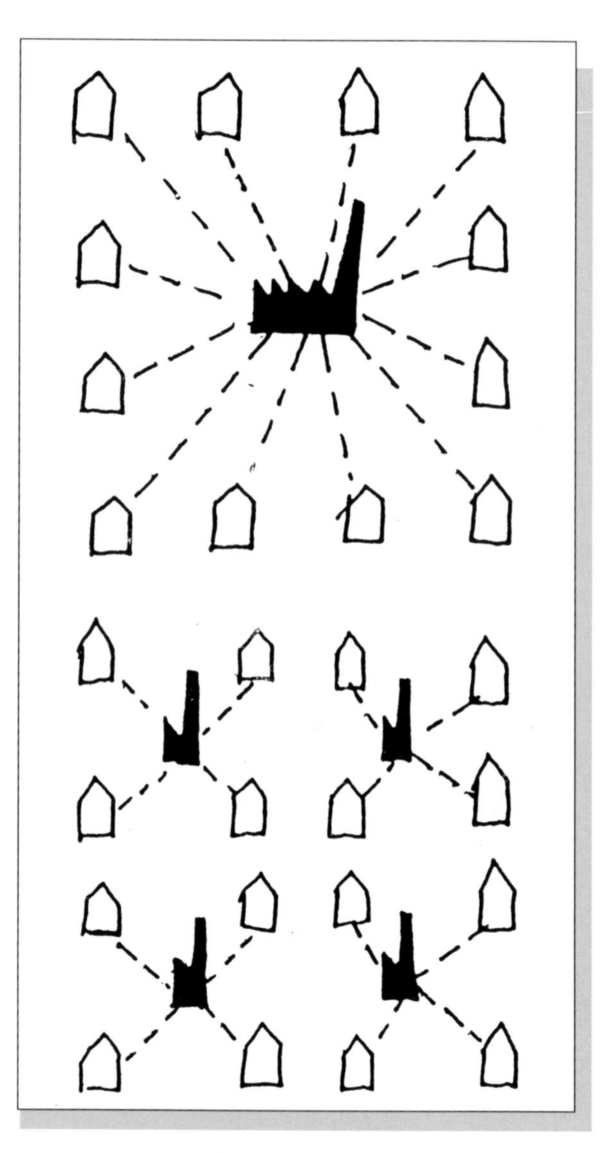

Figure 1.6: Local industries create less need for transport. Source: Plum 1977

Use of highly efficient sources of energy

Electricity produced from oil, coal and nuclear power achieves only 25–30 per cent of the potential energy available. The degree of efficiency is thereby 0.25–0.3, and the rest is lost. Hydro-electricity has an efficiency coefficient of 0.6, which is not particularly impressive either. In many cases it would be better to avoid electricity and use sources of energy within production that use direct mechanical or intensive heat energy – rotational power is an example. The source of energy must have a clear relationship with the manufacturing process used. This principle can be determined in terms of levels of energy quality (see Table 1.2).

Use of local sources of energy

The shorter the distance between the power station and the user, the smaller the amount of energy lost in the network/distribution line. Over larger distances the loss can be as great as 15 per cent. Small local power stations have shown definite economical advantages over recent years.

Other energy saving changes

It is possible to reduce energy consumption in certain industrial processes by using efficient heat recovery and improved production techniques. Cement

Table 1.2: Renewable sources and the levels of energy

	Mechanical energy	*Electricity*	*Production temperatures* Above 600°C	*Between 200–600°C*	*Between 100–200°C*	*Temperatures under 100°C (room temperatures and hot water)*
Sun	(x)	(x)	(x)	(x)	x	x
Water/wind/waves	(x)	x	x[1]	x[1]	x[1]	x
Wood and peat		(x)	x	x	x	x
District heating						x

Notes:
x: commercially available industrially, 1991
(x): under development
x[1]: from electricity

burning in shaft furnaces needs 10–40 per cent less energy than traditional rotational furnaces. In the steel industry one could reduce the use of energy by 50 per cent by changing from open blast-furnaces to arc furnaces.

Energy saving during the building process

Local materials
The use of local materials means less transport requirements.

Low energy materials
Give priority to materials that have a low primary energy consumption and are durable.

Labour intensive processes
The energy needed to keep a worker and his family is so small that it has little effect in the total energy calculation. Labour intensive processes are almost without exception energy-saving processes.

Natural drying out of the building
There is a lot to be gained by choosing quick drying materials – brick rather than concrete, for example – and by letting the building dry out naturally.

Building techniques that favour re-use and recycling
Most building materials have used a great deal of energy during manufacture. By re-using seven bricks, a litre of oil is saved! Recycling metals can save between

Table 1.3: Effects on resources

Material	*Technical properties*			*Material resources*		*Energy resources*			*Water*
						Primary energy consumption			
	Weight (kg/m³)	*Durability*	*Loss factor[1] (%)*	*Statistical number of years left as reserves*	*Raw material (see Table 1.1) R = renewable*	*North Europe (MJ/kg)*	*Central Europe (MJ/kg)*	*Combustion value[(2)] (MJ/kg)*	*Use of water (litres/kg)*
Aluminium, 50% recycled	2700	high	21	220	3	58	184	–	29 000
Cast iron, from iron ore	7200	high		119	12		13	–	
Steel:									
100% recycled	8000	high		–	–	6	10	–	
galvanized from ore	7500		21	21	12–24	12	25	–	3400
stainless steel from ore	7800		21	21	12–24	12	25	–	3400
Lead from ore	11 300	high	21	20	13		22	–	1900
Copper from ore	8930	very high	16	35	8		70	–	15 900
Concrete with Portland cement:									
structure	2400	high	16	–	14	0.6	1	–	170
roof tiles	2200	medium	4	–	14		2	–	
fibre reinforced slabs	1200	medium	20	–	14		7	–	450
mortar	1900	high	10	–	14	1	1	–	170
Aerated concrete blocks and prefab units	500	medium	5	220	3–14–18		4	–	300
Light aggregate concrete blocks and prefab units	750	medium	6	–	14–7	2	4	–	190
Lime sandstone	1600	medium	11	–	14–18		1	–	50
Lime mortar	1700	medium	10	–	14		1	–	
Calcium silicate sheeting	875	medium	20	–	14–18		2	–	
Plasterboard	900	medium	25	–	11	5	5	–	240
Perlite, expanded:									
without bitumen	80	high	1	–	17		8	–	
with bitumen	85		1	40	27–17		8	–	
with silicone	80		1	40	27–17		8	–	

Glass:	2400	high	3	–	18–15–14	7	8	–	680
with a tinoxide layer	2400		3	–	22–18–15–14		–		
Foam glass:									
slabs	115	high		–	18–15–14		11		
granulated, 100% recycled		high		–	–				1300
Mineral wool:									
rockwool	30	medium	6	390	25-14-15	11	16	–	1360
glasswool	20	medium	6	390	25–18–15	20	18	–	1360
Stone:									
structural	2700	very high		–	20		0.1	–	10
slate	2700	very high	6	–	20		0.1	–	10
Earth, stamped structure	2000	high	1	–	9		0.1	–	10
Bentonite clay	1800	high							
Fired clay:									
bricks	1800	very high	10	–	7	2	3	–	520
roof tiles	1800	medium	3	–	7		3	–	640
Ceramic tiles	2000	very high	18	–	7	8	8	–	400
Fired clay pellets	450	very high	1	–	7	2		–	
Bitumen	1000	low/medium		40	27		5		
Polyethylene (PE)	940	low/medium	11	40	27		67	(44)	
Polypropylene (PP)		low/medium	11	40	27		71	(44)	
Expanded polystyrene:									
EPS	23	low/medium	11	40	27	75	75	(20)	
XPS	23	Medium	11	40	27	72		(20)	
Expanded polyurethane (PUR)	35	low/medium	11	40	27	98	110	(76)	18 900
Polyvinyl chloride (PVC)	1380	medium/high	11	40	15–27	56	84	(23)	
Expanded urea-formaldehyde (UF)	12	low/medium		390	25		40		
Polyisobutylene (PIB)		Low/medium		40	27		95		
Polyester (UP)	1220	medium		40	27		78		

continued

Table 1.3: Effects on resources – *continued*

Material	*Technical properties*			*Material resources*		*Energy resources*			*Water*
				Statistical number of years left as reserves	*Raw material (see Table 1.1) R = renewable*	*Primary energy consumption*		*Combustion value*[2] *(MJ/kg)*	*Use of water (litres/kg)*
	Weight (kg/m³)	*Durability*	*Loss factor*[1] *(%)*			*North Europe (MJ/kg)*	*Central Europe (MJ/kg)*		
Styrene butadiene rubber (SBR)	1000	low/medium		40	27		70		
Timber:									
untreated	550	medium/high	20		R	3	3	16	330
pressure impregnated	550	medium/high	20	21	R–6–2			(16)	
laminated timber	550	medium/high		390	R–25	4		16	
Wood fibre insulation	100	medium		–	R				
Cork	70	medium	11	–	R		4		24
Wood fibre board: porous									
without bitumen	300	medium		–	R		16	10	350
porous with butumen	350	medium		40	R–27	18		(10)	
hard without bitumen	900	medium/high	20	–	R	4	15	7	2,500
hard with bitumen	900	medium/high	20	40	R–27			(7)	
Woodwool slabs	230	High	21	–	R–14		20	(7)	
Chipboard	750	medium/high	20	390	R–25	2	4	(14)	1000

Cellulose fibre insulation, 100% recycled and boric salts	60	medium	1	295	R-4	19	21	(17)	10
Cellulose fibre matting (fresh) and boric salts	80	medium	5	–	R-4			(17)	
Cellulose building paper (unbleached): 98% recycled	1200	medium	12	–	R		16	11	
Cardboard sheeting, laminated with polyethylene	750	low/medium	20	40	R–27				
laminated with latex	750	low/medium	20	–	R				
Linenfibre: strips	150	medium/high	1	–	R			12	
Linen matting	16	medium/high	5	–	R				
Linoleum	1200	medium	11	–	R	7	1	10	140
Straw:									
thatch	100	low		–	R				
bound with clay	600	medium		–	R–9				
Coconut fibre, strips	100	medium		–	R				
Jute fibre, strips	100	medium		–	R			12	
Peat slabs	225	medium	5	–	R				
Wool paper	500	medium	12	–	R				
Woollen matting	18	medium	5	–	R				

Notes:

(1) Loss factor is the percentage of material that is usually lost during storing, transporting and mounting of the product.

(2) The figures in brackets under combustion value show the value that is no longer available due to its poisonous character or the structure of the material.

40 per cent and 90 per cent compared with extracting from ore. The ability to recycle locally is a decisive factor, otherwise transport energy costs quickly change the picture from gains to losses.

References

ALTENPOHL D, *Materials in World Perspective*, Berlin/Heidelberg/New York 1980

BITSCH OLSEN E, *Genbrug af materiale og bygningsdele*, NBS seminarrapport, Trondheim 1992

BRAND S, *How Buildings Learn*, Viking Penguin, New York 1994

BROWN LR (ed.), *State of the World*, Washington 1990

CRAWSON P, *Mineral Handbook 1992–93*, Stocton Press, New York 1992

DUFFY F, *Measuring building performance*, Facilities, May 1990

FOSSDAL S, *Energi og miljøregnskap for bygg*, NBI, Oslo 1995

THORMARK C, *Återbygg*, Lunds tekniska högskola, rapp. TABK -95/3028, Lund 1995

THORVALD NO, *Avfallsreduksjon og kildesortering i byggebransjen. Erfaring fra tre gjennomførte prosjekter*, SFT rapp. 94:11, Oslo 1994

WORLD RESOURCE INSTITUTE, *World Resources 1992–93*, Oxford University Press, Oxford 1992

2 *Pollution*

People in all industrialized countries have daily contact with pollution problems: smarting eyes in exhaust-filled streets, decaying marble monuments, murky fishing water, the fact that 80–90 per cent of all cases of cancer are influenced by environmental factors and that the number of allergies are rapidly increasing. In Sweden it has been calculated that 12 000 to 16 000 people die every year because of environmental pollution (Gillberg, 1988). At the same time the rate of extinction of animal and plant species is accelerating. Between 1900 and 1950 one species disappeared annually; in 1990 between one and three species disappeared every hour! Species have always died out and new ones have appeared, but the rate of extinction today is approximately a hundred times greater than the natural rate.

The building industry is directly or indirectly responsible for a great deal of environmental pollution. One example is the damage caused to nature by the over-extensive exploitation of raw materials. Large open limestone, sand or gravel mines, and other open-cast mines, produce visual damage and destroy local plant and animal life as well as polluting ground water.

When talking about pollution, the physical and chemical effects of gaseous and particle pollution, electromagnetic fields and radioactivity primarily come to mind. In these cases, damage to ecosystems tends to be at a lower level than damage to human beings.

The problems can be referred to in terms of 'energy pollution' and 'material pollution'. Energy pollution relates strongly to the primary energy consumption (PEC) and the source of energy used. The sources of energy vary a great deal from country to country. In Scandinavia hydropower and nuclear power are diminishing; in Great Britain and on the Continent the main sources are fossil fuels and nuclear power. Statistics for energy pollution from fossil fuels are as follows:

Energy pollution from fossil fuels in g/MJ

Fossil fuel	CO_2	SO_2	NO_x
Oil for oilfiring	75	0.5	0.15
Natural gas	57	0.01	0.16
Coal, low carbon content	110	0.03	0.16
Coal, high carbon content	93	0.01	0.16

Energy pollution is also caused by the transport of materials. The deciding factors are the type of materials, weight, method of transport and distance travelled.

Energy pollution from different forms of transport (g/ton km)

Type of transport	CO_2	SO_2	NO_x
Diesel: road	120	0.1	1.9
Diesel: water	50	0.3	0.7
Diesel: rail	50	0.05	0.75

(Source: Fossdal, 1995)

Material pollution relates mainly to pollutants in air, earth and water from the material itself and from the constituents of the material when being worked, in use and during decay. The picture becomes quite complex when considering that around 80 000 chemicals are in use in the building industry, and that the number of health-damaging chemicals has quadrupled since 1971. Damage to the ground water system, local ecological systems etc. occurs due to the excavation or dynamiting of raw materials.

Pollution from production, the construction process and completed buildings consists of emissions, dust and radiation from materials that are exposed to chemical or physical activity such as warmth, pressure or damage. In the completed building these activities are relatively small, yet there is evidence of a number of materials emitting gases or dust which can lead to serious health problems for the inhabitants or users; primarily allergies, skin and mucous membrane irritations. The electrostatic properties of different materials also play a role in the internal climate of a building. Surfaces that are heavily negatively charged can create an electrostatic charge and attract a great deal of dust. Electrical conductors such as metals can increase existing magnetic fields. It is also important that materials in the building do not contain radioactive constituents, which can emit the health-damaging gas, radon.

Waste is part of the pollution picture and needs to be discussed, particularly as these materials move beyond the scope of everyday activities and can be overlooked. The percentage by weight of environmentally-damaging material in demolition and building waste is relatively small, but is still a large quantity and has a considerable negative effect on the environment. Waste that has a particularly damaging environmental effect and cannot be recycled is usually burned or dumped.

While some materials can be burned in an ordinary incinerator with no particular purifying treatment, others need incinerators with highly efficient smoke purifiers. Far too few incinerators can do this efficiently – many still emit environmentally-damaging materials such as sulphur dioxide, carbon fumes, hydrogen chloride, heavy metals or dioxides.

Depending on the environmental risk of the materials that are to be dumped, the disposal sites must ensure that there is no seepage of the waste into the water system. This is the most serious type of environmental damage that can occur at such depots when the constituents of the materials are washed out by rain, surface water or groundwater.

The most dangerous materials are those containing heavy metals and other poisons, and also plastics which are slow to decompose and cause problems because of their sheer volume. Organic materials contain enzymes that break down materials, but synthetic materials do not. They take a long time to decompose, so they have to be broken down mechanically before further treatment. Synthetic materials tend to be deposited in the most remote places, and become very difficult to eradicate.

There is an evident relationship between the natural occurrence of a material and its potential to damage the environment. If the amount of a substance is reduced or increased in the environment (in air, earth, water or inside organisms), it can be assumed that this increases the risk of negative effects on the

Table 2.1: Pollution in the material life cycle

Stages of the material life cycle	*Material pollution*	*Energy pollution*
1. Extraction of raw materials	x	x
2. Production process	x	x
3. Building process	x	x^1
4. Transport between stages 1, 2, 3 and 7	x^1	x
5. Materials in use	x	x^2
6. Materials in combustion	x	
7. Materials during demolition	x	

Notes:

x^1: Very small proportions, e.g. accidents during the transport of building materials, though such accidents can lead to leakage of highly dangerous chemicals such as construction glue, which contains phenol.

x^2: Highly polluting building materials give rise to higher use of energy through the increased ventilation required in the building.

Table 2.2: Natural occurrence of elements in the accessible part of the Earth's crust

Amount (g/ton)	*Elements*
Greater than 100 000	O, Si
100 000–10 000	AI, Fe, Ca, Na, K, Mg
10 000–1000	H, Ti, P
1000–100	Mn, F, Ba, Sr, S, C, Zr, V, CI, Cr
100–10	Rb, Ni, Zn, Ce, Cu, Y, La, Nd, Co, Sc, Li, N, Nb, Ga, Pb
10–1	B, Pr, Th, Sm, Gd, Yb, (Cs, Dy, Hf), (Be, Er), Br, (Sn, Ta), (As, U), (Ge, Mo, W), (Eu, Ho)
1–0.1	Tb, (I, Tm, Lu, TI), (Cd, Sb, Bi), In
0.1–0.01	Hg, Ag, Se, (Ru, Pd, Te, Pt)
0.01–0.001	(Rh, Os), Au, (Re, Ir)

Source: Hägg 1984

environment. Table 2.2 shows the natural occurrence of certain elements in the accessible part of the Earth's crust. Elements of approximately the same concentration are placed within brackets in order of their atomic number.

Types of pollution

Environmental poisons

Toxic substances that are heavily decomposible and/or bio-accumulative, which means that they concentrate themselves within nutrient chains. In addition to the heavy metals, it is important to consider organic poisons. Many of these substances are spread by air to the most remote places, and they are in the process of becoming concentrated in ground water in highly-populated areas. Many of them are thought to have environmentally dangerous side effects.

Dust

Dust is produced during the extraction of materials, various industrial processes and through incomplete combustion of solid fuel and oil. It is also caused by building materials such as mineral wool and asbestos. Dust can be chemically neutral or carry environmental poisons.

Substances that reduce the ozone layer

These are mainly the chlorinated fluorocarbons.

Table 2.3: Environmental poisons and ozone-reducing substances in building materials

	Substance	Effects
1.	Acrylonitrile	Carcinogenic; irritates mucous membranes; especially poisonous to water organisms
2.	Aliphatic hydrocarbons (collective name for many organic compounds, naphthenes and paraffins)	Irritates inhalation and oral route and skin; promotes carcinogenic substances
3.	Amines (collective group for different aromatic and aliphatic ammonium compounds)	Irritates inhalation routes; causes allergy; possibly a mutagen
4.	Ammonia	Corrosive; irritates mucous membrane; over-fertilizing effect; strong acidifies water
5.	Aromatic hydrocarbons (collective name for many organic compounds such as benzene, styrene, toluene and xylene)	Carcinogenic and mutagenic; irritate mucous membranes; damage the nervous system
6.	Arsenic and arsenic compounds	Bio-accumulative; can damage foetus; mutagenic; many are carcinogenic
7.	Benzene	Anaesthetizing; carcinogenic; irritates mucous membranes; mutagenic
8.	Bitumen (mixture of aromatic and aliphatic compounds, such as benzolalpyrene)	Contains carcinogenic compounds
9.	Boric salts (collective name for borax and boracic acid)	Slightly poisonous to humans; poisonous to plants and organisms in fresh water in heavy doses
10.	Cadmium	Bio-accumulative; carcinogenic; even in low concentrations can have chronic poisonous effects on many organisms such as liver, kidney and lung damage
11.	Calcium chloride	Irritant; strongly acidifying
12.	Chlorinated hydrocarbons (group of substances including dichloroethane, trichloroethane and chlorinated biphenyls (PCBs))	Carcinogenic; persistent; extremely poisonous to water organisms
13.	Chlorine	Acidifying; strongly irritates mucous membranes
14.	Chlorofluorocarbons (CFCs)	Break down the ozone layer
15.	Chrome and chrome compounds	Allergenic; bio-accumulative; carcinogenic; oxidizing; can cause liver and kidney damage
16.	Copper and copper compounds	Bio-accumulative; poisonous to water organisms

continued

Table 2.3: Environmental poisons and ozone-reducing substances in building materials – *continued*

17. 2-cyano-2-propanol	Extremely poisonous
18. 1,2-dichloroethane (ethylene dichloride)	Carcinogenic; persistent; extremely poisonous to water organisms
19. Dichloromethane (methylene chloride)	Carcinogenic; persistent; extremely poisonous to water organisms
20. Diethyltriamine	Acidifies heavy water; corrosive; strongly irritates mucous membranes
21. Dioxin (2,6dimethyl-dioxan-4yl-acetate)	One of the most toxic materials known: persistent bio-accumulative nerve poison; carcinogenic; extremely poisonous to water organisms
22. Dust	Irritates inhalation routes; forms part of photochemical oxidants
23. Epoxy	Very strong allergen
24. Esters (collective name of buthyl acetates and ethyl acetates)	Irritate mucous membranes; mutagen; medium strength nerve poison
25. Ethene, ethylene	Possibly carcinogenic because it becomes ethylene oxide in the body
26. Ethyl benzene	Strongly irritates mucous membranes; poisonous to water organisms
27. Fluorides	Changes in bone structure; damages forests and water organisms; generally poisonous in varying degrees of accumulation
28. Formaldehyde	Allergenic; carcinogenic; irritates inhalation routes; poisonous to water organisms
29. Fungus (collective name for many micro-organisms including aspergillus, cladosporium and penicillin)	Cause asthma and infections in inhalation routes
30. Hydrochrinon	Allergenic; irritates inhalation routes
31. Hydrogen chloride	Strongly acidifying; corrosive; irritates inhalation routes and mucous membranes
32. Hydrogen fluoride	Corrosive; can cause fluorose; extreme irritant of mucous membranes; extremely damaging to water organisms; poisonous
33. Isocyanates (collective group including TDI, MDI)	Very strong allergenics; irritates mucous membranes and skin

continued

Table 2.3: Environmental poisons and ozone-reducing substances in building materials – *continued*

34.	Ketones (group of substrates including methyl ketone and methyl isobutyl ketone)	Slightly damaging to reproductive organs; generally weak nerve poisons; poisonous to water organisms
35.	Lead and lead compounds	Bio-accumulative; can lead to brain and kidney damage
36.	Mercury and mercury compounds	Allergenic; bio-accumulative; can damage the nervous system and reproductive system; persistent
37.	Nickel and nickel compounds	Allergenic; bio-accumulative; carcinogenic; extremely poisonous to water organisms
38.	Nonyl phenol	Bio-accumulative; environmental oestrogen; persistent; poisonous to water organisms
39.	Organic acidic anhydrides (collective name for substances including PA, HHPA, HA, MA)	Acidifying; irritate the inhalation routes
40.	Organic tin compounds	Bio-accumulative; persistent; extremely poisonous to water organisms
41.	Pentane	Slightly damaging to water organisms
42.	Phenol	Carcinogenic; mutagenic; poisonous to water organisms, alkylphenols and bisphenol A are suspected environmental oestrogens
43.	Phosgene	Extremely poisonous: causes lung damage; breaks down to hydrogen chloride when added to water
44.	Phthalates (collective name for substances including DEHP, DOP, DBP, DEP, DMP, DiBP and BBR)	Environmental oestrogen; damaging to the reproductive system; generally persistent; moderately poisonous to water organisms; certain phthalates are allergenic and carcinogenic
45.	Polycyclical aromatic hydrocarbons (PHHs; group of substances which includes benzo(a)pyrene)	Bio-accumulative; carcinogenic; mutagenic; persistent; particularly damaging to water organisms
46.	Propene	Believed to change to 1,2 propylene oxide in the body, which is carcinogenic
47.	Quartz dust	Carcinogenic
48.	Radon gas (gas that contains radioactive isotopes of polonium, lead and bismuth)	Carcinogenic
49.	Styrene	Irritates inhalation routes – can make them very sensitive; damages reproductive organs
50.	Sulphur	Acidifying

Table 2.3: Environmental poisons and ozone-reducing substances in building materials – *continued*

51.	Synthetic mineral wool fibre (group of substances including glass wool and rock wool)	Slightly carcinogenic; irritates inhalation routes
52.	Thallium	Extremely poisonous
53.	Vinyl acetate	Possibly carcinogenic possibly neurotoxicant, possibly respiratory toxicant; poisonous to water organisms
54.	Vinyl chloride	Carcinogenic; irritates the inhalation routes; narcotic; persistent; poisonous to water organisms
55.	Wood dust	Dust from oak and beech can be carcinogenic; irritates inhalation routes

Greenhouse gases

Of gases that increase the greenhouse effect, the most common is carbon dioxide (CO_2), which is released from most industrial processes, primarily as a result of the burning of fossil fuels. A production equivalent is given as its 'global warming potential' (GWP) in units of carbon dioxide equivalents.

According to the United Nations' climate panel IPCC (Intergovernmental Panel on Climactic Change) there needs to be a 60–70 per cent reduction of the carbon dioxide created by man to stabilize the greenhouse effect.

The burning of all biological substances produces carbon dioxide, but no larger an amount than that by the material created through photosynthesis. Replacing burned wood by replanting trees avoids responsibility for carbon dioxide pollution. Trees and plants absorb carbon dioxide from the air and produce oxygen. A large oak absorbs 10 kg of carbon dioxide in a day. Some of this returns to the atmosphere at night, but over a period of 24 hours a total of 7 kg of carbon dioxide is removed.

Acid substances

Substances that lead to acidification of the natural environment reduce the survival rates of a series of organisms. This group of substances include mainly sulphur dioxide and nitric oxides formed through burning fossil fuels and other industrial processes. Release of hydrogen chloride leads to acidification. The acidifying potential of a product is referred to as its 'acid potential' (AP), in sulphur dioxide equivalents. Nitric oxides, for example, have an AP of 0.7 sulphur dioxide equivalents.

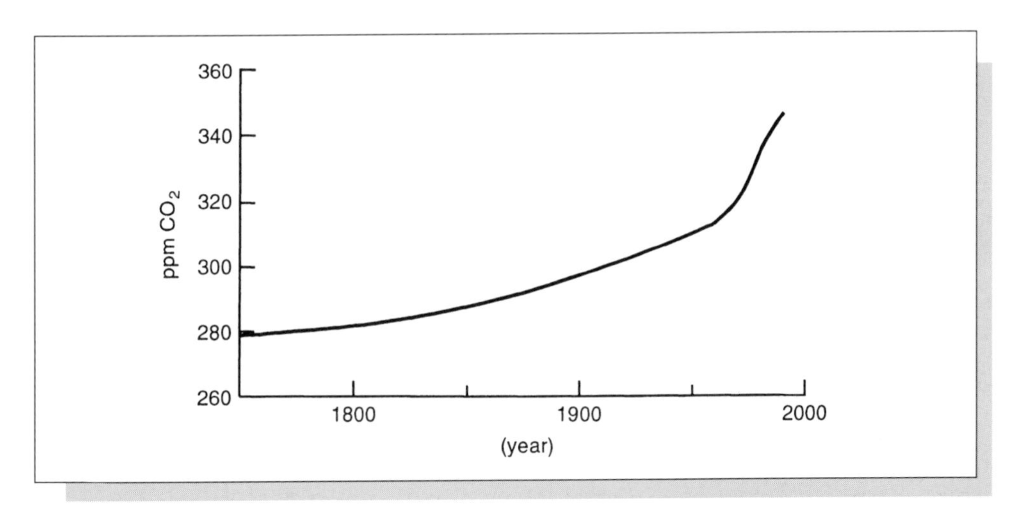

Figure 2.1: The concentration of carbon dioxide in the atmosphere from 1750 until 1988. Source: Mathisen 1990

Substances that form photochemical oxidizing agents, low ozone

Photochemical oxidizing agents are generally very corrosive and are described as smog. They are formed when a mixture of nitrogen oxides from fossil fuels, dust and a few volatile organic compounds like turpentine, are subjected to sunlight. The potential of a product to produce low ozone is referred to as its 'photochemical ozone creation potential' (POCP).

Eutrophicating substances

Over-fertilization and the resulting overgrowth of weeds caused by these substances in water systems is known as 'chemical oxygen depletion' (COD). In the building industry the most critical emission of nitrogen is in the form of nitric oxides from combustion processes. Artificial fertilizers used when producing plant substances can also cause problems. It is important to realize that the effects of eutrophicating substances are dependent upon their location and the type of earth in which they are placed.

Electromagnetic radiation

This includes radioactive radiation and radiation at lower frequencies, which can affect life-processes. Building materials contribute to radioactive pollution through the amount of nuclear powered energy used in their production. During the use of the building some materials can emit small amounts of radioactive radon gas, and materials that are good conductors can strengthen the low frequency magnetic fields in the building. Radioactive radiation can cause cancer. It is also assumed that low frequency radiation can cause sickness, reduction of potency and in some cases, cancer.

Physical encroachment of nature

This leads to a worsening of the quality of life in the area, and a loss of bio-diversity. A variety of species is necessary to maintain the ecosystem. At the moment, we know very little about the interdependence of these factors. The 'hindsight' principle is often used, only to find that what seemed to be a small encroachment has had disastrous effects. Most assaults on nature are in conjunction with efforts to obtain raw materials.

Genetic pollution

Genetically manipulated plant species are now being used in agriculture and forestry to increase production and improve resistance to cold, mould and insects. The goals are often environmentally legitimate, e.g. in order to reduce the use of pesticides. But this must still be regarded as hazardous. We know that generally every change that occurs in a natural species which gives them a defensive advantage also affects that species' environment.

Reduction of pollution in the production stage

Reduction in the use of fossil fuels during extraction and production processing of building materials. This also means a reduction in transport. The possibilities of using renewable energy sources such as solar-wind, hydro-power and biomass should be investigated, and priority given to manufacturing processes and materials which put these principles into practice. As far as the heating of furnaces and operations involving pressure are concerned, in combination with mechanical processes it should be possible to work without electricity.

Careful use of natural resources

An increased use of materials that involve less environmentally-damaging methods of extraction and production would entail an increased use of renewable resources and recycled materials.

More efficient purification of industrial waste

There are plenty of possibilities in this area. It is even possible, in some cases, to reprocess waste for the manufacture of new products. Sulphur can be removed from oil by treatment with hydrogen. In the actual combustion process, the main compounds can be precipitated by adding lime:

$$CaO + SO_2 = CaSO_4 \quad (1)$$

The amount of nitric oxide emitted can be reduced a great deal by reducing the combustion temperature to 1000°C. The emission will also be less if the amount of oxygen within the process is reduced. Nitrogen oxides can also be removed from the emissions by adding ammonia (NH_3); the resulting products are nitrogen and water:

$$NH_3 + NO_x = N_2 + H_2O \quad (2)$$

Combustion over 1000°C greatly reduces the amount of polycyclical aromatic hydrocarbons (PAHs), but at the same time increases the amounts of nitrogen oxides. The PAH substances are otherwise difficult to remove. It is possible to clean out the heavy metals from the smoke by using a highly efficient filter.

Reduction of pollution during building use

This use of local materials would reduce transport-related pollution.

Table 2.4: Energy sources and pollution

Energy source	*CO_2*	*CO*	*NO_x*	*SO_x*	*Heavy metals*	*Dust*	*PAH*	*Radio-activity*
Solar power								
Wind power								
Hydro-power								
Wave power								
Wood burning (dry and efficient)	(x)[1]	x				x	(x)[2]	
Peat burning	(x)[1]	x				x	(x)[2]	
Coal burning	x	x	x	x	x	x	x	
Natural gas burning	x	x	x				x	
Oil burning	x	x	x	x	x	x	x	
Nuclear power								x

Notes:
(x)[1]: see p. 32
(x)[2]: small amount by effective combustion

Table 2.5: Effects of pollution

Material	*Environmental poisons and ozone reducing substance (numbers refer to substances in Table 2.3)*			
	Health		*External environment*	
	Working environment	*Interior environment*	*Exclusive of waste*	*From demolition waste*
Aluminium, 50% recycled	45–22	–	45–22–27	–
Cast iron, from iron ore	22	–	22	–
Steel: 100% recycled	22	–	22–10–6–27	–
galvanized from ore	22	–	2–5–15–10–27	–
stainless steel from ore	22	–	2–5–15–10–27–37	–
Lead, from ore	35	–	35	35
Copper, from ore	22	–	16	–
Concrete with Portland cement:				
structure	22–15	–	22–15–52	–
roof tiles	22–15	–	22–15–52	–
fibre reinforced slabs	22–15	–	22–15–52	–
mortar	22–15	–	22–15–52	–
Aerated concrete, blocks and prefab. units	22–15–45	–	22–15–52–45	–
Light aggregate concrete, blocks and prefab. units	22–15	–	22–15–52	
Lime sandstone	47	–	–	–
Lime mortar	22	–	–	–
Calcium silicate sheeting	47	–	–	–
Plasterboard	(20)	–	–	50
Perlite, exapanded:				
without bitumen	22	–	–	–
with bitumen	8–22–2–5	(8)	8	8
with silicone	22–(19)	–	(19)	–
Glass:	47	–	47–11	
with a tinoxide layer	47	–	47–11–31–32–40	
Foam glass:				
slabs	47	–	47–11	
granulated, 100% recycled	–	–	–	
Mineral wool:				
rockwool	42–28–51–7	(2–51–29)	42–28–51–7–4	42
glasswool	42–28–51–7	(2–51–29)	42–28–51–7–4	42
Stone:				
structural	22–(47)	48	–	–
slate	22–(47)	–	–	–
Earth, rammed structure	22	–	–	–
Bentonite clay	–	–	–	–
Fired clay:				
bricks	22	–	27–50	–
roof tiles	22	–	27–50	–

Dominating air pollution						*Waste from the production process*		*Building and demolition waste*
Scandinavian peninsular			*European continent*					
GWP(1) *(g/kg)*	*AP*(2) *(g/kg)*	*COD*(3) *POCP*(4) *(g/kg)*	*GWP*(1) *(g/kg)*	*AP*(2) *(g/kg)*	*COD*(3) *POCP*(4) *(g/kg)*	*g/kg product*	*Percentage taken to special dumps*	*Waste category*(5)
1900	13	3	11 102	60	119	715	20	D
			771	6	5			D
250	2	1	557	3	4			D
1000	4	1	2230	10	840	601	5	D
1000	4	1	2230	10				D
			1137	10	63	265	5	E
1200	5	6	5234	140	64	2410	84	D
120	0.5	0.4	65	1	0.3	32	–	C
			131	1	1			C
			434	2	3	81	10	C
180	0.5	0.6	98	0.8	11	17	10	C
			280	2	30	49	12	C
230	1	0.4	307	2	38	58	13	C
			68	0.6	0.4	2	–	C
						17		C
			130	1	1			C
330	5	5	265	3	2	8	10	D
			871	2	1			C
								E
								D
600	4	4	569	44	2		C	
								D
								C
								C
770	3	2	1076	6	5	320	5	D
880	8	9	1210	7	6	90	5	D
			8	0	0			C
			8	0	0			C
			8	0	0			C
							–	C
160	2	3	190	2	17	87	15	C
			190	2	17	95	10	C

Table 2.5: Effects of pollution – *continued*

	Environmental poisons and ozone reducing substance (numbers refer to substances in Table 2.3)			
	Health		*External environment*	
Material	*Working environment*	*Interior environment*	*Exclusive of waste*	*From demolition waste*
Ceramic tiles	22	–	27–50	(10–37)
Fired clay pellets	22	–	–	–
Bitumen	2–5–8	2–5–8	2–5–8	2–5–8
Polyethylene (PE)	25–2–5	–	2–5	
Polypropylene (PP)	46–2–5	–	2–5	
Expanded polystyrene: EPS	7–25–49–2–5	(49)	7–26–49–41–2–5	(49)
XPS	7–25–49–2–5	(49)	14–7–26–49–2–5	(49)
Expanded polyurethane (PUR)	33–2–5	(33)	14–33–2–5	(33)
Polyvinyl chloride (PVC)	18–13–54–44–2–5	44–34–(54)	18–44–54–13–21–3	31–(54)
Expanded ureaformaldehyde (UF)	28–3	28–3	28	
Polyisobutylene (PIB)	44	44	44	44
Polyester (UP)	49–12	49	12	
Styrene butadiene rubber (SBR)	49–5	49	5	
Timber:				
untreated	22	–	–	–
pressure impregnated	22–6–27–15	(6–27)	6–15–27	6–15
laminated timber	22–42–2–5	–	42–2–5	–
Wood fibre insulation	22	–	–	–
Cork	22	–	–	–
Wood fibre board:				
porous without bitumen	22	–	–	–
porous with bitumen	22–8–2–5	(8)	8–2–5	8
hard without bitumen	22	–	–	–
hard with bitumen	22–8–2–5	–	8–2–5	8
Woodwool slabs	22–15	–	22–15–52	–
Chipboard	22–42–28–2–5	28	42–28–2–5	(42)
Cellulose fibre insulations, 100% recycled and boric salts	22–9	–	22–9	9
Cellulose fibre matting (fresh), and boric salts	22–9	–	22	9
Cellulose building paper (unbleached); 98% recycled	22	–	22	–
Cardboard sheeting				
laminated with polyethylene	22–25–2–5	–	25–2–5	–
laminated with latex	22–5	–	24–5	–
Linenfibre, strips	22	–	–	–
Linen matting	22	–	–	–
Linoleum	22–24–5	–	24–5	–

Dominating air pollution						*Waste from the production process*		*Building and demolition waste*
Scandinavian peninsular			*European continent*					
GWP[1] *(g/kg)*	*AP*[2] *(g/kg)*	*COD*[3] *POCP*[4] *(g/kg)*	*GWP*[1] *(g/kg)*	*AP*[2] *(g/kg)*	*COD*[3] *POCP*[4] *(g/kg)*	*g/kg per product*	*Percentage taken to special dumps*	*Waste category*[5]
			571	4	51	9	–	C
120	0.2	0						C
			489	4		3	–	B/D
			751	9	0.1			B/D
			900	7	0.1			B/D
2000	14		1650	11	0.2			B/D
2200	15							B/D
4800	38	14	3900	30	42	486	7	B/D
700	13		1400	13	0.5			D
								D
								B/D
								B/D
40	0.6	0.8	116	1	1	25	–	A/D
								E
60								B/D
							–	A/D
			277	–	1			A/D
						81	5	A/D
120	2	1						B/E
			766	3	8	80		A/D
								B/E
			980	4	11	79	5	D
20	0.3	1	69	1	102	40	2	B/D
140	2	2	160	3	3			E
								E
							–	A/D
								B/D
								B/D
								A/D
							–	A/D
1000	4	4				2	–	B/D

Table 2.5: Effects of pollution – *continued*

Material	*Environmental poisons and ozone reducing substance (numbers refer to substances in Table 2.3)*			
	Health		*External environment*	
	Working environment	*Interior environment*	*Exclusive of waste*	*From demolition waste*
Straw: thatch	22	–	–	–
bound with clay	22	–	–	–
Coconut fibre, strips	22	–	–	–
Jute fibre, strips	22	–	–	–
Peat slabs	22	–	–	–
Wool paper	–	–	–	–
Woollen matting	–	–	–	–

Notes:
The first four columns only give the potential problems that can arise from these materials, so it is not possible to use them as a basis for any quantitative comparison. Figures in brackets show pollution that is rare or only occurs in small doses – means that there are no known pollution problems. Open space means that there is no available information

(1) GWP = Global Warming Potential in grams CO_2 equivalents.
(2) AP = Acid Potential in grams SO_2 equivalents.
(3) COD = Chemical Oxygen Depletion in grams NO_x.
(4) POCP = Photochemical Ozone Creation Potential in grams NO_x
(5) Waste categories:
A: Burning without purification
B: Burning with purification
C: Landfill
D: Ordinary local authority tip
E: Special tip
F: Strictly controlled tip
(Sources: Fossdal, 1995; Hansen, 1996; Kohler, 1993; Suter, 1993; Weibel, 1995)

Reduced use of materials which emit harmful gases, dust or radiation

Gases, dust and radiation can emanate from the building or from waste. Alternative materials are now available.

Increased use of timber and other 'living' resources in long-term products

Products made from plants function as absorbers of carbon, and therefore reduce emission of the greenhouse gas carbon dioxide.

Increased recycling

Through recycling, energy-use and the use of resources can be reduced, which also reduces pollution.

Dominating air pollution						*Waste from the production process*		*Building and demolition waste*
Scandinavian peninsular			*European continent*					
GWP[1] *(g/kg)*	*AP*[2] *(g/kg)*	*COD*[3] *POCP*[4] *(g/kg)*	*GWP*[1] *(g/kg)*	*AP*[2] *(g/kg)*	*COD*[3] *POCP*[4] *(g/kg)*	*g/kg product*	*Percentage taken to special dumps*	*Waste category*[5]
							–	A/D
							–	A/D
								A/D
								A/D
							–	A/D
							–	A/D
							–	A/D

References

FOSSDAL S, *Energi og miljøregnskap for bygg*, NBI, Oslo 1995

GILLBERG B O *et al*, *Mord med statlig tilstånd. Hur miljöpolitiken förkortar våra liv*, Uppsala 1988

HÄGG G, *Allmän och oorganisk kemi*, Stockholm 1984

HANSEN K *et al*, *Miljøriktig prosjektering*, Miljøstyrelsen, Copenhagen 1996

KOHLER N *et al*, *Energi- und Stoffflussbilanzen von Gebäuden während ihrer Lebensdauer*, EPFL-LESO/ifib Universität Karlsruhe, Bern 1994

MATHISEN G, *Varm framtid*, Universitetsforlaget Oslo 1990

PLUM NM, *Økologisk handbog*, Christian Ejlers Forlag, København 1977

SUTER P *et al*, *Ökoinventare für Energisysteme*, ETH, Zürich 1993

WEIBEL T *et al*, *Ökoinventare und Wirkungsbilanzen von Baumaterialen*, ETH, Zürich 1995

3 Local production and the human ecological aspect

There are basically three different ways of manufacturing a product:

- It can be manufactured by the user, based on personal needs or on the local cultural heritage.
- It can be manufactured by a craftsman who has developed a method of manufacture through experience.
- It can be manufactured by an engineer who directly or indirectly, through electronics, tells the worker which steps to take.

The first two methods share a common factor – the spirit of the product and the hand that produces it belong to the same person.

Up to the earliest Egyptian dynasties around the year 3000 BC, it is assumed that the dominant form of manufacture was 'home production'. Everybody knew how a good hunting weapon should be made, or how to make a roof watertight. Certain people were more adept and inventive than others, but they shared their experience. Knowledge was transferred to the next generation and, through time, became part of the cultural heritage. Home production has been the dominant form of manufacture until relatively recently, especially in village communities. On isolated farms, clothes, buildings and food have been home produced late into this century.

Today there are not many forms of serious craft production left. Herb gardens have had a small renaissance and small handicraft companies still survive. The fact is that the division of production into different units is the most common model in all the major manufacturing industries today.

Craftsmen have existed for at least 5000 years. During the Middle Ages the guilds were formed; apprentices learned from their masters and further developed their own knowledge and experience. In this way, they became masters in their own trade. The potter, lacking advanced measuring instruments relied on his own judgement to know when the pottery had reached the right temperature in the kiln. This judgement consisted of his experience of the colour, smell and consistency of the material. And as long as he manufactured products that satisfied his customers, he could decide how the product was manufactured. The method of production was not split up into different parts – the craftsman followed the product through the whole process.

The working situation of a quarry worker was such that all his senses took part in his work. The quality of the stone was decided by how 'it stuck to the tongue', the resonance of it when struck, the creaking when pressure was applied, the smell when it was scraped or breathed on, or the colour of the stone and the lustre given by scraping it with a knife or nail.

This form of manufacture, where manual labour was the main resource, stretched a long way into the industrial revolution. In the American steel industry of the nineteenth century the workers themselves controlled the production. They led the work and were responsible for engaging new workers. This principle became a contractual agreement between workers and their employers in 1889, giving them control of all the different parts of production. The factory-owner Cyrus McCormick II soon became tired of this system. He came up with the idea that if he invested in machinery he would be able 'to weed out the bad elements among the men' (Winner, 1986), i.e. the active union members. He took on a large number of engineers and invested in machinery, which he manned with non-union men. As a result, production went down and the machines became obsolete after three years. But by this time McCormick had achieved what he set out to achieve – the destruction of the unions. Together with the engineers he took full control of production.

McCormick introduced the third form of manufacture, today the established mode of production, controlled by the engineer. From the beginning the engineer situated himself on the side of the capitalist. In this way the worker lost control of the manufacturing process. His experience and sensitivity were replaced by electronic instruments and automation.

The traditional use of timber as a joint material disappeared during this period, partly because of the standardization regulations that came into power. They were replaced by steel jointing materials, bolts and nails. Steel components of a certain dimension always have the same properties. The properties of timber joints are complex and often verified through experiment and experience rather than calculation. After the restructuring of the steel industry took place, many heavy industries in the newly-industrialized world followed suit.

The car industry transferred to engineer-run production after just two years. The paint and paper industries soon followed. In certain other areas, expert-controlled production came later. The largest bakeries were already under expert control during the 1930s, whereas the timber and brick industries were not controlled by engineers until after the Second World War.

Does it matter which method of manufacture a product undergoes? Adam Smith, one of capitalism's first ideologists, states in his book from the beginning of the industrial revolution, *The Wealth of Nations* (1876):

> 'A man is moulded by the work he does. If one gives him mundane work to do, he becomes a mundane person. But to be reduced to a totally mundane worker is the destiny of the great majority in all progressive societies.'

It is understood that work is here to fulfil our needs – after all, most of our life is filled with different types of work. Most will agree that work is not just a means to an end, but an important means, a process of research, a process of discovery where one learns more about the material one is working with, about oneself and about the world. In many situations today, professionalism has transformed work from self-development to mere 'doing'.

The production process, product quality and the quality of work

The relationship between producer and consumer in worker-controlled production is called a 'primary relationship'. Engineer-controlled production is a 'secondary relationship'. In the latter case, contact between customer and producer never occurs; at the most the customer is aware of the country in which the last process of production took place. The name of the company gives very few clues. However, in the primary relationship the client and the manufacturer often have a very close relationship with each other.

Aspects of the primary relationship

The primary relationship has positive effects for the consumer, the manufacturer and the worker.

For the consumer

Better product

It is quite normal today to have built-in weaknesses in most products manufactured by engineer-controlled methods. There are examples in the USA where frustrated production-line workers have taken secret revenge by comprimisng the quality of cars and other products that have rolled past them.

It is doubtful that a skilled worker in a primary relationship would replace a light bulb which only has limited life left, partly because of professional pride and partly for fear of being reprimanded openly. In this way there is a guarantee in the primary relationship.

Responsible use of resources and less pollution
It is doubtful that a small industry manufacturing products for the local community would bury barrels of poisonous waste in the area. A small industry, based on local resources, would most likely have a much longer perspective in planning the use of resources than a larger also firm with a much broader base.

For the manufacturer and the consumer

Less bureaucracy
In most cases there is a feeling of solidarity in the primary relationship. In the secondary relationship solidarity is replaced by laws, rules, production standards etc., and expensive and inefficient bureaucracy.

Flexibility
Possibilities for spontaneity in the production process, e.g. to change a door handle or re-style a suit, are much greater in the primary relationship. This has to do with the use of imagination, which we can assume is appreciated by both the manufacturer and the consumer.

For the worker

Safer places of work
Worker-controlled industries limit their own size and will remain local. People living in such an area realize that by buying local products they are supporting the local industries, and that everyone is dependent upon everyone else. People are also aware of any unemployment. This also creates solidarity.

Meaningful work
There is a big difference in the scope and challenge of the work of a carpenter who builds a complete house and the carpenter who fits the windows into a prefabricated house. The latter misses two important aspects of his identity as a builder: a relationship to the completed house and to the client. Instead of this, he forms a relationship to many houses and many clients which is abstract and not so meaningful. Close contact between worker and client increases the possibilities for a more personal touch in the product.

E. F. Schumacher sums it up like this: 'What one does for oneself and for friends will always be more important than what one does for strangers' (McRobie, 1981).

With the continuous division of industry into separate skills, there has also been a geographical division of work in the direction of forming small communities around these specialized industries. There are now communities whose inhabitants work only for an aluminium factory, for example. Opportunities for different experiences become less and less and the communities become less exciting to live in. Just as with the division of work, the geographical division of

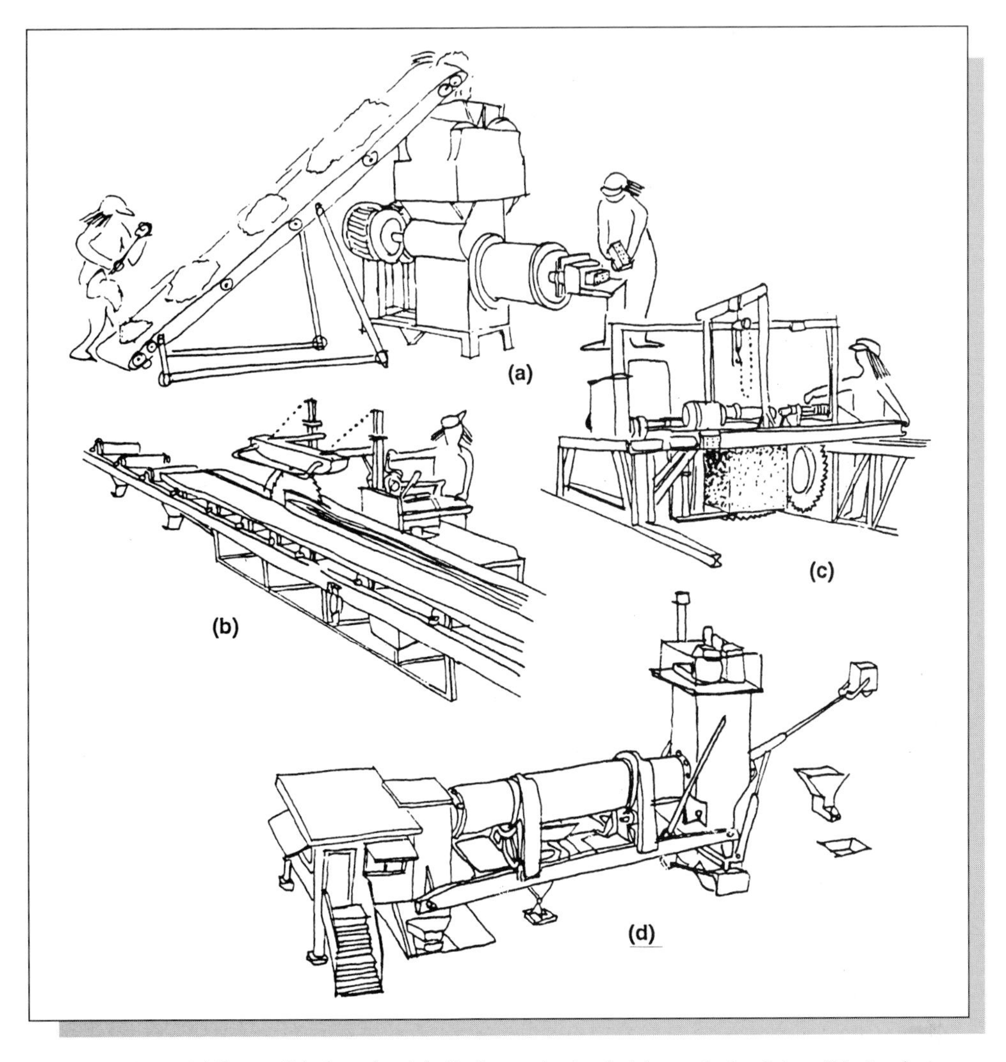

Figure 3.1: Mobile small industries: (a) die for producing bricks ready for firing; (b) circular saw for timber; (c) circular saw for sandstone and limestone and (d) a rotating kiln for the production of calcinated lime, cement and light expanded clay.

specific skills or industrial processes has a power aspect. A community of specialized workers can easily become the victims of internal negotiations which take place totally outside their own sphere of activity.

Technology

Schumacher rejects any mechanization that takes away the joy of creating from people. He demands that work fulfils at least three different functions:

- To give every person the possibility to use and develop their skills
- To make it possible for people to overcome their egoism by doing things together
- To produce articles that are necessary for everyday life.

Ivan Illich focuses even more on the role that power plays: 'We must develop and use tools that guarantee man's right to work efficiently without being controlled by others, and thus eliminate the need for slaves and masters' (Illich, 1978). Through choosing a technology one is also deciding a quality of life for those who are going to serve that technology. Today's society is ruled by a high degree of technological determinism. It is taken for granted that technological development has its own momentum, which cannot be hindered in any way. The technological philosopher Langdon Winner maintains that 'much could have been left undone'. His colleague Jonas follows with the statement: 'One shall only do a part of everything one is actually capable of doing', and thereby introduces a new categorical imperative (Apel, 1988).

There is a mechanism in traditional development theory which is called 'phaedomorphosis'. This means that development can take one step back to an earlier and less specialized phase, in order to take a new line of development later. Progress is not always achieved by taking a step forward.

The following questions then arise: Why did development carry on as it did? Were there actually any alternatives? Why weren't these chosen? And in what way can we now re-evaluate the choices that have been made?

There is a tendency to regard technology as neutral and to believe that the political aspect becomes important when technology first comes into use. The use of a knife can illustrate this view: it can be used to cut bread or to kill someone. But, for example, when a robot becomes part of a work force, it does not only increase productivity but also defines the whole concept of work at that production site. It has been discovered that, within the building industry, apparently small changes in the use of materials can have far-reaching consequences. Until about 1930 all mortar used was a lime mortar, and bricks could only be laid a metre at a time as the mortar needed to carbonize. The bricklayers had to take a break and use that time to design or do detail work. With the introduction of Portland cement this drastically changed the whole situation. Within a few years architects completely took over detail design, which had been the bricklayer's task for centuries.

When describing the development of technology and new products one seldom questions the quality of work for the individual. Usually, discussions centre around the profitability, the economic efficiency or the ergonomic relationships. The only limiting factor of any consequence in technological development is *the risk factor*. That risk we are often willing to take. Seven million people have been killed in car accidents and a hundred million have become invalids since the introduction of the car, and everyone feels that the car is well worth it. What is so striking about the history of modern technology is that the new and innovative ideas become part of everyday life so quickly, totally accepted by everybody. At the same time, there will often be another available technology which can reduce risk, or at least the risk of a catastrophe. These technologies are becoming more and more important. At the moment a whole new industry is growing – environmental technology. It seems that these technologies can only offer solutions to problems they have created themselves. This pattern of production is moving in the direction of pure technophilia.

As early as the 1960s, Lewis Mumford stated:

> 'From late Neolithic times in the Near East, right down to our own day, two technologies have recurrently existed side by side: one authoritarian, the other democratic, the first system-centred, immensely powerful, but inherently unstable, the other man-centred, relatively weak, but resourceful and durable.' (Mumford, 1964)

Economy and efficiency

Principles for an ecological building industry include the following:

- The technological realm is moved closer to the worker and user, and manufacturing takes place in smaller units near to the area where the products will be used.

  Paul Goodman gives the following definition: 'Decentralizing is increasing the number of centres of decision-making and the number of initiators of policy, increasing the awareness of the whole function in which they are involved, and establishing as much face-to-face association with decision-makers as possible.' (Goodman, 1968)

- The use of raw materials is based on renewable resources or rich reserves, products are easily recycled and are economic in terms of materials during construction.

- Priority is given to production methods that use less energy and more sustainable materials, and transport distances are reduced to a minimum.

- Polluting industrial processes and materials are avoided, and energy based on fossil fuels reduced to a minimum.

This can be summed up by saying that the optimal ecological building industry is a cottage industry, which responds to local needs and resources. We will be moving into deep water when comparing this with the European reality. It should be made clear here that we are discussing precepts, not an attainable situation. Different regions have, amongst other things, varying amounts of natural resources. Certain places have plenty of fish while others have an abundance of iron ore. An exchange of goods is self-evident, and is to everyone's advantage. However, during the last 100 years, right up to the present moment, development has followed a path of extreme centralization.

It is the same situation in the whole of the building industry. Many say that this centralization has been necessary. 'Large is efficient' is the refrain that resounds in our ears. But this is not the case if we bring in ecology as a condition.

Efficiency is the increase in production related to the cost of production: wages, devaluation of machinery and costs related to energy and raw materials. The tendency in this century has been a strong increase in the proportion of wages, while the cost of raw materials and energy has been left behind. The gap between these two curves has increased so much that from 1960 to 1970 wages increased fourfold compared to the sum of all other production costs. This development has been compensated for by increased mechanization. Only the larger organizations could cope with the immense investment needed; smaller ones fell by the wayside one by one. Through this expansionist industrial growth, industry became immensely vulnerable to the smallest changes in market forces, with minimal flexibility because of over-specialized production technology.

Then the energy crisis arrived at the beginning of the 1970s, and suddenly the cost of energy became a much more important parameter. Apart from the fact that energy-intensive industries experienced problems, the greatest factor was the increase in transport costs. Today energy prices have stabilized at a lower level.

Godfrey Boyle, a researcher at the Open University, has confirmed that an industry can just as easily be too large as too small, and has concluded that for many industries the most efficient level of production lies in the region of having 10 000 users (Boyle, 1978). In Sweden they have discovered that the optimal size of a farm with cattle and pigs is the family-based farm. Shipping companies are changing from very large to medium-sized ships. Bakeries are closing down large bread factories in favour of local bakeries.

At the same time, though it cannot be denied that we do not really know the true relationship between size and efficiency, at least it can be said that it has very much to do with the actual product. For example, there is no limit to how large an egg farm can be in order to optimize its efficiency. The Norwegian social scientist Johan Galtung has an interesting view on the problem:

> 'High productivity does not necessarily mean something positive. We can already see that efficiency is too high; newly completed articles have to be burned, weaknesses are built in so that the product does not last too long. There is an increasingly wound-up cycle of fashion-oriented products, which age quickly and then have to be replaced by the next fashion, leading to the time when articles are obsolete the moment they are released on the market!' (Galtung, 1980).

Galtung's solution: 'Reduce productivity. The market cannot absorb all the products it manufactures.'

In many EU countries, on the economic front, it is necessary to take into account the fact that a great deal of industry is heavily subsidized by the taxpayer. In addition, there are also subsidies for energy and road building. Even the polluting industries are subsidized, where account must be taken of the extra costs of inspection and control of pollution and any health implications. The most important factor is the cost to nature, which is difficult to calculate financially, but is, nevertheless, a debt which coming generations will have to pay. Besides measurable pollution, other factors must be included, such as the lost wood fuel from a well-balanced forest which has to be sacrificed to reach the iron ore in a mountainous area. Such a calculation is very complex and one that is preferably avoided.

The price tag in the shop is therefore anything but realistic. The price difference between a solid board of timber and a cheap sheet of chipboard coated with a plastic laminate has probably already been paid for by the customer before he enters the shop.

Benjamin Franklin claimed that activity and money are virtues. Industrial economy is a flowing economy. Society devours virgin materials, consumes them in the production process, often with a very low level of recycling, and leaves the waste to nature. The industrial culture of flowing economy is the complete opposite of nature's diligence based on restricted resources. Nature's method is that of integrated optimization, ecological systems tend towards an optimal solution for the natural environment as a whole. Efficiency is based on the greatest variety of species where each has its own special place. There is a continuous interplay between all the different species.

When the Dutch mission came to Labrador in 1771, the Eskimos lived in large family groups in houses of stone and peat. The rooms were small and warmed by lamps fuelled by blubber. One of the first things that the new settlers did was to introduce a new form of house. They built a series of timber houses with large rooms heated by wood-fired iron stoves. This had a radical effect on the whole of the Eskimo society. They had earlier obtained fuel oil from seals by hunting. The meat provided food and the hides could be used for clothes and boats.

The change of house made fetching wood a very important task for them. The forest was a long way away and the sleigh dogs needed to eat more meat to manage the transport, so seal hunting had to increase as well as wood gathering. The need for wood became so great during winter that it took longer than all the other tasks put together. Despite all their efforts, it became clear that the new timber houses could not give the same warmth and comfort as the original earth houses (Arne Martin Claussen).

The goal of this book is to show alternatives to the herrnhutic way of thinking, which diminishes a greater proportion of today's building industry, whether we believe it or not.

References

BOYLE G, Community Technology: Scale versus Efficiency, *Undercurrents* No. 35

GALTUNG J *et al*, *Norge i 1980–årene*, Oslo 1980

GOODMAN P, *People or Personnel: decentralising and the mixed systems the moral ambiguity of America is like a Conquered Province;* Vintage, New York 1967

ILLICH I, *The right to useful unemployment and its professional enemies*, Marion Boyars, London 1978

MCROBIE G, *Small is Possible*, London 1981

MUMFORD L, *Authoritarian and Democratic Technics*, Technology and Culture No. 5/1964

WINNER L, *The whale and the reactor. A search for limits in the age of high technology*, University of Chicago Press, Chicago and London 1986

4 The chemical and physical properties of building materials

Materials are produced in different dimensions and forms. A block is usually defined as a building stone that can be lifted with two hands, while a brick can be lifted with one. Two people are needed to carry a sheet. During recent years a new category has come into play: the building element, which can only be moved and positioned by machines. Each group of materials creates its own particular form of working.

Properties of materials are divided into chemical and physical. Chemistry gives a picture of a substance's elemental contents, while physics gives a picture of its form and structure. As far as chemistry is concerned, it does not matter whether limestone, for example, is powder or a whole stone – in both cases the material's chemical composition is calcium carbonate. In the same way, physical properties such as insulation value, strength etc. are regarded independently of chemical composition.

In traditional building it is usually the physical properties that are considered, and it is almost entirely these properties that decide what the material can and should be used for – its potential. Exceptions where the chemical properties are also an important factor happen in cases where the material will be exposed to different chemicals. Determining the resistance of a material to exposure to moisture, oxygen or gases will include chemical analysis. This is much more necessary nowadays with increased air pollution, which contains various highly reactive aggressive pollutants.

An ecological evaluation of the production of certain building materials requires a knowledge of which substances have been part of the manufacturing process, and how these react with each other. This gives a picture of the possible pollutants within the material, and what the ecological risks are when the material is dumped in the natural environment. Increased attention to the quality of indoor climates creates a greater need for chemical analyses. In many cases problems are caused by emissions from materials in the building. How these react with the mucous membranes is also a question of chemistry. It has been shown that certain materials react with each other, and can thus affect each other's durability, pollution potential, etc.

A small introduction to the chemistry of building materials

There are a total of 89 different chemical elements in nature. Each element is represented by a single letter or two letters, e.g. H for hydrogen or Au for gold. Chemistry is mainly concerned with the way these elements combine to form compounds.

Materials usually consist of several compounds, and when a product consists of several materials the picture can become rather complex. A telephone can contain as many as 42 of the 89 elements (Altenpohl, 1980). Materials exist as solids, liquids or gases. The same chemical compound can exist in any of these three states, depending on temperature and pressure. Water (H_2O) freezes at 0°C and boils, or evaporates, at 100°C without changing its chemical composition.

The smallest unit a material can break down into is a molecule. Every molecule consists of a certain number of atoms. These atoms represent the different elements and can be obtained through chemical reactions.

Relative atomic weight

Each of the 89 elements has its own characteristic atomic structure, mainly described by its weight: the relative atomic weight. Hydrogen has the lowest relative atomic weight, 1, while oxygen has a relative atomic weight of 16.

The molecular weight of water is found through adding up the different atomic weights:

$$H_2O = H + H + O = 1 + 1 + 16 = 18 \quad (1)$$

Calcium carbonate ($CaCO_3$) consists of calcium (Ca), with a relative atomic weight of 40, carbon (C) with a relative atomic weight of 12, and oxygen (O) with 16. The relative molecular weight is therefore:

$$CaCO_3 = Ca + C + O + O + O = 40 + 12 + 16 + 16 + 16 = 100 \quad (2)$$

The relative atomic weights of the different elements are given to two decimal places in the periodic table (see Figure 4.1). The elements are also given a ranking in the table of 1–89. The number of the elements in the ranking order is equivalent to the number of protons in the nucleus of the atom.

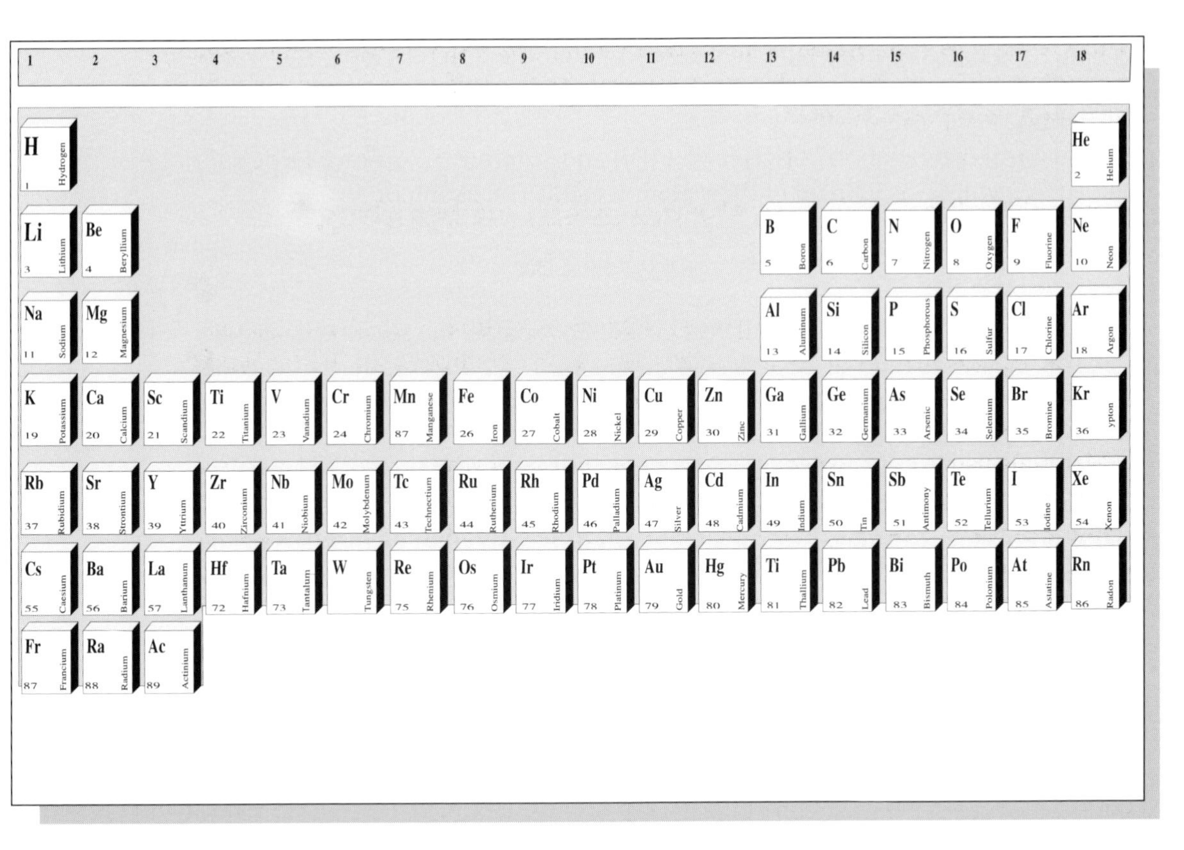

Figure 4.1: The periodic table.

Radioactivity

In the largest atoms there is often a large inner tension. They want to be radioactive and thereby emit radiation into their surroundings. There are three different forms of radiation: alpha, beta and gamma radiation. Gamma radiation is pure electromagnetic radiation and is part of the nucleus of the atom. It can penetrate most materials in just the same way as X-rays. Alpha and beta radiation come from particles and are caused by the atom breaking down, reducing the size of the nucleus. Radium (Ra) with the atomic number 88, will go through a great number of changes and finally become lead (Pb) with the atomic number 82. This process takes thousands of years.

Weights of the different substances in a chemical reaction

For a chemical reaction to take place, substances must have the necessary affinity with each other, and be mixed in specified proportions. Only certain substances react together in certain circumstances, and the different molecular combinations that result always have the same proportion of elements as the original substances.

A chemical combination between iron (Fe) and sulphur (S) making ferric sulphide (FeS) will be as a result of their atom weights consisting of:

$$56 \text{ g Fe} + 32 \text{ g S} = 88 \text{ g FeS} \tag{3}$$

If we begin with 60 g Fe, there will be 4 g Fe left over after the reaction has taken place. In the production of polymers, the remaining products from the reaction are called residual monomers. These by-products usually follow the plastics in the process as a sort of parasite, even though they are not chemically bound to them. This physical combination is very unreliable and can lead to problematic emissions in the indoor climate.

It is possible to calculate how much of each of the different elements is needed to produce a particular substance. In the same way we can, for example, calculate how much carbon dioxide (CO_2) is released when limestone is heated up:

$$CaCO_3 \rightarrow CaO + CO_2 \tag{4}$$

$CaCO_2$ has the following weight, through adding the relative atomic weights:

$$40 + 12 + 16 + 16 + 16 = 100 \text{ g} \tag{5}$$

$$CaO \text{ is } 40 + 16 = 56 \text{ g}$$

$$CO_2 \text{ is } 12 + 16 + 16 = 44 \text{ g}$$

This means that 44 g of CO_2 are given off when 100 g of limestone is burned.

Supply of energy and release of energy in chemical reactions

The conditions governing how a chemical reaction takes place are decided by the physical state of the substances. There are three different states: the *solid* state which is characterized by solid form, defined size and strong molecular cohesion; the gaseous state which has no form and very weak molecular cohesion; and the *liquid* state, which is somewhere between the two other states.

When heated most substances go from the solid state, through the liquid state and to the *gaseous* state. In a few cases there is no transitional liquid state, and the substance goes direct from the solid to the gaseous state. As the molecular cohesion is weakened in the higher states, we can assume that the majority of chemical reactions need a supply of heat. The amount of energy supplied is dependent upon the temperature needed to make the substances transform into the higher state, i.e. the substance's boiling point:

Classification of volatility for organic substances

Type	*Boiling point*
VOC: Volatile organic compounds	Above 250°C
SVOC: Semi-volatile organic compounds	250–380°C
POM: Particle-bound organic compounds	Below 380°C

There are also chemical reactions which emit energy. When water is mixed with unslaked lime (CaO) slaked lime ($Ca(OH)_2$) is formed by the release of a great deal of heat. If slaked lime is then burned, unslaked lime will form and water will be given off in the form of steam. The energy supply in this reaction is exactly the same as the amount of energy released in the first reaction.

Each substance has a given energy content, known as the element's cohesive energy. If the energy content in the original substances of a chemical reaction is greater than the energy content of the resultant substances, then energy is released, mainly in the form of warmth. This is called an exothermic reaction. In an endothermic reaction, energy must be supplied to the reaction. Exothermic reactions usually occur in nature; endothermic reactions are usual in all forms of industrial processes.

It is not only energy in the form of warmth that can stimulate chemical reactions: radioactivity, electricity and light can also have an effect. Sunlight is an example of light that can initiate a number of chemical processes in different materials. One of the most important rules in chemistry is: 'Within a chemical reaction the sum of the mass energy is constant'.

Other conditions for chemical processes

Other factors also affect the reactions process. The solidifying process of chalk ($CaCO_3$) is an example:

$$Ca(OH)_2 + H_2O + CO_2 \rightarrow CaCO_3 + 2H_2O \qquad (6)$$

Note that the solidifying is reduced with lower temperature; it can also be accelerated with larger amounts of carbon dioxide. A higher concentration of carbon dioxide accelerates the chemical reaction, even if not all of it is used in the reaction or is part of the final product.

The size of the particles also plays a part. The finer the particles and the greater the surface of the materials, the quicker the reaction is. Fine cements therefore

Table 4.1: The most common elements

Element	*Chemical symbol*	*% of Earth's crust*
Oxygen	O	49.4
Silicon	Si	25.8
Aluminium	Al	7.5
Iron	Fe	4.7
Calcium	Ca	3.4
Magnesium	Mg	1.9
Sodium	Na	2.6
Potassium	Ka	2.4
Hydrogen	H	0.9
Titanium	Ti	0.6
Total		99.2

have a shorter setting time. In a few chemical reactions with gases, air pressure plays an important role – pressure decides the weight of the gases.

In chemistry there are also catalysts, which increase the rate of the reaction without actually 'chemically' taking part in it. In an animal's digestion system catalysts are known as vitamins and play a vital role in a whole series of processes.

A chemical reaction can in principle be reversed, and must be seen as a reaction in equilibrium. Chemical compounds can be stable, metastable and unstable. Life would not have been possible without metastable systems.

The different elements

Ninety-nine per cent of the Earth's crust consists of ten elements. The other 1 per cent consists of, amongst other elements, carbon, which is a condition for biological processes.

There is a difference between organic and inorganic compounds. Carbon is the basic element in all life, and is in all organic compounds even lifeless compounds, such as oil and limestone, created from hundreds of decomposed organisms.

There are 500 000 carbon compounds. They include many compounds found in nature, e.g. plastics. Inorganic compounds number approximately 80 000.

Important factors in the physics of building materials

In every building project it is very important to have a clear picture of a material's physical properties. There are different demands on the different groups of

Table 4.2: The physics of building materials

	Structural materials	*Climatic materials*	*Surface materials*	*Surface treatment*
Weight	x	x	x	
Compressive strength	x	(x)	(x)	
Tensile strength	x	(x)	(x)	
Thermal conductivity	(x)	x	(x)	(x)
Thermal capacity	(x)	x	(x)	
Air permeability	(x)	x	(x)	
Vapour permeability	(x)	x	(x)	(x)

Notes:
x: primary function
(x): secondary function

materials. The following technical specification can be of great help (see also Table 4.2):

- *Weight* indicates what structural loading can be anticipated in the building, which building techniques can be used, etc.
- *Compressive strength* is an expression of how much pressure the material tolerates before collapsing, and is of particular importance in the design of columns and other vertical structural elements.
- *Tensile strength* expresses how much a material can be stretched before collapsing. This is important for the calculation of horizontal structural elements and suspended structures.
- *Thermal conductivity* describes a material's ability to conduct heat. It describes the insulating properties that can be expected of this material as a layer within an external wall, for example. The conductivity of a material is dependent upon the weight of the material, the temperature, its moisture content and structure.
- *Heat capacity* of a material is its ability to store warmth, which tends to even out the temperature in a building and also in many cases reduces energy consumption. Heat capacity is strongly related to a material's weight.
- *Air permeability* indicates how much air is allowed through a material under different pressures. It depends upon a material's porosity, the size and the structure of its pores. The moisture content of the material also plays an important role, as water in the pores will prevent air passing through. The right specification of material is particularly important when making a building airtight.

- *Vapour permeability* gives the equivalent picture of water vapour penetration under different pressures. This can vary according to the material's moisture content and temperature, and is a decisive factor in the prevention of damage caused by damp.

In the third section of this book, the primary technical specifications are presented in tabular form. Secondary specifications are not discussed further, even though they are often a decisive factor in the choice between alternative materials. Other physical properties such as bending strength, elasticity, expansion, porosity, etc., will only be sporadically discussed.

References

ALTENPOHL D, *Materials in World Perspective*, Berlin/Heidelberg/New York 1980

KARSTEN R, *Bauchemie*, Verlag C.F. Müller, Karlsruhe 1989

Section 1: Further reading

ABBE S *et al*, *Methodik für Oekobilanzen auf der Basis ökologischer Optimirung*, BUWAL Schriftenreihe Umwelt no.133, Bern 1990

ALSBERG T *et al*, *Långlivade organiska ämnen och miljön*, Naturvårdsverket, Solna 1993

ANDERSON J, *Tyskland, Återvinningskvoterna växer*, Byggforskning 93:6, Stockholm 1993

APEL K-O, *Diskurs und Verantwortung. Das Problem des Übergangs zur postkonventionellen Moral*, Frankfurt 1988

BAKKE J V, *Overfølsomhet i luftveiene og kjemiske stoffer*, Arbeidstilsynet, Oslo 1993

BERGE B, *De siste syke hus*, Universitetsforlaget, Oslo 1989

BERGE B, *Bygningsmaterialer for en bærekraftig utvikling*, NKB rapp. 1995:07, Nordic Ministry, Helsingfors 1995

BERGE B, *Byggesystem for ombruk*, Eikstein Forlag, Marnardal 1996

BERGE B, *ADISA-structures. Principles for Re-usable Building Construction*, PLEA Proceedings Vol. 2, Kushiro 1997

BERGE B, *Nedbrytingsdyktige Konstruksjoner*, Landbrukets Utviklingsfond Pnr. 2-0350, Oslo 1997

BOKALDERS V *et al*, *Byggekologi*, 1–4, Svensk Byggtjänst, Stockholm 1997

BREEAM Building Research Establishment, *Environmental Assessment Method*, BRE 1991 New Homes Version 3/91

British Petroleum Corporate Communication Services, *BP Statistical Review of the World Energy*, London 1993

Curwell S *et al*, *Buildings and Health*, RIBA Publications, London 1990

ERIKSEN TB, *Briste eller bære*, Universitetsforlaget, Oslo 1990

FLYVHOLM M-A *et al*, *Afprøvning og diskussion af forslag til kriterier for kemiske stoffers evne til at fremkalde allergi og overfølsomhed i hud og nedre luftveje*, NKB rapp. 1994:03, Helsingfors 1994

GRUNAU E B, *Lebenswartung von Baustoffen*, Vieweg, Braunschweig/Wiesbaden 1980

GUSTAFSSON H, *Kemisk emission från byggnadsmaterial*, Statens Provningsanstalt, Borås 1990

HÄRIG S, *Technologie der Baustoffe*, C.F. Müller, Karlsruhe 1990

HOLDSWORTH B *et al*, *Healthy Buildings*, Longman Group, London 1992

IVL, *The EPS Enviro-accounting method*, IVL, Report B 1080:92

KARSTEN R, *Bauchemie*, C.F. Müller, Karlsruhe 1989

KASSER U, *Grundlagen und Daten zur Materialökologie*, Büro für Umweltchemie, Zürich 1994

KOHLER N *et al*, *Energi- und Stoffflussbilanzen von Gebäuden während ihrer Lebensdauer*, EPFL-LESO/ifib Universität Karlsruhe, Bern 1994

KÖNIG H L, *Unsichtbare Umwelt. Der mensch im Speilfeld Elektromagnetischer Feltkräfte*, München 1986

LIDDELL H *et al*, *New Housing from Recycled and Reclaimed Components*, Scottish Homes Research Project, Edinburgh 1994

LÖFFLAD H *et al*, *Das recycling-fähige Haus*, Katalyse, Köln 1993

NÆSS A, *Anklagene mot vitenskapen*, Universitetsforlaget, Oslo 1980

NYBAKKEN Ø *et al*, *Miljøskadelige stoffer i bygg- og anleggsavfall*, Hjellnes Cowi, Oslo 1993

PAPANEK V *et al*, *How things don't work*, Pantheon Books, New York 1977

PERSSON J, *Hus igen*, CTH, Göteborg 1993

SAX I, *Dangerous Properties of Industrial Materials*, Van Nostrand Company, New York 1990

SOLBJØR O, *Miljøbelastning forårsaket av fyllinger*, SFT rapp. 92:23, Oslo 1992

STANG G *et al*, *Historiske studier i teknologi og samfunn*, Tapir, Trondheim 1984

STOKLUND LARSEN E, *Service life prediction and cementious components*, SBI report 221, Hörsholm 1992

STRUNGE *et al*, *Nedsiving fra Byggeaffald*, Miljøstyrelsen, Copenhagen 1990

TILLMANN A *et al*, *Livscykelanalys av golvmaterial*, Byggforskningsrådet R:30, Stockholm 1994

TÖRSLÖV J *et al*, *Forbrug og fororening med arsen, chrom, cobalt og nikkel*, Miljøstyrelsen, Copenhagen 1985

TURIEL I, *Indoor Air Quality and Human Health*, Stanford University Press, Stanford 1985

VALE B & R, *Green Architecture*, Thames & Hudson, London 1991

section 2

The flower, iron and ocean

Raw materials and basic materials

5 *Water and air*

Water and air are needed for all life and therefore for all animal and vegetable products; they are the constituents of many materials. Water can dissolve more chemical compounds than any other solvent and is used a great deal in the painting industry. When casting concrete, water is always part of the mixture, even if it evaporates as part of the setting process. Air is also an important component in the chemical processes required for the setting of concrete. The majority of industrial processes also use great amounts of water for cooling, cleaning etc.

Clean air and pure water are very limited resources in many places, especially dense industrial areas. During recent years large areas of the European continent have experienced drastic disturbances in the ground water situation, including widespread pollution of ground water.

Water

Water is seldom just water. It nearly always contains other substances to some degree such as calcium, humus, aluminium, nitrates etc. The quality of water is important, not only for drinking, but also as a constituent in building materials. Water with a high humus content produces bad concrete, for example, as the humus acids corrode the concrete.

The terms 'hard' and 'soft' water are well known. Hard water contains larger amounts of calcium and magnesium, 180–300 mg/l, than soft water, which contains approximately 40–80 mg/l. Very soft water will have a dissolving effect on concrete.

Water also has different levels of acidity which is expressed in a so-called pH-scale with values from 0–14. The lower the pH value, the more acidic the water. A pH value of 6.5–5.5 has a slightly aggressive effect on concrete and materials containing lime, while a pH value under 4.5 is very aggressive. Marsh water contains large amounts of sulphuric acid and is therefore unsuitable for use. Free carbonic acid, found in most water, attacks lime and corrodes iron. Sulphates in water, especially magnesium sulphate in salt water, is also corrosive and attacks lime.

Improving colloidal properties

Energized water, E-water, is water which has been produced in a levitation machine. The machine is a hyperbolic cylinder where the water is spun in a powerful and accelerated spiral movement. The process was developed by Wilfred Hacheney in Germany in 1976. When the water is used in cement, for example, it has been found that the material has an amorphous mineral structure as opposed to the ordinary crystalline concrete. This is probably due to the increased colloidal properties, i.e. a reduced tension in the water which increases contact between the water and the particles in it. The practical consequences are better compressive and tensile strength and a higher chemical stability, e.g. against air pollution. According to research the level of tolerance can drop to pH2, and at the same time the proportion of water and the setting time can be reduced. More conventional ways of increasing the colloidal properties usually entail mixing in small quantities of waterglass, natron and/or soda.

Ice and snow

Ice is a building material of interest in colder climates. The former Soviet Republics have a special category of engineering, engineering of 'glasology': the design of ice structures such as roads and bridges in areas of permafrost. Snow's potential as an insulating material against walls and on roofs has been used in the north throughout recorded history. One of the main reasons for having a grass roof is that in appropriate climates it retains snow for longer.

Air

In the lower level of the atmosphere the percentage by weight of the different gases is oxygen (O_2) 23.1 per cent, nitrogen (N_2), 75.6 per cent, carbon dioxide (CO_2) 0.046 per cent, hydrogen (H_2) 0.000 003 5 per cent, argon (Ar) 1285 per cent plus smaller amounts of neon (Ne), helium (He), krypton (Kr) and xenon (Xe). Water vapour and different pollutants also occur.

At very low temperatures air becomes a slightly blue liquid. From this state oxygen and nitrogen can be extracted through warming. Nitrogen is used for the production of ammonia (NH^3) by warming hydrogen and nitrogen up to 500°–600°C under a pressure of 200 atmospheres and passing it over a catalyst, usually iron filings. Amongst other things, ammonia is used in the production of glass blocks, glass wool and waterglass via soda, and as the main raw material for the production of ammonium salts, which are used to a certain extent as fire-preventing agents in insulation products. By reacting with hydrocarbons it forms amines which can be used in the production of a whole series of plastics.

When a material oxidizes, it forms a chemical compound with oxygen. This is an exothermic reaction which is automatic. In the building field, this is a very common occurrence with metals, more commonly known as rust or corrosion. The process is electrolytic. In many cases this oxidization is not a welcome process – metals are often coated with a protective sheath.

Other compounds in the air can also break down building materials, including natural carbon dioxide and air pollutants, such as sulphur dioxide and soot.

6 *Minerals*

The majority of the planet on which we live consists of inorganic, mineral materials. Stone consists of minerals in the form of crystals, and in general it is estimated that there is 4000 times as much solid rock on the earth as there is water in all the oceans put together.

There are thousands of different minerals. They can be characterized by colour, lustre, translucence, weight, hardness and their ability to split, and also by chemical formulae, because all types of crystal have their own unique chemical structure. In normal rock species there are only a few hundred different minerals, and in a simple species there are seldom more than four or five different minerals. Granite is made up of the minerals quartz, felspar and mica, the latter contributing sparkle. In a few cases minerals can be found in a pure state.

The first use of minerals can be traced back to Africa in the production of colour pigments. These were retrieved from the earth through a simple form of mining.

In chemistry minerals are divided up according to their chemical composition. The most important groups include pure elements: sulphides, oxides, carbonates and silicates. The most widespread of these is the silicates, while oxides and sulphides are mostly used as ore for the extraction of metals. In order to simplify the picture one can reduce minerals into two groups: *metals* and *non-metals*.

The occurrence of minerals is most often quite local. The purer the mineral when extracted, the easier it is to use. However, most minerals are extracted from conglomerate rocks or different types of loose materials.

Certain minerals have a tendency to occur together in the natural environment. When looking for a certain mineral, it is usually straightforward to work out where to find it.

Metallic minerals

Some minerals have a chemical composition which makes it possible to extract metals from them. These minerals are usually mixed with other minerals in the

Table 6.1: Metals, their ores and their use in building

Metal	*Ore*	*Use in building*
Iron (Fe)	Hematite, magnetite	The most important constituent in alloy steels; balconies; industrial floors; pigment (red); ingredient in timber impregnation
Aluminium (Al)	Bauxite, nepheline, kaolin	Light structures; roof sheeting; wall cladding; window frames; door; foil in reflective sheeting and vapour-proof barriers; window and door furniture; guttering; additive in lightweight concrete
Manganese (Mn)	Braunite, manganite, pyrolusite	Part of alloy steel; pigment (manganese blue); siccatives
Copper (Cu)	Chalcocite, chalcopyrite	The most important constituent in bronze; roof-covering; door and window furniture; guttering; ingredient in timber impregnation
Lead (Pb)	Galena	Roof covering; flashing; pigment (lead white); siccatives; additives in concrete
Zinc (Zn)	Sphalerite	Zincing/galvanizing of steel; roof covering; pigment (zinc white); ingredient in timber impregnation; additive in concrete
Cadmium (Cd)	Polluted sphalerite	Pigment (cadmium red and cadmium yellow); stabilizer in PVC; alloys
Chrome (Cr)	Chromite	One of the alloys in stainless steel; pigment (chrome yellow and chrome green); ingredient in timber impregnation
Nickel (Ni)	Pentlandite	One of the alloys in stainless steel; galvanizing of steel; pigment (yellow, green and grey)
Titanium (Ti)	Ilmenite, rutile	Pigment (titanium white)
Cobalt (Co)	Cobaltite	Pigment (cobalt white); siccatives
Antimony (Sb)	Stibnite	Pigment (yellow)
Gold (Au)	Gold ore	Colouring of glass; vapourized onto windows as a special protective coating
Tin (Sn)	Casseterite	Stabilizer in PVC; colouring agent in glazing for ceramics; ingredient in timber impregnation; catalyst in the production of silicone and alkyd
Arsenic (As)	Arsenopyrite	Ingredient in timber impregnation
Zirkonium (Zr)	Zircon	Siccatives

Table 6.1: Metals, their ores and their use in building – *continued*

Metal	*Ore*	*Use in building*
Metal alloys:	**Constituents:**	
Steel	Iron (85–98%) Manganese (0.1–0.5%) Nickel (1–10%) Silicon (0.5–1.0%)	Structure for floors, walls and roofs; roof covering; reinforcement in concrete; wall cladding; guttering; door and window furniture; nails and bolts (galvanized or zinced)
Bronze	Copper (more than 75%) Tin (less than 25%)	Roof covering

ores (see Table 6.1). The most common ore from which aluminium is extracted is bauxite, which contains iron as well as aluminium oxides.

In earlier times metals were worth a great deal because they were often inaccessible and required complicated working techniques. At first they were used for weapons and tools. During the industrial revolution great changes occurred in production techniques, and metals became more essential in the building industry, which mainly uses steel and aluminium, followed by copper and zinc. The areas of use are spread over a wide spectrum, from roof-laying and window frames to structures, nails, impregnation materials and colours in plastic, ceramics and paints.

In general metals can be replaced with other materials such as timber, cement products, etc. The exceptions are mechanical jointing elements such as nails and bolts.

During the extraction of ore, the mountains of slag and dust produced from breaking up and grinding cause environmental problems. Extraction can also create huge scars in the landscape which require filling and planting to restore afterwards. This is especially the case with shallow opencast mines. Even after much work it can be difficult or even impossible to rehabilitate or re-establish the local flora and fauna and an acceptable water table level. All industries that deal with metal extraction or smelting are environmental polluters. This is partly through the usual energy pollution from burning fossil fuels and partly through material pollution from the smelting process. Amongst other things the ores often contain sulphur, and during smelting huge amounts of sulphur dioxide are released. It is usual for this to be extracted and used in the production of sulphuric acid.

The consumption of energy for the extraction of metals from ore is far too high. All metals can in principle be recycled and through recycling of steel, copper, zinc and lead from waste the energy consumption can be reduced by 20–40 per cent and for aluminium by 40–70 per cent. The metal industry has good potential as far as excess heat is concerned, which can be recycled and distributed as district heating or for heating industrial premises.

Table 6.2: Potential pollution in the production phase

Metals	*Boiling point (°C)*	*Potential process pollution*
Cast iron	up to 3000	SO_2, CO_2, dust, Ar (when smelting scrap iron)
Steel	1535	Pb, Hg, Cd
Aluminium	2057	PAH, Al, F, CO_2, SO_2, dust
Chrome	2200	Cr
Cadmium	767	Cd, SO_2
Nickel	2900	Ni, SO_2
Zinc	907	Pb, Hg, Cd, SO_2
Lead	1620	Pb, Cd, SO_2
Copper	2310	SO_2, Cd
Zincing		Cr, Fl, phosphates, cyanides, organic solvents
Galvanizing		Cr, Fl, phosphates, cyanides, organic solvents

Note: The boiling point indicates the risk of vapourizing during different processes, such as when making alloys

The usage cycle of metals in buildings causes relatively few environmental problems, except for particles that are washed off the surface when exposed to different weather conditions. Lead roofing and flashings and metallic salts used in the impregnation of timber can lead to the pollution of local wells or soil. Large amounts of metal, as in reinforcement for example, can lead to a stronger electromagnetic field in the building.

Figure 6.1: Heavy extraction of minerals can cause damage and destroy the local biotopes and the quality of the groundwater.

In waste products, metals that are exposed to running water release metallic particles into soil and water which can damage many different organisms, depending upon the amount and degree of poison contained in them. It is important to note that pollution due to metals is irreversible. Metals left in the natural environment will always be there – they do not decompose. Even if the amount of metals released is reduced, the total amount of metals ending up in the environment will still be increasing. The possibilities of recycling metals, however good, only postpone the inevitable pollution.

Iron, aluminium, magnesium and titanium can be considered relatively 'benign' metals, even if the environmental consequences of their extraction and production are quite severe. They have a relatively good base as a raw material and their recycling potential is also high. They are not particularly poisonous and are abundant in the Earth's crust (see Table 2.2 in section 1).

Chrome, nickel, copper and zinc, however, should be used very sparingly, or not at all. The use of mercury, cadmium and lead should be banned. All metals in the long term should be kept within closed cycles, in order to maximize their re-use and minimize their loss during production or the life of the building.

Raw materials

Metals are the most limited reserves. On current statistical predictions, iron reserves will last 119 more years (from 1992), aluminium 220 years, copper 36 years and zinc 21 years (Crawson, 1992). These statistics do not take into account a possible increase in the consumption of metals. The use of aluminium in countries with low and medium industrialization increased by 460 per cent between 1960 and 1969, and is still increasing.

The production of aluminium is based on the ore bauxite, which contains 40–60 per cent aluminium oxide. Ninety per cent of the bauxite reserves are in countries with low and medium industrialization, while the same proportion of extracted aluminium is used in highly-industrialized countries. There are also other sources of aluminium such as kaolin, nephelin and ordinary clay. In the former Soviet Republics there are low reserves of bauxite, so aluminium oxide is extracted from nephelin, although it is much more expensive to extract aluminium from these minerals than from bauxite.

Primary use of energy for some metals

Metals/alloys	*From the ore (MJ/kg)*	*50% recycling*	*100% recycling*
Aluminium	165–260	95	30
Copper	80–127	55	
Steel	21–25	18	6–10
Zinc	47–87		

Probes are now being made to find new sources of iron ore, and have resulted in the discovery of interesting sources on the ocean floor – the so-called iron nodules. These also contain a large amount of manganese. Extraction of iron from bog-ore is now being considered. A more systematic recycling of scrap metal is in fact the most sensible method of obtaining iron. It is also possible to use alternative metals. There are in fact alternatives for all metals/alloys except for chrome, which is a part of stainless steel.

Recycling

Metal materials corrode, and 16–20 per cent of the total iron content effectively disappears. Chemical corrosion occurs mainly in the presence of water and oxygen; it is an oxidation process. Copper, aluminium and chrome are relatively resistant to corrosion. Metals are also attacked by acids: carbonic acid from carbon dioxide and water, and sulphuric acid. Iron, aluminium and magnesium are the metals most commonly affected. Base materials such as lime solution and concrete can attack metals, particularly aluminium, zinc and lead. Electro-corrosion can occur with certain combinations of metals.

The remaining metals can in theory be recycled or re-used.

Pure steel structures in heavy sections are usually easy to remove; as they are standardized, they are quite easy to re-use. In reinforced concrete, where the steel content can be up to 20 per cent, recycling is the only alternative, even if the process is relatively difficult.

A differentiation must be made between industrial and domestic waste. Industrial waste is usually pure and can be recycled without difficulty, whereas domestic waste may contain a whole variety of substances and therefore can cause problems. Copper in the electric cables of old cars and tin from tin cans make it impossible to recycle the steel in these products. Another problem is that waste metal often has a surface treatment, which can lead to complications.

All metals and metal alloys used in the building industry can be melted down and recycled. The metal can be added to new processes in varying proportions, from 10–100 per cent depending upon the end product and its quality requirements. Steel and aluminium alloys can only be used for similar alloy products, whereas copper, nickel and tin can be completely reclaimed from alloys in which they are the main component. Copper, for example, can be removed from brass through an electrolytic process.

The technology for smelting is relatively simple. A normal forge is all that is necessary. Breaking down alloys electrolytically and further refining, casting or rolling techniques, require much more complex machinery.

Metals in building

Iron and steel

Iron was used in prehistoric times. Pure iron has been found in meteorites and could be used without any refining. Smelting iron from iron ore has been carried out for at least 5000 years. Iron was not used in building until the eighteenth century, and then it was used for balustrades, balconies, furniture and various decorative items. The first structural iron girder was manufactured by Charles Bage in 1796, in England, and was used in a five-storey linen mill.

While cast iron contains a large proportion of carbon, steel is an iron alloy with a carbon content of less than 2 per cent. Towards the end of the nineteenth century steel became a serious rival to, and gradually replaced, brittle cast-iron. Buildings with a steel structure started to appear just before the turn of the century. Today steel is the only iron-based material used in the building industry. It is possible to use about 20 different alloys in steel and up to 10 can be used in the same steel. Normal building steel such as reinforcement, structural steel and most wall and roof sheeting does not usually contain any alloy. A particularly strong steel quality is formed through alloying it with small amounts of nitrogen, aluminium, niobium, titanium and vanadium. Sheeting products are protected against corrosion by a protective layer of aluminium or zinc. Facing panels in aggressive environments are often made of stainless steel which is 18 per cent chrome alloy and 8 per cent nickel. By adding 2 per cent molybdenum alloy an acid-resistant steel can be produced.

Ninety-five per cent of the cast iron manufactured is used in the production of steel. Even if materials are known as iron reinforcement, iron beams, ironmongery etc., they are all basically steel products.

As a resource iron is a very democratic material. Iron ore occurs spread evenly over the surface of the earth, and is extracted in over 50 countries. But the consumption of iron in certain parts of the world is so high that there are very high transport costs, from Australia, India or Brazil to Japan, from West Africa and Brazil to Europe and from Venezuela to USA. Rapidly diminishing iron ore reserves are also a problem, and the alloy metals required (nickel and zinc) also have very limited reserves.

Together with iron resources carbon is also an important element, and is generally a prerequisite for the production of cast iron from iron ore. The exception to the rule, where the reduction process uses natural gas, requires ore with a very high iron content. Rock iron ore is normally extracted by mining.

Bog iron ore lies in the soil and is much more easily accessible: it was the dominant source in earlier times. It lies in loose agglomerations in swamps or bogs. To find it, the bog is probed with a spear or pole. Where there is resistance to the spear, it can be assumed that there is ore. There may even be small traces of iron filings when the pole is removed.

Extraction of iron ore usually occurs in open quarries and extends over large areas, which means that the groundwater situation can change and the local ecosystem can be damaged. A large amount of waste is produced, usually about 5–6 tons for 1 ton of iron ore. Extraction of coal takes place either in open quarries or mines and causes the same environmental damage as the extraction of iron ore.

Extraction of iron from iron ore can be simple or complicated, on a small scale or large scale. There is quite a clear correlation today between size and efficiency in the metal industry. The fact that, 250 years ago, there were handbooks on extraction of iron for domestic needs proves that times and technologies have changed.

The conversion of iron from ore to steel requires a long series of processes. They begin with the breaking up of the ore, then cleaning, followed by sintering. The iron is smelted out and reduced in a large blast furnace at 1700–1800°C. A large, modern blast furnace can produce 1000 tons of pig iron every 24 hours. The amount of air needed is four million cubic metres, and the cooling water is equivalent to the amount a small town would use. It takes 440–600 tons of coal to produce 1 ton of iron (either charcoal or coal can be used). The amount needed can be reduced by half if an oil spray is injected into the furnace. Carbon is used in the process to remove oxygen from the ore by forming carbon dioxide, leaving the iron behind. Earth kilns were once used to smelt out the iron. The ore was filled in from above with layers of charcoal. In newer methods ore is mixed with lime and sand. The function of the lime is to bind ash, silica, manganese, phosphorous, sulphur and other compounds. The lime and other substances become slag from the blast furnace, which can be used as pozzolana in the production of cement.

Steel can be made of pig iron and steel scrap. Most of the carbon in the iron is released through different methods, e.g. oxidizing. This is done in blast furnaces or electric arc furnaces. The latter consumes far less energy and is today used in 30–40 per cent of the world's production.

Finally the steel is rolled out to produce stanchions, beams, pipes, sheeting and nails.

Iron and steel products that are not exposed to corrosive environments usually last for very long periods. Robust products can be recycled locally with a little cleaning up. All steel products are well suited for recycling.

Large amounts of sulphur dioxide and dust can come from the production of iron, while steel production releases large amounts of the greenhouse gas carbon dioxide, as well as dust, cadmium and fluorine compounds, into the air and water. This pollution is reduced when producing steel from waste. When producing steel from stainless steel, there will be a release of nickel and chrome.

Arsenic is a common pollutant of iron. It is well bound in the ore, but with a second smelting of steel scrap a good deal is released. Steel scrap is virtually inert, but ions from iron and other metal alloys can leak into water and the earth and damage various organisms.

Protection against corrosion

When ordinary steel is exposed to damp air, water, acids or salt solutions, it rusts. This is hindered by coating it with zinc, tin, aluminium, cadmium, chrome or nickel through zinc coating or galvanizing.

For zinc coating, metal is dipped into molten zinc at a temperature of at least 450°C. Zinc and iron bind with each other giving a solution which forms a hard alloy layer. Galvanizing is an electrolytic process. The metal to be coated acts as a cathode, and the material which coats the metal acts as an anode. A thin metal layer is formed on all the free surfaces without any chemical reaction.

These two processes, zinc coating and galvanizing, are considered serious environmental polluters. In both cases there is an emission of organic solvents, cyanides, chrome, phosphates, fluorides etc., mainly in the rinse water. These pollutants could be precipitated in a sludge by relatively simple means, which requires treatment as a special waste. Most of the galvanizing industries do not take advantage of these possibilities. Processes do exist that do not produce waste water, or have a completely closed system.

One method for relatively pollution-free galvanizing is a process making use of the natural occurrence of magnesium and calcium in sea-water. The technique was patented in 1936 and quite simply involves dropping the negatively charged iron into the sea-water and switching on the electricity. The method has proved effective for underwater sea structures. It is, however, not known whether this technique gives lasting protection from rust for metal components that are later exposed to conditions on land.

Treating surfaces of steel and metals with a ceramic coating would also give a better result environmentally. These methods are currently only used on materials in specialized structures.

Steel reinforcement is not galvanized. Concrete provides adequate protection against corrosion. But even concrete disintegrates in time and the reinforcement is then exposed. Correct casting of concrete should give a functional life span of at least 50 years. The most corrosive environment for galvanized iron and reinforced concrete structures is sea air and the air surrounding industrial plants and car traffic.

Aluminium

Aluminium is one of the newcomers amongst metals, and was produced for the first time in 1850. It is used in light construction and as roof and wall cladding. The use of aluminium in the building industry is increasing rapidly.

Aluminium is usually extracted from the ore bauxite. The Norwegian companies Elkem and Hydro import their bauxite from Brazil, Surinam and Venezuela, which are important rainforest areas. Extraction occurs mainly in opencast quarries after clearing the vegetation, which causes a great deal of damage to the local ecosystems. Production of aluminium entails a highly technological process of which electrolysis is an integral part. Building efficient production plants requires high capital investment, and countries with low and medium industrialization with large reserves of bauxite have mostly been forced to export the ore rather than refine it themselves. This is, of course, also because of the enormous amount of energy which is required to produce aluminium. As far as future development is concerned, it is safe to assume that there will be an expansion of energy resources and expertise in these countries, and that today's large aluminium producing plants in USA, Canada and Northern Europe are just an intermezzo.

Aluminium is produced from bauxite in two stages after extraction of the bauxite ore. Aluminium oxide is first extracted from the ore by heating it to between 1100°C and 1300°C with an increased air flow. This is called calcination. The oxide is then broken down in an electrolytic bath at around 950°C with sodium and fluorides. The pure aluminium is deposited on the negative pole, the cathode. On the positive pole, the anode, oxygen is released which combines with carbon monoxide, (CO) and carbon dioxide (CO_2). The anode consists of a paste mixture of powdered coal and tar – for every kilo of aluminium, half a kilo of paste is required. A huge amount of water is used.

The processes in the aluminium industry release huge amounts of carbon dioxide and acidic sulphur dioxide, along with polyaromatic hydrocarbons (PAHs), flourine and dust. These pollutants are washed off with water and then rinsed out into the sea or water courses without treatment. Some sulphur dioxide, hydrocarbons and fluorine escape the washing down with water and come out as air pollutants instead. Emissions into both air and water can have very negative consequences for the local environment and its human population. PAH substances, fluorine and aluminium ions remain in the sludge and slag from the production processes. This causes problems in the ground water when deposits have to be stored on site.

The amount of energy needed for the process from ore to aluminium is very high. Aluminium produced from bauxite ore is used to produce sheeting. Recycled, it can be used a great deal in cast products (known as downcycling). Aluminium waste is recycled by smelting in a chloride salt bath at 650°C, which at best only requires 7 per cent of the energy needed for production from ore. The waste aluminium has to be pure, not mixed with other materials. Recycling of aluminium requires a great deal of transport because of its centralized production. Most aluminium goods are relatively thin and easily damaged during demolition or removal, so local re-use is seldom practical. Aluminium is susceptible to corrosion, but less so than steel.

Copper

Copper was most likely the first metal used by mankind. The oldest copper articles that have been found were made about 7000 years ago in Mesopotamia. An early development was the invention of bronze, produced by adding tin to create a harder metal.

There are many examples of bronze being used in building relatively early, especially as a roof material. The roof on the Pantheon in Rome was covered in bronze sheeting. This was subsequently removed and transported to Constantinople. Copper has always been an exclusive material. It is found mainly in churches and larger buildings.

The most important alloy, brass, consists of 55 per cent copper and 5–45 per cent zinc, occasionally combined with other metals. It is commonly used in light fittings and a variety of timber impregnation treatments.

Copper ore is extracted from quarries and mines in the Congo, Zimbabwe, Canada, USA and Chile and entails a heavy assault on the natural environment. The natural reserves are very limited. Large quantities of sulphur dioxide are emitted during traditional copper smelting. Modern plants resolve this problem by dissolving the ore in sulphuric acid, then extracting pure copper by electrolysis. Copper is poisonous and can be washed out of waste. It can accumulate in animals and plants living in water, but unlike many other heavy metals it does not accumulate in the food chain. Copper has a very high durability but is expensive. Most copper in Western Europe is recycled. Some, however, is re-used locally, such as thick copper sheeting.

Zinc

Zinc is the fourth most common building metal in Scandinavia. It probably came into use around 500 BC. It has commonly been used as roofing material and later to galvanize steel to increase corrosion resistance. It is also used as a pigment in paint and a poison against mould in impregnation treatments. Zinc is part of brass alloy. Extraction of zinc causes the release of small amounts of cadmium. Zinc is susceptible to aggressive fumes. In ordinary air conditions one can assume a life span of 100 years for normal coating but only a few years in sea air, damp town air or industrial air. There are very restricted reserves of zinc. It was estimated at 21 years in 1992, and ought to be greatly restricted in its use. When zinc is broken down, the zinc particles are absorbed in earth and water. In higher concentrations, zinc is considered poisonous to organisms living in water. It can be recycled.

Secondary building metals

The following metals collectively represent a very small percentage of the use of metal in the building industry.

Lead

Lead has been in use for 4000–5000 years. It is not found freely in the natural environment but has to be extracted, usually from the mineral galena – lead sulphide

(PbS). The most common use of lead has been for roofing material and for detailing, but it has also been used for pipes, in Rome and Pompeii for example. Danish churches have a total of 30 000–50 000 tons of lead covering their roofs. The paint pigment, lead white, was also very common until recently, when its poisonous effect on humans was discovered. Useful lead resources are very limited.

Lead is mostly used nowadays in flashing for chimneys and for dormers on roofs etc. It is very durable, but can still be broken down in aggressive climates. When lead is exposed to rain, small, highly poisonous lead particles are washed out into the ground water. Lead has a tendency to biological amplification.

Cadmium

Cadmium does not occur naturally in a pure form, but in the compound cadmium sulphide (CdS) which is often found with zinc sulphide (ZnS). The metal was discovered in Germany in 1817, and is used as a stabilizer in many polyvinyl chloride (PVC) products. It is also used as a pigment in painting, ceramic tiles, glazes and plastics. Colours such as cadmium yellow or cadmium green are well known. The metal is usually extracted as a by-product of zinc or lead ores. Cadmium has a relatively low boiling point, 767°C, which is why it often occurs as a waste gas product in industrial processes, house fires and incinerators. Accessible reserves are very limited. Cadmium particles are washed out of waste containing cadmium. Cadmium has a tendency to biological amplification, and in small doses can cause chronic poisoning to several organisms.

Nickel

Nickel is used in steel alloys to increase strength. It is also an important part of stainless steel. It is used as a colour pigment in certain yellow, green and grey colours, for colouring ceramic tiles, plastics and paint. Nickel has very few accessible sources. During production of nickel large amounts of metal are liberated. Nickel has the property of biological amplification and is particularly poisonous for organisms living in water. In the former Soviet Union a connection has been registered between nickel in the soil and the death of forests (Törslöv, 1985).

Manganese

Manganese is a necessity for the production of steel. Between 7 and 9 kg are required per ton of steel. It is also used as an alloy of aluminium, copper and magnesium. Manganese is also a pigment – manganese blue. Manganese can cause damage to the nervous system.

Chrome

Chrome is used for the impregnation of timber and in stainless steel. There is no alternative to its use in stainless steel, so chrome is very valuable. Chrome compounds have the property of biological amplification and are very poisonous.

Arsenic

Arsenic is usually produced from arsenopyrite (FeAsS). Its main use is in timber impregnation, where it is mixed with copper or chrome. Accessible sources of arsenic are very limited. Arsenic has been the most popular poison used for murder for many centuries! The metal has a tendency to biological amplification and is extremely toxic.

Magnesium

Magnesium is not used very much. It is a light metal which in many ways can replace aluminium. It is extracted from dolomite and sea-water and is thus the only metal with large accessible reserves. Magnesium is not considered toxic.

Titanium

Titanium is the tenth most common element in the Earth's crust, even if the accessible reserves are very few. The metal has been given a positive prognosis as extraction costs for the other metals are increasing, but it is relatively difficult to extract and requires high energy levels to do so. Titanium dioxide is produced from ore of ilmenite ($FeTiO_3$), and 92 per cent is used as the pigment titanium white, usually for paints and plastics. Production of titanium oxide is highly polluting, whereas the finished article causes no problems.

Cobalt

Cobalt is a metal used as a pigment and drying agent in the painting industry and also as an important part of various steel alloys. Cobalt is slightly poisonous for plants, but very little is known about how it affects organisms in water.

Gold

Gold has a very limited use in the building industry. The most important use is the application of a thin layer on windows to restrict the amount of sun and warmth coming into a building, and to colour glass used for lanterns in yellow and red. Of the 80 000 tons of gold calculated to have been mined since the beginning of its use, most is still around, partly because gold does not oxidize or break down and partly because of its value. The gold used in window construction is considered to be taken out of circulation, but this only represents a small quantity.

Non-metallic minerals

The most important non-metallic minerals in the building industry are lime and silicious acid.

Table 6.3: Non-metallic minerals in the building industry

Mineral	*Areas of use*
Anhydrite, $CaSO_4$	Render; mortars; binders on building sites
Asbestos, $Mg_3Si_2O_5(OH)_4$	Thermal insulation; reinforcement in concrete; render; mortars; plaster and plastics
Borax, $Na_2B_4O_7.10H_2O$	Impregnation; fire retardant
Boric acid, $B(OH)_3$	Impregnation; fire retardant
Dolomite, $CaMg(CO_3)_2$	Filler in plastics and paint; production of magnesium oxide (MgO), glass and fibreglass
Gypsum, $CaSO_4.2H_2O$	Portland cement; gypsum cement
Graphite, C	Additive in sulphur concrete; oven lining; absorption layer for solar energy
Limestone, $CaCO_3$	Cements; binder; constituent in rockwool; mineral paints; ingredient in boards; filler; varnish and paint; glass and fibreglass; source of slag in the metal industries
Potassium chloride/sylvite, KCl	Used to obtain potash and soda for the production of glass
Various calcium silicate minerals	Glass and glazing on ceramics
Kaolin, $Al_2Si_2O_5(OH)_4$	Filler in plastics and paint
Magnesium oxide/periclase, MgO	Cement floor covering
Montmorillonite, $Al_4Si_8O_{20}(OH)_4+H_2O$	Waterproofing
Sodium chloride/halite, NaCl	Soda for the production of glass and waterglass; base for hydrochloric acid used in the plastics industry
Olivine, $(Mg, Fe)Si_2O_4$	Moulds for casting; filler in plastics
Silicon, SiO_2: as quartz	Glass; Portland cement; glasswool; rockwool; surface finish on roofing felt; aggregate; bricks; filler in paint and plastics
as fossil meal	Pozzolana; thermal insulation; filler
as perlite	Expanded for thermal insulation
Mica, different types	Fireproof glass (as in stove windows); expands to become vermiculite
Sulphur, S	Constituent in concrete and render
Talc, $Mg_3Si_4O_{10}(OH)_2$	Filler in plastic materials
Barite, $BaSO_4$	Colour pigment (lithopone)
Ilmenite, $FeTiO_3$	Colour pigment (titanium white); filler

Quartz is almost pure silicic dioxide and the hardest of the ordinary minerals. It is the main constituent of glass and silica and an important ingredient in Portland cement. Pure quartz is as clear as water and is known as rock crystal. Normal quartz is unclear and white or grey, and is a part of granite, sandstone or quartzite, or the sand of these rock types.

Pure limestone is a monomineral rock type of the mineral calcite. Accessible sources of limestone appear as veins or formations in many different types of rocks of different ages.

Limestone is used in a variety of products – it is one of the most important construction materials in the world after sand, gravel and crushed stone. The largest consumer of limestone is the cement industry. Cement nowadays means Portland cement, which is produced from a mixture of two thirds ground limestone, clay, iron oxide and a little quartz, heated to 1500°C. Gypsum is added to the mixture and then it is ground to a fine cement.

Limestone is an important filler in industries producing plastics, paint, varnish, rubber and paper. Some limestone is used in the production of glass and fibreglass to make the material stronger. In the metal industry, limestone is used to produce slag.

As well as quartz and limestone, there are many non-metallic minerals of rather more limited use. Important minerals are gypsum, used in plasterboard and certain cements, potassium chloride and sodium chloride, which form the base of a whole series of building chemicals, partly in the plastics industry, and kaolin, used as a filler in plastic materials and paints. Asbestos, which was widely used earlier this century, is now more or less redundant as a result of its health damaging properties.

Generally, the energy consumption and polluting potential of non-metallic minerals are much lower than in the metal industries, and their resources are generally richer.

Extraction of the minerals usually takes place in a quarry, where stones with the lowest impurity content are cut out as blocks, broken down and ground. In a few cases, the minerals can be found lying on the surface. One important example of this is quartz sand.

Extraction uses large quantities of material, causing large scars on the landscape. As with the metallic ores, serious damage can be caused to local ecosystems and ground water which can be quite difficult to restore later. Certain minerals such as lime and magnesium can be extracted by electrolysis from the sea, where the direct environmental impact is somewhat less.

Minerals from the sea

Apart from H_2O, the main constituents of sea water are the following (in g/kg water): chlorine (Cl) 19.0, sodium (Na) 10.5, sulphate (SO_4) 2.6, magnesium (Mg) 1.3, calcium (Ca) 0.4 and potassium (K) 0.4. Blood has a somewhat similar collection of minerals.

Table 6.4: Base materials

Material	*Main constituents*	*Areas of use*
Cements:	Lime Quartz Gypsum Sulphur Magnesium oxide Fossil meal Ground bricks Fly ash Clay Blast furnace slag	Structural concrete; concrete roof tiles; render; mortar; fillers; foamed up as a thermal insulation
Glass:	Quartz Lime Dolomite Calcium silicate Soda Potash	Openings for light in doors and windows; glasswool or foamglass as thermal insulation; external cladding
Sodium water glass:	Soda Quartz	Surface treatment on timber as a fire retardant
Potassium water glass:	Potash Quartz	Silicate paint

The main material in a snail's shell and in coral is lime. The formation of these structures happens electrolytically by negatively charged organisms, such as snails, precipitating natural lime and magnesium in salt water.

These processes can be performed artificially using electrolysis. The method is effectively the same as that used in galvanizing. A good conductor, usually a metal mesh which can also be used for reinforcement in the structure to be repaired, is dropped in the sea and given a negative charge. This is the cathode. A positively charged conductor, an anode, of carbon or graphite is put into the sea close by. As the magnesium and calcium minerals are positively charged from the beginning, they are precipitated on the metallic mesh. When the coating is thick enough, the mesh is retrieved and transported to the building site. The mesh or cathode can have any form and the possibilities are infinite.

There are many experiments nowadays around such sea-water based industries, even using solar panels as sources of energy. There is evidence that this is an environmentally acceptable method for the production of lime-based structures (Ortega, 1989).

In the continued working of raw materials, high process temperatures and fossil fuels are often used. Depending on the temperature level there is also a chance that impurities can evaporate into the air, such as the heavy metals nickel, thalium and cadmium. The environment is usually exposed to large amounts of dust of different types and colours.

Pollution due to the production of base materials

Material	*Potential pollution*
Calcined lime	SO_2, CO_2, unspecified dust
Natural gypsum	SO_2
Portland cement	SO_2, PAH, NO_x, Tl, Ni, quartz dust, unspecified dust
Glass	SO_2, CaCl, CO_2, unspecified dust

Many forms of silica dioxide (SiO_2), have to be seen as risks for the working climate. The problem is dust from quartz; overexposure to quartz can lead to silicosis. Dust from quartz can be emitted from several sources such as bricks containing quartz, or the production of stone, cement, concrete, rockwool, glass, glass wool, ceiling paper (where the paper is coated with grains of quartz), paint, plastics and glue. Olivine sand is not dangerous and can be used instead of quartz sand at foundries. Quartz sand can be replaced by materials such as perlite and dolomite as a filling for plastics. Silica dioxide dusts in the form of fossil meal and perlite are amorphous compounds and harmless apart from an irritatant effect.

Primary use of energy for the production of base materials

Material	*MJ/kg*	*Temperature (°C)*
Calcined lime	4.5	900–1100
Calcined natural gypsum	1.4	200
Portland cement	4.0	1400–1500
Glass from raw materials	10.0	1400
Glass, 50% recycled (varies according to the type of glass and its purity)	7.0	1200

When producing cements and lime binders workers are exposed to many different risks, depending upon the type of product, such as the heavy rate of work, high noise levels, vibrations and dust that can lead to allergies. Large amounts of the greenhouse gas carbon dioxide and acidifying sulphur dioxide are released.

Once in the building, the materials are relatively harmless, and as waste they are considered inert. The exceptions to this are asbestos and boron substances which have a pollution risk during their entire life span.

The non-metallic minerals are usually impossible or difficult to recycle as they are usually in the form of new chemical compounds in the final material. They nearly always have to be extracted from their raw state, sulphur though is an exception, which can be smelted out easily.

All glass can be recycled by smelting. But smelting of coloured glass has been found to be impractical. Also, used glass must be cleaned of all impurities for smelting.

The most important non-metallic mineral raw materials in the building industry

Lime

Lime is the starting point for the production of pure lime binders, as well as cements. It is also an important ingredient in glass. In the production of aluminium from nephelin, a great deal of lime is used, which becomes Portland cement as a by-product.

Most places on the Earth have deposits of lime, either as chalk deposits or coral and sand formed from disintegrated seashells. The purity of the lime is the decisive factor as far as the end product is concerned. For pure lime binders there has to be a purity of 90 per cent, preferably 97 per cent. Lime in Portland cement can be less pure. Chalk is a white or light grey lime originating from the shell of *Foraminifera* organisms.

The production of lime binder from lime ore starts with a burning process, usually called calcination:

$$CaCO_3 = CaO + CO_2 - 165.8 \text{ kJ} \quad (1)$$

This dividing reaction is endothermic and continues as long as the energy keeps the temperature 800–1000°C. Calcination is usually performed by breaking up the limestone into pieces of 2–8 cm which are then burned in kilns at 900–1200°C. There are a number of kiln constructions in use. Many are simple both to build and use, and production rates of 30–150 tons per 24 hours can be reached locally. There are mobile variations that can be used on very small lime deposits. Wood is the best fuel, as the flames are long-lasting and create a more even burning of the limestone than other fuels.

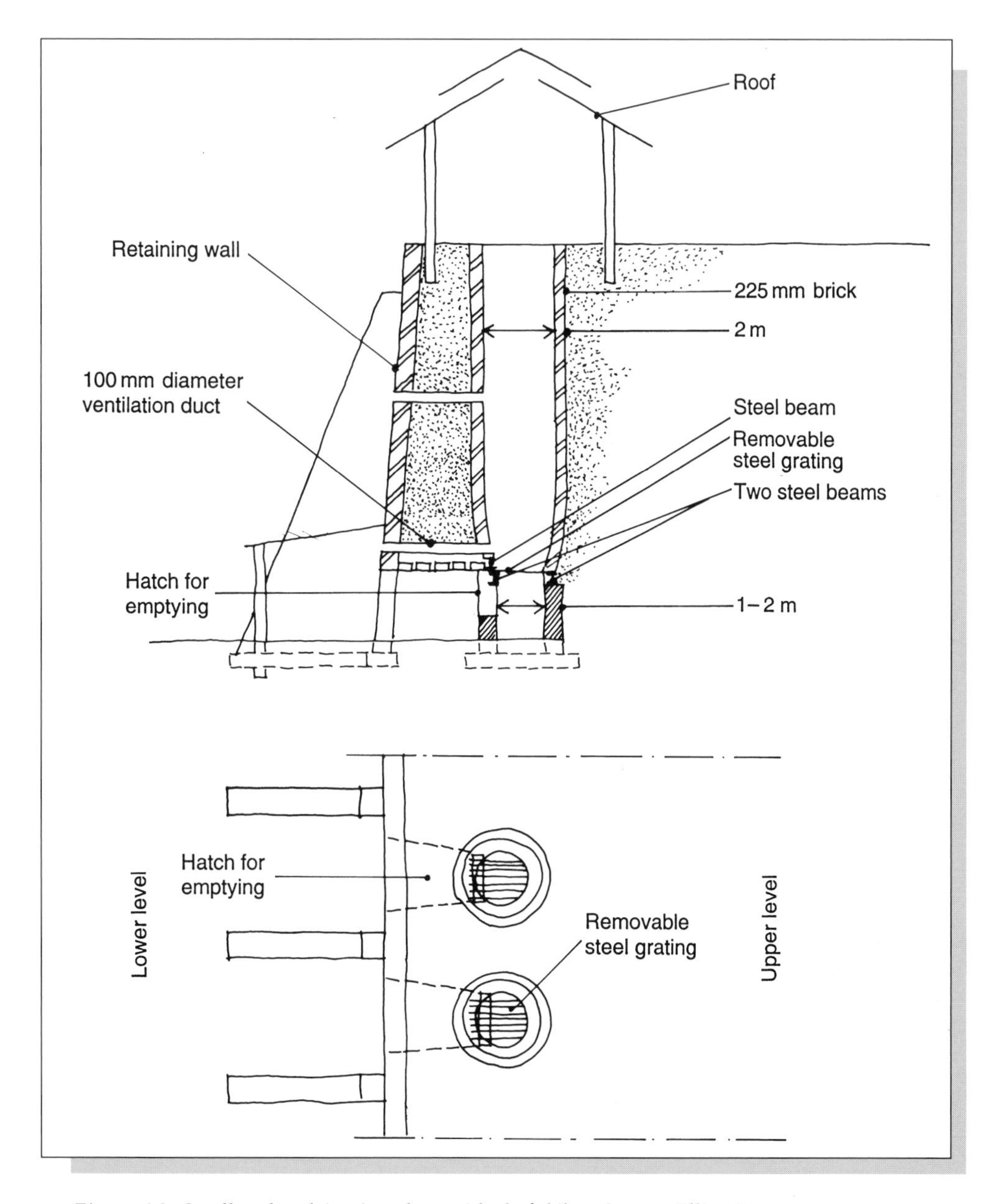

Figure 6.2: Small scale calcination plant with shaft kilns. Source: Ellis 1974

Calcined lime can be used directly to make lime sandstone (see Table 13.2) and pozzolana cements. During the production of Portland cement calcining occurs after the necessary extra constituents are added.

Lime has to be slaked so that it can be used, without introducing any additives, for render, mortar and concrete. The slaking process starts by adding water to the

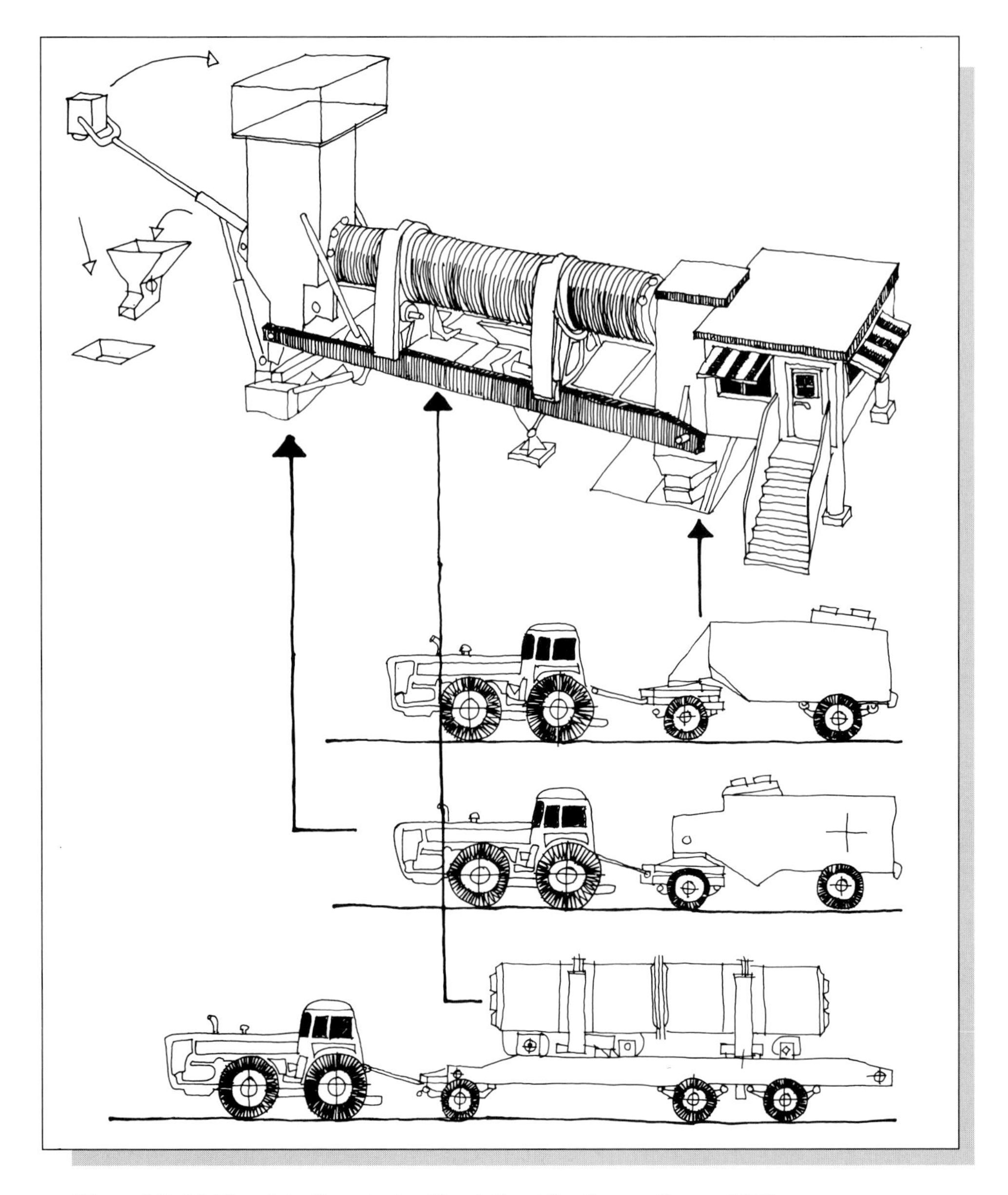

Figure 6.3: Mobile calcination plant with rotating kiln. Source: Spence 1976

lime on a slaking bench. Figure 6.4 shows a very simple slaking bench. The principles are the same regardless of the size of the system. The reaction is exothermic:

$$CaO + H_2O = Ca(OH)_2 + 65.3\ kJ \qquad (2)$$

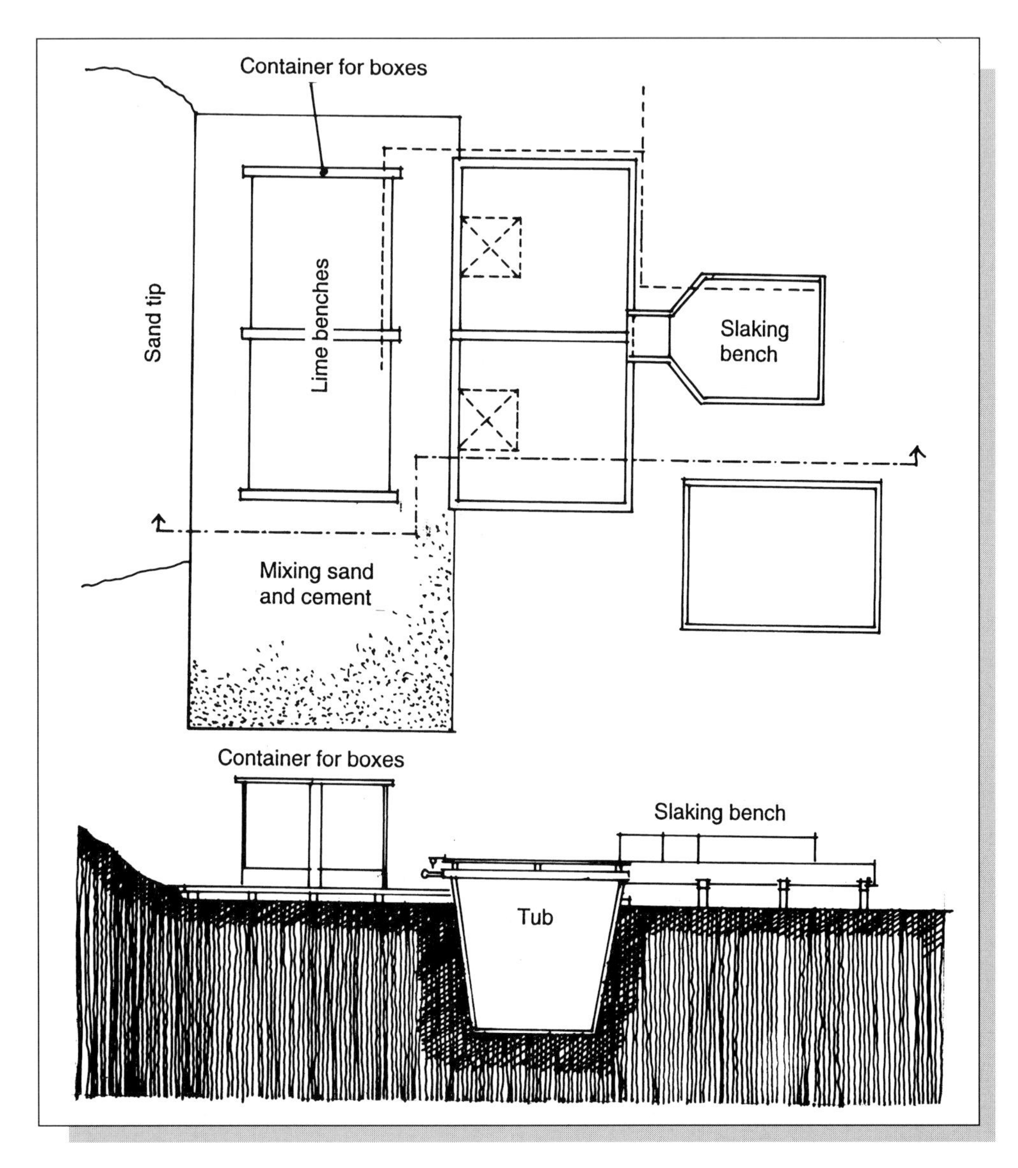

Figure 6.4: Small slaking bench. Source: Jessen 1980

A part of the energy needed for combustion is now released as heat. The lime swells up quickly and breaks up during a strong 'explosion' of heat. The lime milk is drained into a hollow and covered with sand. The lime is reslaked and after a week it is usable as mortar, while lime for rendering needs two to three months storage in the hollow.

The quality of lime gets stronger and harder if the Earth's moisture performs the slaking process. In this case, the storage has to take place from three to seven years, anaerobically, at a depth below frost level.

The technique of dry slaking has become more widespread recently. The is an industrial process where the exact amount of water needed is added. The product is called 'hydrated lime'. While ordinary slaked lime is usually mixed with sand and water, hydrated lime is in powder form. This has the advantage of lower transport costs and easier handling on site, where it is mixed with sand. Waste from demolition does not cause any problems. Lime products can, in principle, be recycled by burning.

Dolomite

Dolomite usually has a finer grain than lime, but otherwise has similar properties. The content of magnesium is too high for use in Portland cement, but it has a certain potential as an alternative to lime in pozzolana cement. The methods for calcination and slaking are approximately the same as for lime.

Gypsum

This is an aqueous calcium sulphate which is a natural part of stone salt deposits, precipitated in seawater or in lakes. Anhydrite is a white, translucent material which forms gypsum when water is added. Anhydrite and gypsum are used in the production of plasterboard, sheeting, mortars and as constituents in Portland cement. During recent years industrial gypsum by-products have made up a large proportion of the total volume of gypsum produced (see 'Industrial gypsum', p. 183).

In order to cast moulds with gypsum, the raw material has to be calcined, preferably in lime kilns. A temperature of 200°C is needed, which entails a relatively low energy consumption. The burning is complete when the vapour smells like rotten eggs.

Waste from demolition and building sites can develop sulphurous pollution from the breaking down of microbes, but this can be avoided by adding lime to the waste. Waste gypsum can be recycled, but these products are heavy and therefore need high energy in terms of transport.

Silicium dioxide

This is usually used in the form of quartz sand. It has an important role in several cements and in the production of glass and silicon.

Silicone is the only plastic that is not based on carbon. The molecule consists of silicium and oxygen atoms, but needs hydrocarbons and copper to initiate the process. Silicium is extracted through the reaction of quartz sand in electric furnaces.

Fossil meal is a type of earth which is rich in silicium dioxide. It consists of petrified and closed shells from silicious algae. Fossil meal is used as pozzolana, or as insulation for high temperatures, alone or mixed with brick or mortar.

Perlite is a volcanic type of earth with a high content of silicium dioxide is usually expanded for use as insulation. The deposits in Iceland are the largest in the world. In most types of clay there is usually a high concentration of silicium.

Potassium chloride and sodium chloride

These are extracted from salt water and used to produce two important base materials, potash and soda, which in turn are the starting point for the manufacture of glass and waterglass.

Potassium waterglass is produced by smelting potash and quartz at a temperature of more than 1710°C. Potash, K_2CO_3, was once produced from the ash of deciduous trees. It is now more common to produce it from potassium chloride.

Sodium waterglass is produced by allowing soda to replace potash in a combination with quartz. The soda is made by passing carbon dioxide and ammonia through a concentrated solution of sodium chloride.

Chlorine is produced electrolytically from a solution of sodium chloride. This substance is very important in the production of chlorinated hydrocarbons for the plastics industry. Hydrochloric acid is made industrially by igniting hydrogen and chlorine gas and is used in the production of PVC.

Sulphur

Sulphur occurs in its natural state and can be used independently for casting by smelting and then pouring into a mould. It is most practical to use it when it is an industrial by-product (see 'Sulphur', p. 184) or it occurs naturally, as in Iceland.

Mica

This consists of aluminium silicates and is used in windows of oven doors. Vermiculite is also a form of mica which can be expanded to make an insulation material for high temperatures.

Montmorillonite

This is found mainly in bentonite clay. Its most important use is as a waterproofer or watertight membrane. By adding water, the clay expands up to twenty times its own volume. There are many sources on the European continent, but the USA is the main producer.

Borax

Borax is extracted mainly from kernite which contains boron. Boracic acid is produced through a reaction with sulphuric acid. Sources of borax are relatively common. Borax and boric acid are used as fungicide and fire retardants, in insulation made of cellulose fibre and for timber impregnation. Boron substances are slightly poisonous, but in larger concentrations they affect plants and fish in freshwater.

Asbestos

This fibrous material was used as a reinforcement for ceramics as early as the Stone Age. As a building material it was widely used during the middle of this century and reached its peak around 1965. It has been used as reinforcement in different types of concrete, plastic and plaster products, and as insulation. It has became very clear that asbestos is carcinogenic. Products containing asbestos are now banned in most European countries, and elsewhere their use has been minimized.

Non-metallic minerals in building

The basic materials for which non-metallic minerals are used are mineral binders and glass.

Cements and limes

Cement is a collective name for mineral binders in powder form, which set to become solid when mixed with water. Pure lime binders are not usually considered cements. The main difference is that lime solidifies when it reacts chemically with air, while cement reacts with water in a hydrating process. While lime is a binder reacting in air, cement is a hydraulic binder which can also be used under water.

For use within a building, a material should not take longer than seven days to set, though this depends upon where the material is going to be used.

The cement most usually used in building is Portland cement, but there are plenty of other cements that have been used through the ages. In many cases, pure lime products can replace cement. The high energy consumption during production of Portland cement and the functional advantages of alternatives have recently led to experiments with alternative cements.

Cements can have three basic building functions: as render, mortar or concrete. The consistency depends on the number and size of the constituents, whether sand or stone, and the proportion of water and any additives.

History

The use of lime-based materials for casting goes back a long time. Excavation of Neolithic dwellings in Jericho in the Middle East has revealed an extensive use of concrete as a floor material. The concrete is almost completely made of lime, used as both the cast material and the fill. The technical quality can be compared with modern concrete in relation to its absorption of water and compressive strength, and it is so widespread that there must have been a relatively well-developed production technique using high-temperature kilns (Malinowski, 1987).

In Egypt there are solid structures that are 5000 years old and have gypsum as the main constituent in the mortar, while Greece used lime mortar. In Mychae on the Greek mainland, exposed lime mortar 3000 years old is still intact. The mortar was made the 'modern' way by mixing burnt and slaked lime with sand in the proportions 1:1 or 1:2.

The Romans mixed finely ground volcanic stone with their lime mortar 2000 years ago. They thereby produced a hydraulic mortar, which could withstand both saltwater and freshwater. The volcanic stone was fetched from Pozzuoli, and named pozzolana. The Romans later discovered other mineral substances which could be used as pozzolana, e.g. ground bricks and pottery.

The introduction of different pozzolanas revolutionized the building of inner walls and stronger arches and vaults. The Pantheon in Rome has a cassette vault cast in pozzolana cement. These pozzolanas were also used to make baths, water pipes and aqueducts watertight, and as a jointing material between roof tiles.

During the Dark Ages after the fall of the Roman Empire, the pozzolana technique seems to have been forgotten. With very few exceptions, such as the Sophiysky Cathedral in Kiev (1000–1100), builders returned to slaked lime. Certain places managed with clay, for example the stone churches of Greenland (1100–1400), but this was rather disappointing for future archaeologists – when the roofs had disintegrated, the rain washed the clay away, leaving only a pile of stones!

During this period there were several efforts to put oxblood, casein and protein into lime. This produced watertight, more elastic mortars with quicker setting times. The pozzolana mixture turned up again in England during the sixteenth century, and around 1800 James Parker from Northfleet made 'Roman cement' – a somewhat misguiding nomenclature – by firing broken up argillaceous limestone, which contains small amounts of fossil meal found along the banks of the Thames.

In 1824 an Englishman by the name of Aspedin patented what he called Portland cement, because it resembled rock quarried on the Portland peninsula in the south of England. In 20 years it was developed into the mixture still in use today. Many more cements similar to Portland cement have been developed since then, in which Portland cement is often an important ingredient. These cements have different expanding, elastic or quick-drying qualities.

In northern Europe there are approximately 35 different types of cement on the market. In the industrial countries its use is of the order of 1.7 m^3/year/per person; in countres with low and middle industrialization it is approximately 0.3 m^3.

Apart from the usual problems associated with centralized industry, such as vulnerability to market forces and distance from the user, the cement industry also has high transport costs because of the weight of the cement and extra care is required because of cement's sensitivity to moisture.

The alternative is a cement industry based on medium- or small-sized businesses. Setting up takes little time, and investment is small enough to be covered by local demand. These smaller plants can be placed where the cement is to be used and the raw materials extracted. The local infrastructure should be able to support them, and as changes in market forces will be local, they will be less devastating. The technology is relatively straightforward and could be adequately served by local small workshops and services.

Hydraulic binders

Hydraulic binders include lime pozzolana cements, hydraulic lime, Portland cement, Portland pozzolana cement and mortar cement – a mixture of lime and Portland cement.

A hydraulic binder can harden with dampness, even under water, but it must contain an acid. The most suitable are silicium dioxide and aluminium silicates, which are plentiful in clay. Argillaceous ingredients, pozzolanas such as broken up brick, can be added with other silicium-rich additives such as fossil meal and volcanic earths. Ashes from silica plants can also be used, (see 'Silicates', p.185). The hardening reaction is:

$$2(2CaO \times SiO_2)+4H_2O = 3CaO \times 2SiO_2 \times 3H_2O + Ca(OH)_2 \qquad (3)$$

At the outset one may think that quartz sand, which is almost pure SiO_2, would be usable. However, quartz sand in principle cannot form silicic acid under normal pressure and temperature conditions. It can in a damp, warm atmosphere and under pressure – a method used in the manufacture of lime sandstone. In many of the castles of the Middle Ages on the European continent a mixture of lime and quartz sand was used as a cold mix: we must assume that the silicic acid has been released from the sand, thus forming a durable binder, as these buildings are still solid today.

Pozzolana cements are low energy because the pozzolana undergo only a moderate warming. For the same reason there is very little gaseous pollution during production. Heavy metals such as nickel and thallium need a much higher temperature for vaporizing. Pozzolana cements can also be produced more economically than Portland cement, but they are often weaker. A ton of Portland cement is equivalent to 1.7 tons of lime pozzolana cement.

The following hydraulic binders are the most common.

Lime pozzolana cements

Fossil meal/slaked lime

Fossil meal is an earth rich in SiO_2 which consists of shells of petrified silica algae. Pure fossil meal reacts with slaked lime in its natural state even in weak frost, while fossil meal mixed with clay needs to be fired to a temperature of 600°C to be mixed with slaked lime. Higher temperatures reduce the reactivity of the lime. Very few experiments have been undertaken with this cement.

Calcined clay/calcined lime

Most clays react with lime after they are calcined. Clays to be used as pozzolana must be calcined to sintering level, which is usually around 550–650°C. Firing time is about half an hour, but the reactivity and viability of different types of clay varies. All ceramic clays are suitable for pozzolana. Clay and lime cements are used today in several parts of Asia. In India this cement is called Surkhi, and consists of lime ground with pulverized brick. It is weaker than Portland cement, but has better waterproof properties and has been used widely in dam building.

Blast furnace slag/calcined lime

The starting point for a reactive blast furnace slag is granulation. The glowing slag is tipped into a vessel filled with cold water. It is then ground into powder and mixed with calcined lime. An alternative is a mixture with dolomite calcined at 800°–900°C which also works well. The strength of slag and lime cements is good, but the mixture cannot be stored for long periods and must therefore be used shortly after production.

Hydraulic lime

Hydraulic lime is produced from natural limestone containing 6–20 per cent clay impurities. The firing is done in the same way as with lime. After hydraulic lime is mixed with water, it begins to set in the air. It will also eventually set underwater, and can be used for casting underwater in the same way as hydraulic cement.

The strength in this concrete is from about half to two-thirds that of normal Portland cement.

Portland cement

The main constituent of Portland cement is lime, which is 1.7–2.2 parts for each part of the other substances. The limestone is broken up and ground with quartz sand and clay or just clay:

$$CaO + SiO_2 + Al_2O_2 + Fe_2O_2 \qquad (4)$$

The content of sulphur compounds must not be more than 3 per cent. Water is added during grinding so that it becomes a slimy gruel. Next it is is fired in kilns at 1400°–1500°C and sintered to small pellets called cement clinker.

Vertical shaft kilns or rotating kilns can be used, but the rotating kiln is dominant in the industry. Rotating kilns, at their most efficient, yield 300–3000 tons a day; shaft kilns produce 1–200 tons a day. Modern shaft kilns have a higher efficiency and certain functional advantages, such as low energy consumption (Spence, 1980).

After firing, the mass is ground again and usually a little finely ground glass or gypsum is added to regulate setting. Pure Portland cement is seldom used today – it is usually mixed with lime or pozzolana.

Portland pozzolana cements

Pozzolanas also react with lime in Portland cement, resulting in cements that not only use less energy in production but also have higher strength and elasticity. In fossil meal/Portland cement, fossil meal is mixed in a proportion of 20–30 per cent. In calcined clay/Portland cement, clay is mixed-in, in a proportion of 25–40 per cent.

Industrial pozzolanas can also be used. For the production of blast furnace slag/Portland cement, the slag is granulated and ground with Portland cement in a proportion of 1–85 per cent. So-called Trief-cement consists of 60 per cent slag, 30 per cent Portland cement and 2 per cent cooking salt. It is usually recommended to use far less slag – preferably under 15 per cent. Fly ash/Portland cement has about 30 per cent ground in fly ash. The same proportions are used if mixing with industrial silicate dust, microsilica.

Blast furnace slag often slightly increases radioactive radiation from the material. Particles of poisonous beryllium can be emitted from fly ash, and easily-soluble sulphates can leach out from pollute waste and the ground water.

Lime/cement mortar

Lime/Portland cement is made by grinding larger or smaller amounts of slaked lime or hydrated lime into Portland cement. This mixing can also take place on the building site. The mix has a better elasticity than normal Portland cement, both during use and in the completed brickwork.

Non-hydraulic binders

Lime

Lime reacts as a binder with carbon dioxide in the air to form a stable compound.

$$Ca(OH)_2 + CO_2 = CaCO_3 + H_2O \quad (5)$$

This reaction is exothermic in the same way as slaking, in that the energy used in firing is now released. It takes a long time for the lime to set, and the process is slower at low temperatures. During setting, moisture escapes, which needs to be ventilated.

Gypsum

Calcined gypsum is a widely-used binder. It is usual to grind the calcined substance with larger or smaller parts of lime or dolomite, which act as catalysts for setting. The calcined gypsum can even be used as plaster of Paris as it is.

In Germany, a plaster cement which can compete functionally with Portland cement is developed. This is a hydraulic product.

Additives in cement

Cement is often complemented with additives, either while dry or during mixing when water and other mineral constituents are added. The first additives were used as early as 1920, but only in small amounts. During the 1960s and 1970s the amounts grew. In Denmark there are now additives in 60–70 per cent of all concrete (Strunge, 1990). The actual amounts vary, but the additives seldom form more than 1 per cent of the weight of the cement. Amongst the most important additives are:

- *Airing agents*, used to increase the workability, reduce the need for water, etc. These additives are benzene-compounds and phenolaldehyde condensates.
- *Water reducing agents* up to 5–10 per cent by weight which reduce the surface tension of water. Examples are waterglass, sodium and soda.
- *Accelerators*, which increase the rate of setting. Calcium chloride at 1.5 per cent by weight. Different amounts of sodium, potassium, lithium or ammonia salts can also be used. Triethanolamine, waterglass, soda and aluminium compounds can be used.
- *Retarders*, which delay setting during transport. These contain sugar, petrol, etc.
- *Water-repellents*, which make the substance more waterproof. Metal salts from stearic acid such as zinc stearate and silicone are used.
- *Adhesive agents*, which increase the tensile strength and ability of the cement to adhere to other materials, such as polyvinyl acetate and polyvinyl proprionate.

Cement products and pollution

To produce Portland cement in rotary kilns requires the use of energy sources such as coal, heavy oil or gas. Effluent from combustion, therefore, is the same as

Table 6.5: Additives in cement and concrete

Additive	*Contents*
Anti-freeze	Alcohol, glycol, inorganic salts
Expander	Iron powder, sulphur-aluminate cement
Water repellent	Stearic acid, oleic acid, fats, butyl stearate, wax emulsions, calcium stearate, aluminium stearate, bitumen, silicone, artificial resins
Permeability reducer	Bentonite clay, lime, fossil meal
To improve pumping	Alginates, polyethylene oxides, cellulose ethers
To reduce reactions with alkali–silica compounds	Lithium- and barium salts, pozzolanas
To reduce corrosion	Sulphites, nitrites, benzoates
Fungicide	Copper salts, dieldrin, polyhalogenized compounds
To reduce foaming	Polyphosphates, polyphthalates, silicones, alcohol
Aerating	Hydrogen peroxide, aluminium powder, magnesium, zinc, maleic acid-anhydride
To increase adhesion	Silicones, artificial resins such as PVA, PVP and acryl, epoxy, polyurethane, styrene and butiadiene compounds
To mix in air	Natural timber resins, fatty acids and oils, lignosulphonates, alkyl sulphonates or sulphates (e.g. ethylene ether sulphate, sodium dodecyl sulphate, tetradecyl sulphate, cetyl sulphate, oleoyl sulphate, phenol etoxylates, sulphonated naphthalenes) tensides, plastic pellets
To reduce the amount of water	Ligno sulphonates, polyhyroxy-carboxyl-acids and salts, polyethylene glycol, melamine formaldehyde sulphonates, naphthalene formaldehyde sulphonates, aliphatic amines, sodium silicate, sodium carbonate
Accelerators	Calcium chloride, other calcium salts (e.g. bromide, iodine, formiate, nitrite, nitrate, sulphate, oxolate, hydroxide, fluate), the equivalent salts of sodium, potassium, lithium and ammonium, triethanolamine, sodium silicate (waterglass), sodium carbonate (soda), aluminates
Retardants	Carbohydrates (sugar, starch), heptonates, phosphates, borates, silicon fluoride, lead and zinc salts, hydroxy-carboxyl acids and salts (e.g. gluconates), calcium sulphate dihydrate (gypsum)

(Source: U. Kjær *et al*, 1982)

for other production methods that use fossil fuel. The temperatures in the firing zones are so high, around 2000°C, that it must be assumed that nitrogen oxides are also emitted. This is not removed from the effluent today, although the technical facilities exist, e.g. by catalytic reduction. Shaft kilns can be fired with wood. The raw materials in cement also emit large amounts of acidifying sulphur dioxide and the greenhouse gas carbon dioxide.

Sulphur dioxide can, in principle, be cleaned out by adding lime to the exhaust gases. This is more difficult with the carbon dioxide which results from the calcination of limestone. This amount of carbon dioxide is a much larger proportion of the total carbon dioxide emissions from cement production than that caused by the firing processes, even though coal is the main fuel. The extremely high temperatures suggest that heavy metals are also emitted.

The problem of dust has previously received the most attention in connection with cement production. Today the dust problem is often much reduced as a result of closed systems for handling the clinker, more efficient dust filters, etc.

A similar pollution situation arises when calcining ordinary lime in charcoal-kilns, even though the temperatures are somewhat lower and the use of wood as an energy source gives a lower level of energy pollution.

The most effective step towards reducing pollution in the production of cements lies in the increased use of pozzolana mixtures in both hydraulic lime and Portland cements. In this way the amount of lime can be reduced, with a reduced emission of sulphur dioxide and carbon dioxide as a result.

On building sites the use of cements can produce dust problems. Wet Portland cement can cause skin allergies. In the construction process, cement products are relatively free of problems, though if setting is not effective, chemical reactions can occur between it and neighbouring materials, e.g. with PVC floor coverings. As waste, cement products are relatively inert.

Cement production and energy use

Energy consumption in cement production varies according to the type, but is mainly somewhere between the energy consumption levels of timber and steel production. Portland cement has a relatively high energy consumption, largely due to the high temperatures needed for production (up to 2000°C in the firing zone). The cement industry is usually very centralized, and the use of energy for transport is high.

It would be a significant achievement to reduce energy consumption in both production and transport. A decentralizing of cement production could save a great deal of energy, not only in transport, but also because smaller plants can be as efficient as larger plants. Today rotary kilns are used, but smaller, more efficient, modern shaft kilns could reduce energy consumption by 10–40 per cent. Rotary kilns are very specialized – shaft kilns have a greater variety of possibilities. They can be used for both calcination and sintering of most cement materials.

There are many ways of utilizing the heat loss, e.g. by production of steam, electricity or district heating. It is also possible to preheat the clinker in a pre-calcination. This process has been developed in Japan and has saved energy in the process.

Another step in the right direction is pozzolana mixing, which is now standard in many European factories, but this requires a local resource of pozzolana.

The greatest gains can be achieved through developing cements requiring less energy in production, where lower temperatures are required. The most profitable cements with the greatest potential are probably the lime pozzolana mixtures.

Glass

Glass surfaces bring in views, light and solar warmth. However, like the rest of the wall, they must protect the inhabitants against rain, cold, heat and noise. Few materials can satisfy these different demands at the same time. There have been many alternatives throughout history: shell, horn, parchment, alabaster, oiled textiles, crystalline, gypsum (selenite) and thin sheets of marble. Eskimos have used the skin of intestines. In Siberia mica is cut into sheets for windows. This is known as Russian glass.

None of these seriously rival glass, and the only alternative commonly in use is rice paper, used in Japan for letting light pass from room to room internally. More recently, plastic alternatives have been developed, such as plexiglass.

Normal clear glass lets about 85–90 per cent of daylight through. There are many other types of glass on the market: diffuse, coloured, metal-coated, reinforced, etc. Glass has also been developed to perform other functions, e.g. as insulation, such as foamglass and glasswool, the latter having a very large proportion of the insulation market nowadays.

History

The Phoenicians were probably the first to produce glass, about 7000 years ago. But the oldest known piece of glass is a blue coloured amulet from Egypt. Glass painting began in the Pharaohs' eighteenth dynasty (1580–1350 BC), but it is difficult to say if glass windows were produced during this period.

A broken window measuring 70 × 100 cm and 1.7 mm thick, opaque and probably cast in a mould was excavated from the ruins of Pompeii. It was originally mounted in a bronze frame in a public bathhouse.

Flat glass technology spread very slowly through Europe. Glass craftsmen kept their knowledge close, and only the Church, with a few exceptions, was allowed to share the secrets. Early glass was blue-green or brown, partly because ferrous sand (containing iron) was used as a raw material. Later it was discovered that adding magnesium oxide, 'glassblowers' soap', neutralized the effect.

During the eighteent century glass became affordable for use as windows in all houses. Glass was still very valuable and far into the nineteenth century it was normal to put many small pieces together to make one pane. From 1840 the methods of glass plate production became modernized and glass became even cheaper. The methods of production were still basically manual – glass spheres were blown, then divided.

In Belgium in 1907 the first glass was produced by machine. In 1959, float glass was developed, for the first time giving a completely homogeneous surface without any irregularities.

Different proportions of raw materials can be used to make glass, but it usually consists of 59 per cent silicon dioxide in the form of quartz sand, 18 per cent soda ash, 15 per cent dolomite, 11 per cent limestone, 3 per cent nephelin and 1 per cent sodium sulphate. The formula for the process is:

$$Na_2\,CO_3 + CaCO_3 + SiO_2 = Na_2O \times CaO \times 6SiO_2 + CO_2 \quad (6)$$

This glass, based on natron, is the most common. Replacing the soda ash with potash (K_2CO_3) gives a slightly harder glass. Lead glass is achieved by replacing limestone in the potash glass with lead (Pb).

For glass that needs high translucency for ultraviolet light an important constituent is phosphorous pentoxide (P_2O_5).

Fluorine compound agents decrease the viscosity and melting point of glass mixtures, which can reduce the use of energy. Antimony trioxide (Sb_3O_2) can be added to improve malleability, and arsenic trioxide (As_2O_3) acts as an oxidizing agent to remove air bubbles from the molten glass. Both are added in a proportion of about 1 per cent each. Stabilizers which increase the chemical resistance are often used: CaO, MgO, Al_2O_3, PbO, BaO, ZnO and TiO_2.

Coloured glass contains substances which include metal oxides of tin, gold, iron, chrome, copper, cobalt, nickel and cadmium, mixed in at the molten stage or laid on the completed sheet of glass electrolytically or as vapour. Traditionally, coloured glass has been used for decoration. In modern coloured glass the colouring is very sparse and it can be difficult to differentiate from normal glass. Decorative qualities are less important than the ability of the coloured glass sheet to absorb and/or reflect light and warmth. The aim is to reduce the overheating of spaces or reduce heat loss. Products which achieve this are usually known as energy glass, and have a high energy-saving potential. There are two types: 'absorption glass', which is coloured or laminated with coloured film, and 'reflection glass', which has a metal or metallic oxide applied to it in the form of vapour. Early energy glass reduced the amount of light entering the building by up to 70 per cent; today's is much more translucent, but the area of glass in a room may need to be increased to achieve adequate levels of light.

Production of glass for windows

To produce good quality glass, good quality raw materials with no impurities must be used. The ingredients are ground to a fine powder, mixed and smelted down.

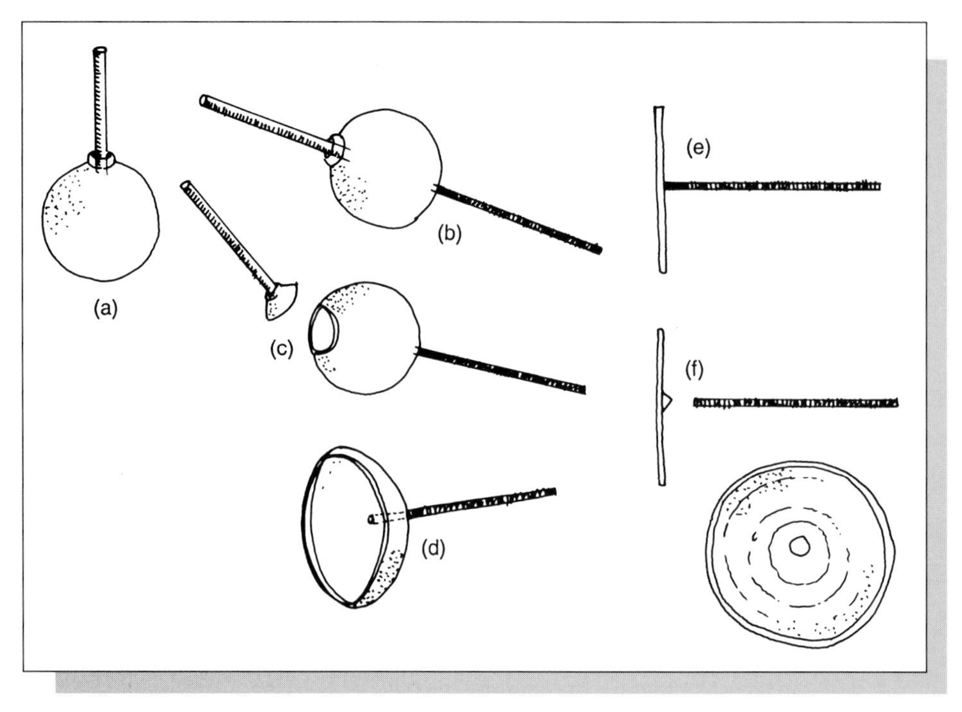

Figure 6.5: The production of crown glass: (a) the glass is blown up into a bubble; (b) an iron rod is fixed to the glass bubble; (c) the blowpipe is removed; (d) the glass bubble opens up after being warmed and rotated; (e) when completely open, the bubble becomes a flat, circular pane of glass; (f) the iron rod is removed. The pane of glass has a thick edge and centre, but is otherwise clear.

Smelting

As early as the Middle Ages, glass-works used 'pot kilns'. The method is comparable to ordinary cooling. The pot is warmed up by a fire or gas flame. Dry glass mix is poured into the pot and heated to 1400–1500°C. Recycled glass only needs 1200°C. When the mass has become even and clear, the temperature is lowered, and the substance removed in small portions and cast into a mould. In theory, the glass is soft and can be worked until the temperature reaches 650°C. The usual working temperature in the production of windows is about 1000°–1200°C. The capacity of a pot kiln is about half a ton per day. They are still used in smaller glass-blowing workshops for glass goods, but not in the production of windows.

In more industrial smelting methods, closed tanks with an inbuilt oil burner or electrical element are used. The tank is made of fireproof stone and has a capacity of 200–300 tons per day. The working temperature etc. is the same as that of the pot kiln. A tank kiln will be worked at full capacity continuously and may only last two to three years. The glass produced can be shaped using a series of different techniques.

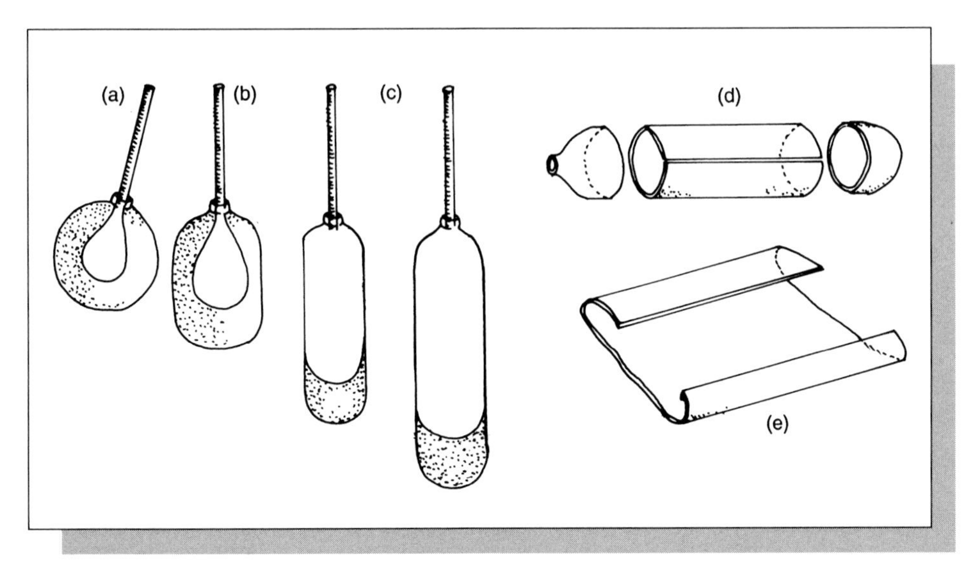

Figure 6.6: The production of table glass: (a)–(c) the glass is blown within a mould into a cylinder; (d) the end pieces are cut off; (e) the cylinder is opened up and divided into the required sizes.

Casting

Casting, most likely the first method for glass plate production, works on the simple principle that the smelted glass mass is poured into smooth moulds and then rolled out. This technique is still used for some types of glass where translucency is less important, e.g. decorative glass, profiled glass and wired glass. Glass bricks are made from two cast half blocks stuck together.

Crown glass

Crown glass was the most usual method up to about 1840. Figure 6.5 shows the production process. The glass is blown up to a bubble, a pin is stuck to the sphere, and the blowpipe removed. The pin is spun while the glass is warmed and the glass bubble opens up, becoming a circular disc up to 1 m in diameter, which can then be cut into panes. The pane in the middle – the bottle glass – is the lowest grade. Crown glass has low optical quality, with bubbles, stripes and uneven thickness. Today it is only used as decoration, or in panes where translucency is not required.

Table glass

Figure 6.6 shows the production process for table glass. The glass mass is blown into an evenly thick cylinder in a mould 2–2.5 m long and 60 cm in diameter cast in the floor. After blowing, the end pieces are removed and the cylinder is opened along the middle. The glass is then warmed and stretched into a large flat sheet. Table glass has a much better optical quality than crown glass. With this method, larger panes of glass can be produced.

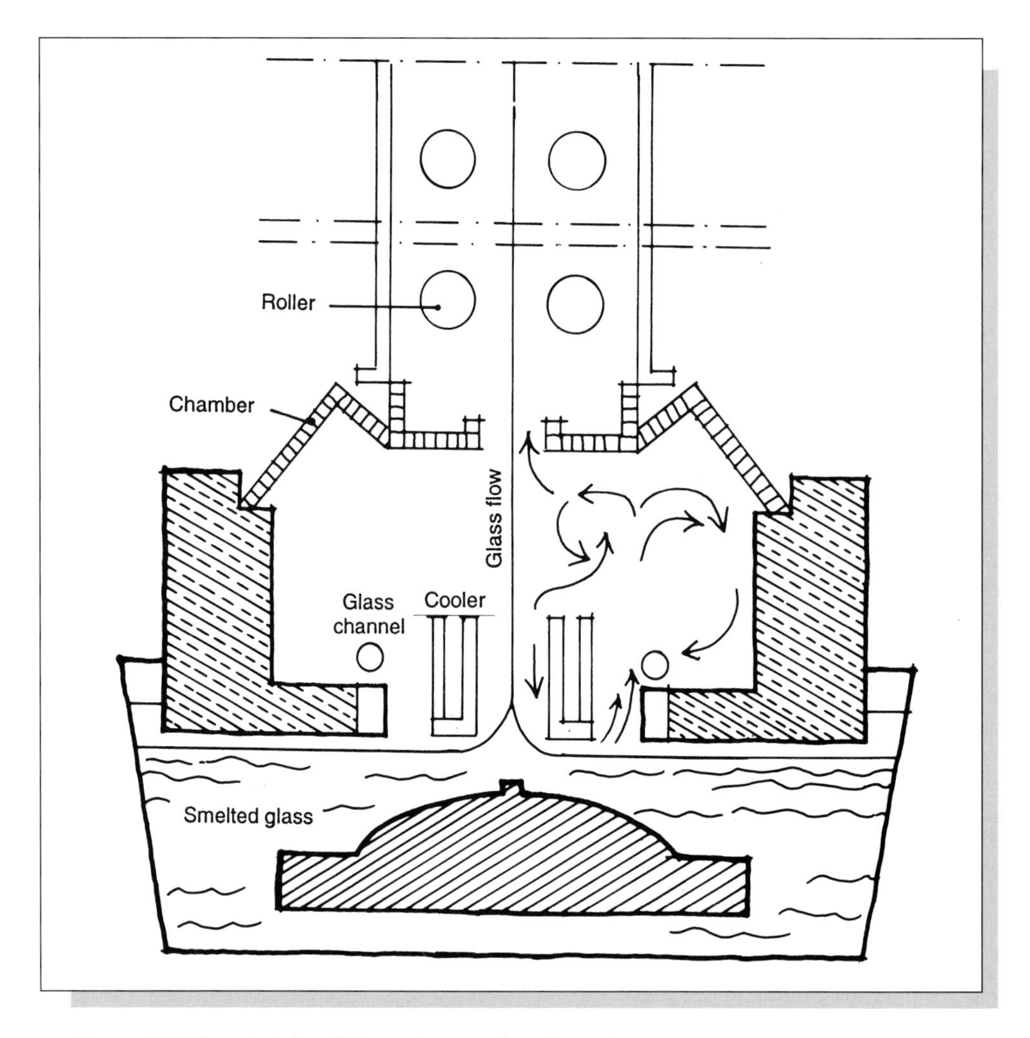

Figure 6.7: The principles of the production of machine glass. Source: Saten 1980

Machine glass

Figure 6.7 shows the production process for machine glass. The glass mass is cooled to 950°C to become a little tougher. It is then drawn through a flat nozzle out of the kiln and vertically up between a set of asbestos rolls in a cooling shaft about 12 m long. At the end of the shaft the glass is cut into the required lengths and slowly cooled.

Float glass

Instead of pulling up the glass substance vertically it is poured out over a bath of floating tin. This produces a totally flat sheet which it is cut and cooled. This is the method used by most glass manufacturers today.

Ecological aspects of glass production

The reserves of raw material for glass production are rich, even if deposits of quartz sand are regionally limited. Accessible reserves of the metallic oxides necessary for colouring or covering energy glass, most often tin and gold, are generally extremely limited. The most important environmental factors are the high primary energy consumption with related energy pollution, and the material pollution. Pollution by quartz dust and calcium chloride can also occur. When tin oxide is applied as a vapour, hydrogen chloride and hydrogen fluoride are emitted, in addition to tin pollution. Gold film emits less pollution than tin.

Glass does not produce pollution when in use, but both antimony trioxide and arsenic trioxide can seep out after disposal, causing environmental pollution. Coloured glass and metal-coated glass may contain heavy metal pigments which can be washed out on a dump, and must be left at a controlled waste-disposal tip.

Clear glass is very well suited for recycling. The production of new glass can in principle use up to 50 per cent returned glass. Recycled glass can also be used in the production of glasswool, foamglass and granulated glass (see 'Foamglass', p. 268) Glass covered with a metal film cannot be recycled.

Production of glass has become sophisticated and technology-dependent, and requires high investment. It is difficult to imagine that a small plant for local production of perhaps, 1 ton in 24 hours could be competitive in both price and quality. For glass with a lower standard of translucency and clarity it should be possible to set up local production based on casting, recycled glass, etc. for products such as glass blocks.

References

ALTENPOHL D, *Materials in World Perspective*, Berlin/Heidelberg/New York 1980

CRAWSON P, *Mineral Handbook 1992–93*, Stockton Press, New York 1992

ELLIS CI, *Small scale lime manufacture in Ghana*, Intermediate Technology, London 1974

JESSEN C, *Byhuset*, SBI, København 1980

MALINOWSKI R *et al*, *9000 år gammel betong med nutida hållsfasthet*, Byggforskning 6: 1987

ORTEGA A, *Basic Technology: Sulphur as a Building Material*, Minamar 31, London 1989

ORTEGA A, *Basic Technology: Mineral Accretion for Shelter. Seawater as Source for Building*, Minamar 32, London 1989

SATEN O, *Bygningsglass*, Oslo 1980

SPENCE RJS, *Small-scale production of cementious materials*, London 1976

STRUNGE *et al*, *Nedsiving fra byggeaffald*, Miljøstyrelsen, København 1990

TÖRSLÖV J *et al*, *Forbrug og forurening med arsen, chrom, cobalt og nikkel*, Miljøstyrelsen, København 1985

7 *Stone*

Many myths compare stones with the bones of Mother Earth. Extraction of minerals in most cultures has been accompanied by complex rituals and rites, undertaken as carefully as possible by, amongst other things, filling up the holes and passages into the mine when the extraction was finished. A Sioux Indian smallholder expressed this spiritual attitude thus:

> 'You ask me to dig in the earth. Do I have to take a knife and plunge it into my Mother's breast? You say that I must dig and take away the stones. Do I have to remove her flesh to reach down to her bones?'

There are three main categories of stone:

- *Igneous stones.* Consolidated pieces of rock which have forced their way up through splits in the crust of the earth. These are the hardest types of rock such as the granites, syenites and dolerites.
- *Sedimentary stones.* Petrified and disintegrated stone which has combined with organic materials. In this group are sandstone, slate and limestone.
- *Metamorphic stones.* Formed by exertion of pressure and the action of high temperatures on igneous or sedimentary rock types, which transforms them into another structure. Examples of these rock-types are crystalline slate and quartzite.

None of these groups can be referred to as the oldest, as the geological processes are in a continuous, rotational process. Sedimentary rock types can be formed through hardening of gravel, sand and clay which originate from the disintegration or breaking down of igneous or metamorphic stones; igneous stones can arise through the smelting of metamorphic and other types of rock and a later consolidation, and metamorphic stones can arise from changes in older sedimentary, igneous or metamorphic stones.

According to Asher Shadmon of the HABITAD centre in Nairobi:

> 'Stone is the building material of the future. We are on our way into a new Stone Age. The resources are limitless and evenly spread over the whole globe. Extraction does not require a lot of energy and does not pollute. And most important of all is that the material is durable' (Shadmon, 1983).

A differentiation is usually made between loose stones and quarry stone. The former are found on beaches or in fields; the latter are deliberately quarried. Stone primarily is used in the form of blocks, cut slabs or sheets, slate or crushed stone. It is used to create the walls of buildings, retaining walls, edging and bridges. Dressed stone and specially made slabs can be used for exterior or interior cladding, framing around doors and windows, fireplaces, floors and stairs. Slate can be used on floors, stairs, fireplaces, as framing around doors and windows, as roof covering and as wall cladding.

Crushed stone or gravel is used as aggregate in various concrete structures.

Stone has a very high compressive strength and a low tensile strength. Consequently, it is therefore possible to build high buildings of solid stone, whereas a stone lintel has a very limited bearing capacity. The Greek Temple shows this very clearly, where dimensions are immense just to achieve small spans. In Roman aqueducts the stones form arches; the compressive strength is thereby used at its maximum, making spans of up to 70 m possible.

The strength of stone varies from type to type. Slate has a higher tensile strength than other stone and is therefore a good floor material on a loose underlay.

The art of building stone walls for protection against the forces of nature goes back to prehistoric times. The earliest remaining stone buildings were built in Egypt and Mesopotamia about 5000 years ago. Stone has been the only building material used almost continuously until modern times, with its apotheosis during the late Middle Ages when a widespread stone industry developed throughout northern Europe.

The stone villages of this period were usually built with a foundation wall and ground floor in stone; the rest of the building was brick. By the beginning of the First World War the stone industry had lost its status, mainly due to the rapid rise in the use of concrete. Large quantities of stone are still quarried and sawn into slabs, mainly as marble in southern Europe, and a reasonable amount of slate extraction still continues, but the dominant use for stone today is crushed stone for concrete aggregate.

Many in the building industry anticipate a renaissance in stone building, even if not quite as optimistically as Asher Shadmon. Façade cladding is seen as the major area of use, because, with the exception of limestone and sandstone, stone is less sensitive to pollution than concrete and related materials. New technology has made it possible to re-open many disused quarries.

Table 7.1: Uses of stone in the building industry

Type of stone	*Minerals*	*Areas of use*
Granite	Feldspar Quartz Mica	Crushed stone; structures; floor finishes; wall cladding
Gabbro	Feldspar Pyroxene	Crushed stone; structures; floor finishes; wall cladding
Diabase	Plagioclase Pyroxene	Rockwool; crushed stone; structures
Sandstone/quartzite	Quartz, possibly lime or feldspar	Ground to quartz sand; smaller structures
Phyllite slate	Quartz Feldpar Mica	Roof covering; wall cladding; floor finishes
Mica slate	Quartz Feldspar Mica	Roof covering; wall cladding; floor finishes
Quartzite slate	Quartz Aluminium silicates Mica	Roof covering; wall cladding; floor finishes
Gneiss	Aluminium silicates Quartz Mica	Crushed stone; structures; floor finishes; wall cladding
Syenite	Aluminium silicates Pyroxene	Crushed stone; structures; floor finishes; wall cladding
Marble	Lime/dolomite	Structures above ground; floor finishes; cladding
Limestone	Lime	Ground to limeflour (cement, lime binder, etc.); smaller structures
Steatite/soapstone	Talc Chlorite Magnesite	Structures above ground; cladding
Serpentine	Serpentine minerals Chlorite Magnesite	Cladding; floor finishes
Clay slate	Clay minerals	Roof covering; floor finishes

Table 7.2: Primary energy consumption in stone production

Final product		*MJ/kg*
Granite:	as blocks	0.3
	as crushed stone	0.2
Marble		0.3
Limestone		0.3
Sandstone		0.3
Slate		Less than 0.3[1]

Note: (1) There are no relevant figures for slate, but we can assume that the use of primary energy is much lower than for a block of stone

Table 7.3: Potential pollution during the working of stone

Final product	*Potential pollution*
Granite/sandstone	Dust containing quartz
Phyllite slate/mica	"
Slate/quartzite	"
Slate/gneiss	"
Diabase/gabbro	Dust containing no quartz
Syenite/marble	"
Limestone/soapstone	"
Serpentine/clay slate	"

The lifespan of stone containing limestone can be prolonged to a certain extent by treating the surface with linseed oil. Epoxy and silicone-based surface treatments are also used. Stone is ubiquitous, even if in short supply in certain regions. Extraction and refining is labour-intensive, consequently the use of primary energy is a lot lower than the equivalent for brick and concrete. Stone is therefore not responsible for any significant energy pollution.

Extraction and stone crushing is usually a mechanical process with no need for high temperatures. Various energy sources can be used, ranging from handpower to wind and waterpower, either directly or as electrically-based technology.

The weight of stone suggests that the distance between quarry and building site should be short. Quarries along the coast have the potential advantage of energy-conserving water transport. Small, travelling extraction plants could be moved to very small quarries near relevant building sites, employing local labourers.

Large quarries spoil the landscape even if they eventually become overgrown and part of the landscape. They can also lead to altered groundwater conditions and damage local ecosystems. To extract granite for use as crushed stone by the 'gloryhole' method involves drilling the mountain or rock from the top and extracting stones by drilling a vertical tunnel which gets wider the deeper it goes. This means less visual disturbance of the landscape.

Stone often contains radioactive elements such as thorium and radium, and a quarry can increase the general level of radiation in a neighbourhood by emitting radon gas. Generally the extraction of slate, limestone, marble and sandstone have very little, if any chance, of causing radiation risks. Extracting volcanic or alum slate requires caution, including the measurement of radiation levels before removing stone for general use.

Environmental hazards of the industry include noise, vibration and dust – quartz stone dust is the most harmful. The more work stone needs, the greater the potential damage. By using undressed stone direct from the field these problems are avoided. If radioactive stone is avoided in construction there will be no problem during the use of the building, and demolition waste will also be inert.

All building stone is recyclable, especially from bridges, steps and other forms of pressed blocks. These second-hand products are usually valuable. Crushed stone has a potential for recycling when concrete is re-used as aggregate for further concrete production.

Production of building stone

Stone quarrying has always been based on a simple and labour-intensive technology which had difficulty in competing with growing industrialization. The work was heavy and could cause physical damage to workers. Developing technology could make the work lighter and should make stone a more competitive material. In many countries with low and medium industrialization stone can cost as little as a quarter of the price of concrete. In highly industrialized countries there are signs of improved competition as part of an aesthetic and qualitative drive. A significant factor which will strengthen the case for using local stone is that in conventional concrete production the amount of energy comprises 25–70 per cent of the price of the product, and is likely to increase.

Stone from fields and beaches lie freely scattered in nature. Throughout time these stones have been used and carefully stored. In Denmark as recently as the twentieth century, the round beach stone was so highly valued that several parts of the coast have been totally emptied! This round stone is particularly suitable for building in or near water, especially for piers. But the possibilities are still relatively limited, as concrete has difficulty bonding to the smooth surface. For larger buildings these loose stones have usually been cut into rectangular blocks for ease of handling.

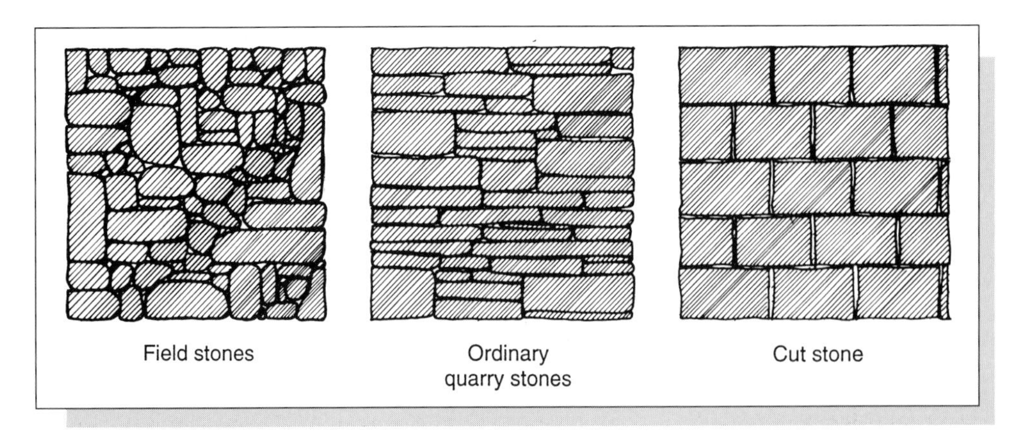

Figure 7.1: The different building stones.

Quarry stone has been extracted since the early Middle Ages. The work has been by pure muscle power, chisels, sledge hammers and pickaxes as late as the twentieth century. The stone quarryman's work is one of the least modernized, despite the introduction of explosives and saws, flame cutting tools and other cutting machinery.

Extraction methods

Extraction methods for various types of stone vary slightly, but the main principles are as follows.

Reconnaissance

The rock is inspected and samples are taken and tested for damp absorption, strength, etc. It is important to split the rock without cracking it or causing it to crumble or disintegrate. Layered and slate-like rock is the least problematic, but the distance between splits should not be too small. Rock of the same structure is usually evaluated by its sound when hit by a hammer, and the splinters or angular forms which split off.

Stone used to go through two further tests – for water absorption and heat resistance. The water test involves leaving the stone in water for several days, and checking that it does not increase in weight. To test heat resistance the rock is placed in glowing coals and must retain its form and structure when raked out afterwards. A good roof slate passes both tests. Another condition is that it must not form a white film on its surface when exposed to air and moisture.

Quarrying

The surface of the rock should be cleared of trees, loose stones, earth and all other organic matter. Holes are drilled for the charges. Placement of these holes is determined by the thickness of the block and the layer formation. The depth of the hole is also important. A 'rimmer' is knocked into the hole. This makes ruts in the wall of the hole along which the block will crack. The hole is then filled with gunpowder, rather than dynamite. Gunpowder has a lower rate of burning and gives a more muted explosion. Dynamite causes microscopic hairline cracks in the blocks which decrease their strength, although for crushed stone this is of no consequence.

Soft stone such as marble, limestone and soapstone can in many cases be removed with a wire saw. This consists of a long line of diamonds which cut 20–40 cm an hour. For rock rich in quartz, e.g. granite, a jet flame can be used. The equipment for the jet flame is a nozzle mounted on a pipe in which there is paraffin or diesel under pressure. The temperature of the flame is about 2400°C, and the speed is very high. A jet flame smelts out about 1–1.5 m^3 stone block per hour.

Dividing and cutting blocks

Stone is seldom used as an unfinished rough block. It is usually divided up into smaller units. This can be done in several ways.

Wedging

Wedging is shown in Figure 7.2. The alignment of the wedges happens in three stages. It requires skill, good knowledge of the nature of the stone and the direction of its layering, and much work.

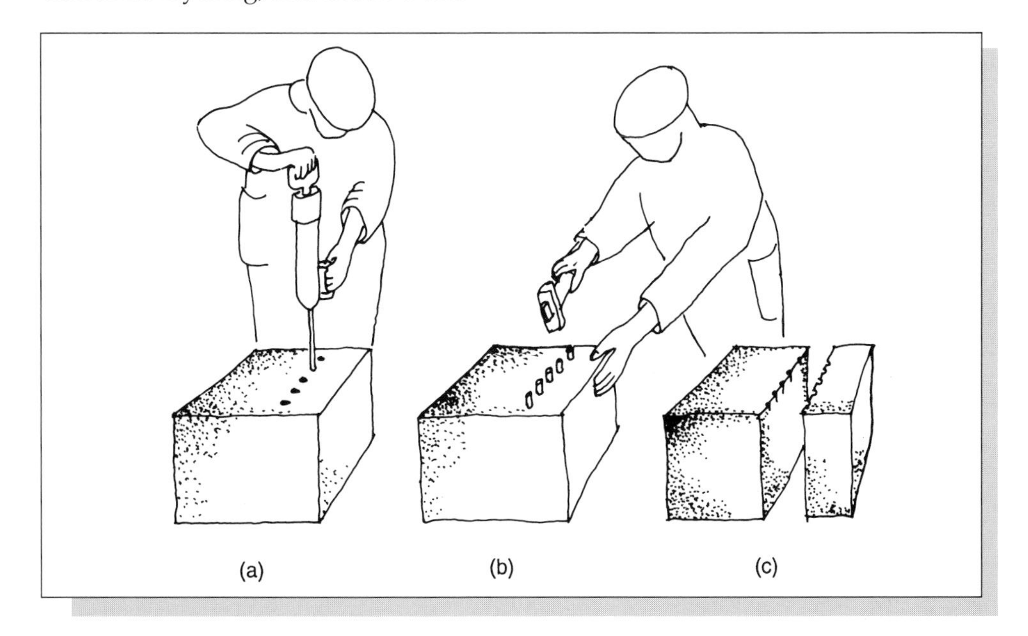

Figure 7.2: Dividing a block with wedges: (a) the seam for the wedge is made; (b) the wedges are knocked in; (c) the block splits.

Guillotining
This is possible for smaller blocks with clear layering. This splits the stone with one blow and is the most labour- and energy-saving technique. It is also the principle upon which modern equipment research and development bases its work. Some methods create an artificial tension within the rock with the help of a strong vice. Fractures then occur, which spread out when the axe falls, and in a single moment maximize the tension in one direction. The maximum size available for a rough block, using modern equipment to split the stone, is up to 250 cm × 50 cm, depending upon the type of stone. Smaller splitting machines can be carried by two men; these can split stone up to 10 cm thick and also work on loose stone.

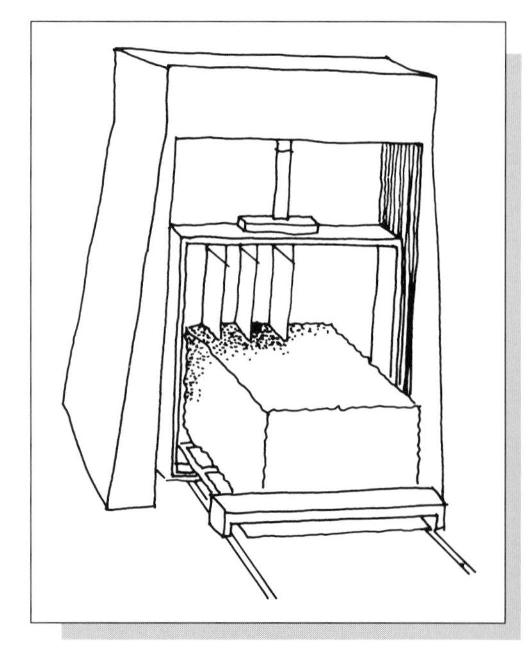

Figure 7.3: The frame saw used for cutting stone blocks.

Sawing
Another common method for dividing the block. A circular saw or frame saw, preferably with a diamond blade, can be used. The frame saw is often used for the production of facing panels. The capacity of a frame saw on hard stone is approximately 30 cm per hour. Circular saws are used for all types of stone and cut considerably faster.

Jetflame
This can be used on quartz stone.

Waterjet
A waterjet has been developed for cutting stone, using a thin spray of water at an extremely high flow speed which cuts stone like butter.

The finishing process
The finishing process is determined by how the stone is to be used. For structural use and foundations the stone does not need much working – the surface can be evened out with a hammer. For cladding panels, tiles, etc., the stone requires planing, grinding and polishing.

Sorting and cutting slate

Every slate quarry has its own characteristics with respect to accessibility, angle of layers and splitting. In particularly favourable locations the layers of rock are

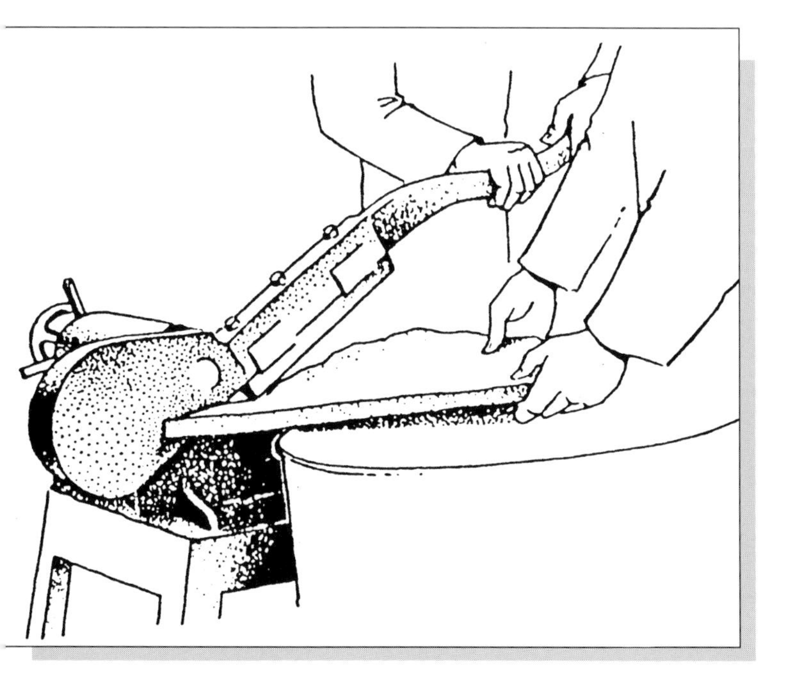

Figure 7.4: Slate 'scissors'. One piece at a time is cut from the edge inwards to the predetermined point. Source: Stenkontoret 1983

separated by a thin fatty layer which makes extraction very simple. In the traditional method, splitting is carried out directly on the exploded shelf within the quarry. In industrial extraction larger pieces are split with a hydraulic hammer and then transported for further splitting.

The secondary working of slate is usually carried out close to its place of extraction. Even at this stage, each slate has its own characteristics and requires its own particular working methods. Slate is typical of a material that requires manual labour; machines are not very useful for processing it.

Generally slates should be no thinner than 6 mm, although this varies with type. Thin slates are easily broken during transport. Once laid on either a floor or a roof, slates will not support high impacts.

If slate is knocked along its natural line, straight or curved, the structure of the stone is crushed to a certain depth inwards, and the stone divides itself. Pouring water over the slate makes the job even easier. This principle was used in manual splitting with a hammer to produce slates. During one working day a craftsman could produce 60 to 80 slates. With the introduction of slate 'scissors' (see Figure 7.4) which dominated production at the turn of the century, the number went up to 400 slates a day. A small wooden block is used to position the notches for the fixing nails, which are knocked out with a pick hammer or cut out with an angle grinder. The working bench is a trestle with slate lying on it. It is possible to knock two slates at the same time.

Crushed stone or stone block

Crushed stone is the only stone used today in foundations and structural work, either as aggregate in concrete or as levelling or loose fill under concrete foundations. In his essay 'Stone Technology and Resource Development' (Shadmon,

1983), Asher Shadmon points out the inconsistency in first crushing stone blocks and then using them in concrete, which in itself is an attempt to copy stone. The extraction and working of stone requires relatively little energy, and at the same time it is a very durable material.

References

ASHURST J, *Stone in Building. Its use and potential today.* London 1977
SHADMON A, *Mineral Structural Materials,* AGID Guide to Mineral Resources Development 1983
STENKONTORET, *Stenhåndboken,* Stavern 1983

8 Loose materials

'Loose materials' is a collective name for fine-particled materials that have originated from mineral and/or organic, decomposed products from animals and plants. In the larger lifecycle these return to a solid form such as rock. During this process, loose materials with a large organic content can form a foundation for the creation of coal or oil. A wide spectrum of raw materials within these states of continuous degradation and regeneration have been used throughout mankind's history for building construction.

Loose materials can be classified according to their origin, e.g. moraine – material originating from a river or sea bed. As well as being the starting point for all of the Earth's food production, they have many different uses in the building process: sand and gravel as aggregate in concrete, clay mixed with earth which can be rammed for solid earth construction and clay for the production of bricks, ceramic tiles and expanded clay pellets.

Table 8.1: Basic building materials from loose materials

Material	*Main constituents*	*Areas of use in building*
Clay bricks, roof tiles	Clay, sand, slag, fly ash, lime, fossil meal	Structures, cladding, floor finishes, roof covering, moisture regulation
Quarry tiles/Terracotta	Substances for colouring	Floor finishes, cladding
Vitrified tiles	Loose materials containing clay, kaolin, substances for colouring, glazing	Floor finishes, cladding
Expanded day	Loose materials containing clay	Thermal insulation, granular fill, sound insulation, aggregate in lightweight concrete products

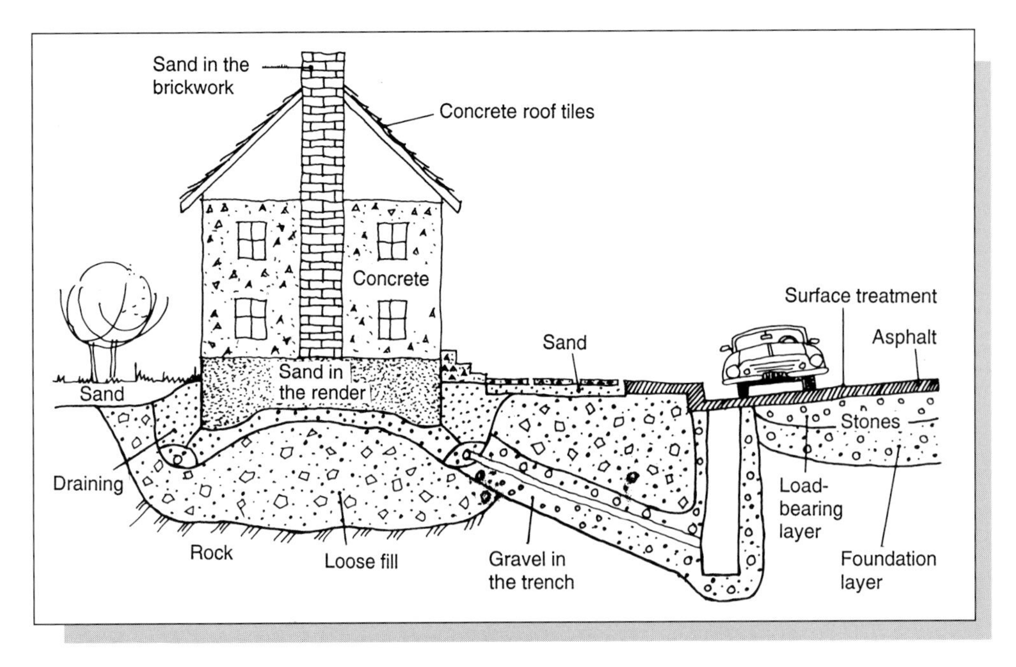

Figure 8.1: The use of sand, gravel and stones for building. Source: Neeb

Loose material	*Use in building*
Clay/silt	Earth construction, bricks
Loose materials	Ceramic tiles, expanded clay pellets, sound insulation in floor structure
Sand	In concrete, brick to decrease 'fattiness', sound insulation
Gravel	In concrete

In contrast to minerals, loose materials are defined by their physical properties rather than their chemical properties. Physical properties include grain size and form.

Material	*Grain size*
Clay:	Less than 0.002 mm
Silt	0.002–0.06 mm
Sand	0.06–2.00 mm
Gravel	2.00–64.00 mm

Different types of earth get their name from the highest percentage of loose material they contain – minimum of 60 per cent. The remaining percentage, if more than 20 per cent, is used to define the quality of that material, e.g. a 'gravelly sand'. They can have quite pure mineral content or they can be a mixture of organic substances such as peat and mud – mostly mould and plant material, known as humus. Loose material that is well suited for cultivation is not suitable for building, as it contains organisms and humus acids which have negative effects on both earth construction and concrete. These materials should be avoided in building construction.

Loose materials in building

Many parts of Europe do not have access to gravel and sand as a building material – not necessarily because the resources are not there, but because extraction would have too much impact on the local environment. Certain types of clay, e.g. clay used for ceramic tiles, can also be limited in certain regions. Otherwise, deposits of argillaceous materials are very large. Their use, however, is very small – in fact this material is an almost unused resource. It will continue to be available as a valuable resource in the future.

Extraction of loose materials for use in the building industry requires very low energy consumption. Drilling into the earth and explosives are unnecessary. It often takes place in quarries, but if these become too large they can damage groundwater and local biotopes. The most suitable clay for the production of bricks and ceramic tiles is usually in the 4–5 m nearest the surface. An annual production of 15 million bricks requires 30 000 cubic metres of clay, which represents 0.6 hectares to a depth of 5 m.

A very large amount of water is used in brickworks and also in the production of expanded clay pellets and ceramic tiles when grinding the clay. The ceramics industry in Italy has developed an efficient re-circulating system with a simple filter for the waste sludge. In this way they have reduced the quantity of water used and kept the sludge effluent to a minimum.

The energy consumption while processing fired clay products is very high. Oil is the usual source of energy, but wood, peat or a combination of electricity and coal can also be used. When oil alone is used, large amounts of greenhouse gas carbon dioxide, acidic sulphur dioxide and nitrogen oxides are released. Emissions are usually much higher than for the equivalent production of concrete.

The brick industry has become increasingly more centralized in Europe. This has resulted in heavy energy consumption in brick transport and distribution, with associated energy pollution.

Heated clay emits pollutants such as sulphur and fluorine compounds. These can be neutralized by adding 15–20 per cent lime to the clay. The red dust resulting from the production of fired clay products does not cause silicosis in workers, but does produce an uncomfortable working atmosphere.

The building of an earth house causes minimal pollution. However, vibrations from the ramming machines (see 'Pisé', p. 212) can cause physical harm to the operator. As far as locally built houses are concerned, there is probably no other technique that can compete with the earth house in terms of the lack of pollution. The most common building technique is to use the earth that is dug out of the ground where the house is going to stand. Transporting earth long distances is not normally economically viable, even though production of pressed earth blocks has now begun in the USA and France at prices much lower than those of brick or concrete.

The use of fired or unfired clay products in building causes no problems. In many cases they can improve the indoor climate by regulating and stabilizing moisture levels.

Clay building waste is inert, and depositing both fired and unfired products has no detrimental effects on the environment. Exceptions are brick or ceramic tiles which are coloured with pigments containing heavy metals, fire-proof bricks that contain soluble chrome and bricks from chimneys which have absorbed large amounts of aromatic hydrocarbons during their life span. These products have to be separated and disposed of at special tips.

Bricks are maintenance free and have an exceptionally high durability. They have also proved to be considerably more effective than concrete in resisting the effects of modern air pollution. Brick can usually be recycled, depending upon the strength of mortar used. Other fired clay products such as ceramic tiles and expanded clay pellets cannot be recycled and are more usually down-graded to become fill. Even roof tiles and bricks can be broken up and used as fill or aggregate in concrete.

When an earth house is demolished, the earth is physically and chemically intact in its original form. It can therefore be easily reinstated as a building material returned to the earth as loose material. To demolish a house of

Table 8.2: Potential pollution by loose materials

Raw materials/base materials	*Potential process pollution*
Sand and gravel	Dust (possibly containing quartz)
Earth for construction purposes	Possible dust
Fired clay products with low lime content	Carbon dioxide, sulphur dioxide, fluorine, possibly chromium, dust
Fired clay products with 15–20% lime content	Carbon dioxide, possibly chromium, dust

Table 8.3: Primary energy consumption during poduction

Raw materials/base products	*MJ/kg*	*Production temperature (°C)*
Sand and gravel	0.1	–
Earth for building, when compressed	0.1	–
High-fired clay	3.5	1050–1300
Well-fired clay	3.0	800–1050
Medium-fired clay	2.5	500–800
Low/light-fired clay	2.0	350–500
Glazed tiles	8.0	1100 (approx.)
Expanded clay	2.0	1150 (approx.)
The zytan block	4.0	1200 (approx.)

rammed earth, either the roof can be taken off and the rain allowed to wash it away, or it can be hosed down with water.

Sand and gravel as aggregate in cement products

Sand, gravel and crushed stone are the main constituents of all concrete. Sand with round or rectangular grains is preferable, with the smallest possible content of humus, mica or sulphur. It is also an advantage if the sand is not too fine – coastal sand is considered to be the best sort. It is possible to use sand dried from the sea, but continual contact with salt water means that it will contain large quantities of chlorine which corrodes steel. This can easily be washed out with fresh water. Sea sand is often very fine, but this can be remedied by adding a coarser sand. High strength is an important quality for aggregate.

Earth as a building material

'From earth you have come, to earth you shall return.'

In 1982 a large exhibition and conference took place at the Pompidou Centre in Paris entitled 'A forgotten building practice for the future'. The theme was earth as a building material. Earth can be used in construction for more than just trenches and potato cellars. It is the second most important building material after bamboo. More than 30 per cent of the world's current population live in earth houses, which once also flourished in Western Europe but have since been forgotten. They are now on the march again, soon at full speed in France, Germany and the USA.

The aspects of earth building that make it popular are:

Figure 8.2: Traditional earth building by the pisé method in Bhutan, 1996. Photo: C. Butters

- It is based on a resource which is abundant in most countries. In many cases the material can be excavated on site
- It requires much less energy, a small percentage of the energy needed for concrete building; if carried out correctly, it also has a long life expectancy
- It has reasonable and simple building methods which make self-building feasible
- The earth buildings create a good indoor climate because of their good moisture-regulating properties
- Buildings can be recycled more easily than those in any other material.

There are two main ways of building earth houses: ramming (pisé) where the earth is rammed between shuttering to make walls, and earth block (adobe) where the earth is first pressed into blocks and dried before use.

Argillaceous marine earth is considered the best raw material for earth building. It is also possible to mix clay with other types of earth. Earth can be used in its natural state, and stabilizers such as cement or bitumen can also be added to increase the cohesion. It can also be mixed with straw, sawdust or light clinker for reinforcement or to increase the insulation value. If it is a good mixture, homogeneous earth construction has strong structural properties. There are examples of German earth houses up to six storeys high. As with other stone and cast materials the tensile strength is poor, and arches or vaults are necessary over openings. Earth structures reach their ultimate strength after a few years. During the first months the walls are soft enough to be chased for electrics and to have holes bored for pipes, niches made, etc. The only enemy of earth construction is damp – very careful design and construction is necessary to avoid damp problems. Even a small detailing error can lead to big problems. Concrete is tougher than earth in such situations.

Earth building is extremely labour-intensive compared to most other methods. In the present economic situation where all the labour must be paid for, building with earth is very expensive. Earth technology is undergoing intensive development on mainland Europe. At present it should be seen as a potential self-build method, mainly in areas where there are earth resources.

History

Earth buildings have probably been around for over 10 000 years. The oldest remains found so far are the ruins of Jericho, estimated at over 9000 years old. In a grave at Mastaba in Egypt there are traces of 5000-year-old cast earth blocks. English archaeologists have found similar 3000-year-old construction techniques in Pakistan. In the Old Testament, references are made to earth blocks made with straw. One of the Pharaohs gives orders that the children of Israel should not be given straw to make their blocks (*Exodus*, Ch. 5, v. 7). Because of its abundance, earth has been used for most of the architecture 'without architects'. There are many historical examples of pure earth towns, from Jericho to Timbuctoo, including temples, churches and palaces. Both the tower of Babylon and the Great Wall of China were partly constructed of earth.

Towns consisting of earth houses are still built in places like the Yemen. These buildings are several storeys high and built in their hundreds, creating the atmosphere of a mud Manhattan!

Figure 8.3: The earth city of Shiban in the south of Yemen. Source: Flemming Abrahamsson

In both Peru and Chile, the Inca Indians knew of these building techniques long before the Europeans came. The Mexican pueblo is the result of a well-developed earth block technique. Earth building can be found in most cultural periods in world history. In Northern Europe they are less common beyond the eleventh century. An old Irish chronicle tells a story of the patron saint, Patrick, building a rectangular church of earth on the Emerald Isle. In the small French village of Montbrisson is a chapel, *La Salle de Diana*, built with earth blocks in the year 1270, which is now the town library.

Figure 8.4: Earth building at Ile d'Abeau in France by architects F. Jourda and G. Perraudin.

Earth building in central Europe flourished from the end of the eighteenth century and continued until the late nineteenth century. The method received a particularly strong following in Denmark, England and Germany. After the First and Second World Wars earth houses became popular again. Towns and villages in Russia destroyed by the fighting were rebuilt in rammed earth, and in Germany around 100 000 earth houses survive from these periods.

Today there is a fresh wave of interest in earth houses. A housing area of 65 earth dwellings has been built in Ile d'Abeau in France, using several different construction techniques (see Figure 8.4). Similar projects are under construction in Toulouse and Rheims. There are professional training courses at universities in both France and Germany for carpenters, engineers and architects who wish to learn earth building techniques. In the southern states of the USA a whole group of contractors now specialize in earth building.

Finding and extracting raw materials

Earth for building should contain as little humus as possible. It must be firm with a good compressive strength and a low response to moisture and workability. The most appropriate earth is found in moraine areas, as the grain size is suitable and the proportion of clay in the earth is within the limits of 10–50 per cent. Clay can also be found in earth originally formed underwater (under the 'marine border', which varies according to geographical location but is usually around 220 m above sea level).

It is said that in Romania, where earth houses have been the most common form of building to the present day, even the children can classify the clay.

Table 8.4: Estimating the clay content of earth

Thickness when rolled	*Percentage weight of clay*
Cannot be rolled out	Less than 2
3–6 mm-thick rolls	2–5
Approx. 3 mm rolls	5–15
Approx. 2 mm rolls	15–25
Approx. 1–1.5 mm rolls	25–40
Approx. 1 mm rolls	40–60
Rolls thinner than 1 mm	More than 60

(Source: Låg, 1979)

Correct perception has become a tradition. The approximate clay content can be estimated through rolling out clay samples and judging their thickness, as shown in Table 8.4.

Deciding technical properties

Many methods have been developed to test the properties of earth. The following is based on a method recommended by the German industrial standard (DIN 18952). There are quicker and simpler methods, but their results are not always reliable. There are also more chemically based methods.

Assessing the binding tensile strength

As with concrete, it is an advantage to have an even proportion of different-sized particles within the earth, no larger than the small stones in shingle. A well-graded clay will bind better as smaller particles fill the gaps left between the larger particles.

There are usually two tests to assess the binding tensile strength – in both tests the moist earth samples are kept under a damp cloth for 6–12 hours before testing:

- *The ball test* tests stiffness. A sample of 200 gm of earth is rolled into a ball, which is then dropped from a height of 2 m over a glass surface. If the diameter of the flattened ball is less than 50 mm after impact, then the earth is good enough.
- *The figure-of-eight* tests the cohesion between the particles. A fracture test is carried out on a piece of earth formed into the shape of a figure eight. This method was once used for testing concrete. The earth is knocked into the figure eight form with a wooden hammer (see Figure 8.5). The mould has specific proportions and can be made of either hardwood or steel. At the narrowest point it has an area of 5 cm^2. The thickness of the mould is 2.23 cm. The hammered piece of earth is taken out and hung in a circular steel ring. It is then loaded with weight in the form of water in a small vessel. An earth with a binding strength of less than 0.050 kp/cm^2 is unusable.

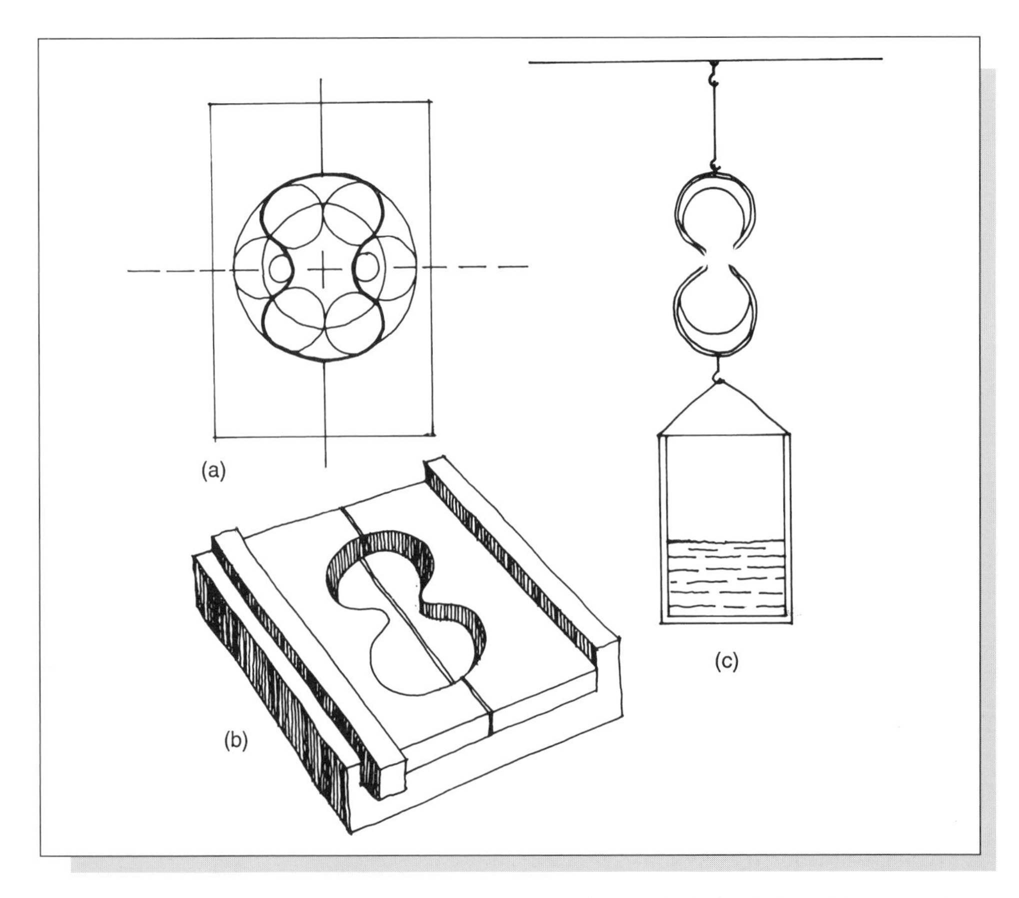

Figure 8.5: Determining the strength of earth using the 'figure-of-eight' technique. (a) Construction of the figure-of-eight mould (DIN 18952). The diameters of the circles from the largest to the smallest are: 78 mm, 52 mm, 26 mm, 10 mm. The distance between the two smallest circles is 22.5 mm. (b) The mould, consisting of two parts, when ready for use. (c) The compressed piece of earths is hung in a steel loop with D = 140 mm; the distance between the claws holding the clay is 75 mm. The width of the claws is the same as the depth of the piece of earth: 223 mm.

Assessing compressive strength

There is a clear connection between the binding tensile strength and the compressive strength. DIN has a standard curve from which the compressive strength can be read as a result of the figure-of-eight tests (see Figure 8.6).

Moisture and shrinkage

Earth that holds its shape has a moisture content of 9.5–23 per cent in its natural state. The more clay it holds, the more moisture it contains. Thoroughly dried walls have a moisture content of 3–5 per cent. This means that earth with

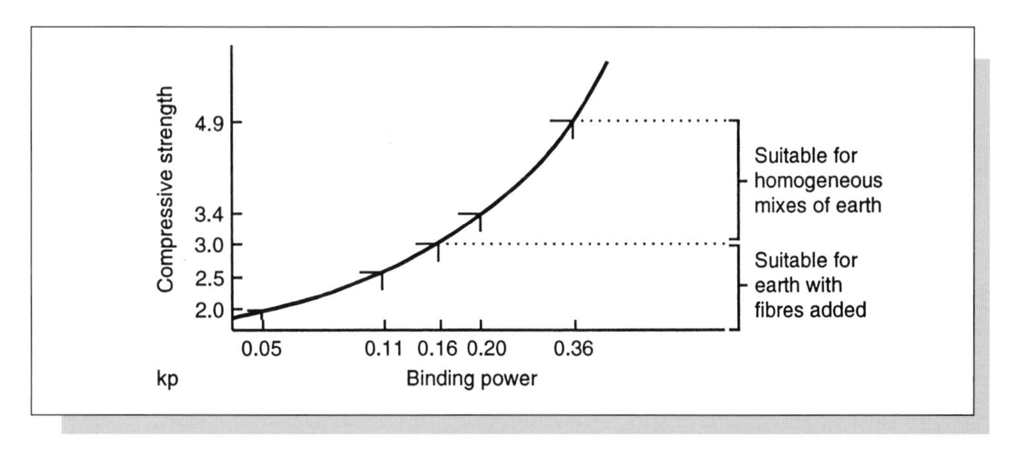

Figure 8.6: Determining compressive strength (according to DIN 18952). The properties of the earth can be read on the right.

a naturally high moisture content will shrink considerably during drying. To assess the moisture content of the earth, a sample of it is weighed, dried out, then re-weighed. The moisture content is the equivalent of the difference between the two weights.

Generally speaking, earth with a high moisture/clay content is best used for an air-dried earth block. Most of the shrinkage will have taken place before the blocks are laid. Through adding plenty of natural fibres, an earth rich in clay can be used for ramming as in the pisé technique.

The preparation of earth

Once the earth has been selected according to the above methods, the topsoil is removed to a depth of 20–30 cm. The earth uncovered is then sieved through a steel net with holes 2.4 cm in size for ramming earth, or 1 cm for the production of earth and clay blocks. If the earth and clay mixture has a variable moisture content, it must be well mixed and stored under a tarpaulin for three to four weeks.

Where necessary, stabilizers or extra amounts of sand or clay can be added either during sieving of the earth or later with an earth grinder. Mixtures containing cement and lime must be used immediately. Others can be stored, but they must be covered with a tarpaulin to preserve the moisture.

Earth structures

Earth is transported straight to the building site without any industrial treatment. Here it is put into casts to make blocks or rammed between shuttering to make walls.

Brick and other fired clay products

'If brick had been discovered today, it would undoubtedly have been the sensation of the century.' (Hoffmann).

Clay is formed by the grinding and disintegration of rock. In a dry state clay has the formula $Al_2O_3.2SiO_2.2H_2O$. By adding water clay becomes workable. The process is reversible.

Clay can be formed and fired up to 1000°C. All water is removed by firing the clay, so the formula becomes $Al_2O_3.2SiO_2$, and this change is irreversible. Water cannot be reintroduced into the clay. It has become a ceramic material, with areas of use that have been the same for thousands of years, in construction, on floors and roofs, as water pipes and tanks. When the temperature in specially-built kilns is increased even more, the clay begins to expand, turning into expanded clay pellets, which in recent years have become an important insulation material and a light aggregate in concrete. If expanded clay is poured into moulds and heated to an even higher temperature, it melts and becomes a highly insulating material called Zytan.

History

'The Chinese invented the compass, gunpowder and the brick' is an old saying amongst brick makers. It could well be true, as archaeologists have unearthed Chinese burnt clay tableaux which can be dated back 6000 years. The first traces of building bricks are from between 1000 and 2000 years later. In Asia there are remains of 4000-year-old brick buildings. In Bombay a brick kiln from about the same period has been found. Between 900 BC and AD 600AD the Babylonians and Assyrians developed a very comprehensive brick-building technique. In Egypt, a pioneering country in many areas, sun-dried bricks were used, except for the occasional use of stone, possibly because of lack of fuel for firing. Brick remains have been found deep in the silt of the Nile, which could mean that there was once brick production even in this area. In Greece, burnt clay probably came into use during the Golden Age of Athens, around 400 BC. The main product was roofing tiles, as used in Italy. The Etruscan walls by Arezzo were built a few years into the Christian epoch and are probably the first brick structures in Italy. The Roman brick industry developed very quickly and produced a whole series of brick elements for both decorative and structural use.

The brick industry in Europe really developed during the eleventh century, and since then it has been the dominant building material in towns. Since 1920 concrete has become a major rival, but brick now seems to be enjoying a renaissance, partly because of its much higher durability.

Brick manufacture

The argillaceous materials used to manufacture bricks must be easily workable and not contain any large hard components or lumps of lime. The latter can cause splitting of the brick when it is exposed to damp. The clay can contain lime, but it has to be evenly distributed. It is an advantage if the clay is well mixed with sand. Clay with too little sand is not easy to shape, but has the advantage of not shrinking so much when drying or being fired. Sand can be added to clays that are too 'fatty'. An idea of the quality of a clay can be found through some simple tests. It must easily form into a ball and keep the prints made by the fine lines of the hand. During drying it must become hard without too many fine cracks.

One thousand square metres of clay can produce about 650 000 bricks per metre of depth. The clay does not usually lie too deep in the ground, so it is relatively easy to extract. This is usually done by first scraping away the soil, then extracting the clay and, after re-planning the area, placing the soil back again.

After the clay has been extracted from the ground, it is covered with water. It then used to be worked by hand with a special hoe or by ramming. The latter method was preferred because it made small stones in the clay obvious. This operation is now carried out by a machine which grinds the clay down to a fine consistency. Additives to reduce its fattiness can be put in the clay and the mixture is then well kneaded. If the clay is stored for between one and three months in an out-house it becomes more workable and produces a better quality final result.

Sand can be used to make the clay leaner, but slag, fly ash and pulverized glass are also suitable. These not only reduce the amount of shrinkage, but make the clay easier to form. The porosity of brick can be increased by adding materials which burn out when the stone is fired, leading to higher insulating values and better moisture regulation. Materials that can be used for this are sawdust, dried peat, chopped straw or pulverized coal. Porosity can also be increased by adding 15–20 per cent of materials that evaporate through heating, such as ground lime, dolomite or marble, which produce carbon dioxide when fired. These additives bind the released sulphur and fluorine into harmless compounds such as gypsum.

Insulating materials such as fossil meal can be added in parts of up to 90 per cent. Fossil meal is a form of earth which consists of air-filled fossils from silica algae. The resulting block has very good insulation value and high porosity. Around Limfjorden in Denmark there is a clay containing fossil meal (about 85 per cent) which occurs naturally. It is called molere, and has a complete brick industry based around it. The resources, however, are very limited.

Forming

Clay needs a water content of approximately 25 per cent in order to be formed. The forming is carried out mechanically by forcing the clay through a die or by

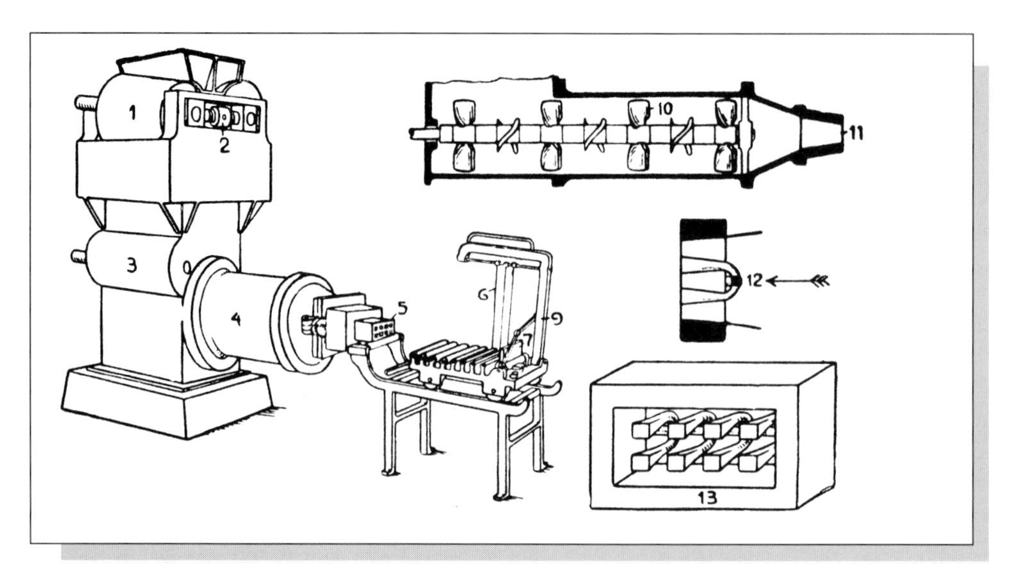

Figure 8.7: The industrial die with mouthpiece.

just knocking the clay by hand into a mould. Mechanical hand presses are also used.

The industrial die presses out the clay through a mouthpiece as a long sausage with a cross-sectional area allowing for shrinkage (see Figure 8.7). Different sizes of mouthpiece and square or round pegs form holes in the clay sausage. Roof tiles can also be produced in this way. The sausage is cut into blocks on a bench. Mobile dies also have equipment to prepare the clay before pressing, and are used where there are smaller deposits of clay.

Handmade bricks are made by placing the clay into wooden or metal moulds in the same way as earth blocks, and striking with a piece of wood (see Figure 8.8). The moulds are sprinkled with sand or dipped in oil or water between strikings. A 'brickstriker' and two assistants can produce 2000 ordinary bricks, 1200 flat roof tiles, or 600 profiled tiles in a day. Even if machine-cut bricks are considerably more economical, the handmade brick with its rustic character is more attractive as a facing brick. As recently as 1973 it was estimated that 99 per cent of all bricks produced in India were handmade (Spence, 1974).

Drying

The unfired brick products are stacked for drying under an open roof for one to two months. For all-year-round manufacturing, bricks need to be stacked inside. This increases energy consumption a good deal, as storage rooms need to be very

Figure 8.8: Wooden mould for handmade bricks.

large. In modern brick factories, special drying houses are kept very hot for two to five days.

Firing

When clay is heated up to boiling point, the water in the pores evaporates, and at 200–300°C the hydrate water evaporates. After this change the clay will not revert to a soft clay with the addition of water, unlike an air dried earth block. Even in the Roman Empire bricks were not fired in temperatures higher than 350–450°C, and this is the case in a great many buildings that still stand today, e.g. the Roman Forum.

If fired at higher temperatures, the particles in the stones are pushed nearer to each other and the brick becomes harder. Between 920 and 1070°C the material begins to sinter. If the temperature is increased even further, the blocks will smelt. Higher temperatures are used in the production of fire-proof bricks and porcelain, using special clay mixtures. To a well trained ear, the temperature at which a brick was fired can be assessed by hitting it with a hammer. The higher and purer the sound, the higher the temperature of the firing. This is especially useful when recycling old bricks.

Clay containing iron turns red when fired, whereas clay containing more than 18 per cent lime turns yellow. There are many different colour variations, determined by the amount of oxygen used during the firing process. Red brick can vary from light red to dark brown.

Chamotte is produced from clay with a low iron and lime content. This can withstand temperatures of up to 1900°C and is classified as fire-proof.

In certain products the brick can be glazed or coloured by the manufacturer using compounds such as oxides of lead, copper, manganese, cadmium, antimony and chromium. To set the glaze onto the brick requires a secondary firing until the glaze smelts. The temperature of this firing should be well under the brick's firing temperature so that it does not lose its form or slide out.

Kilns

Many different types of kiln have been used over the years, but almost all belong to one of three main types: the open charcoal kiln, the circular kiln or the tunnel

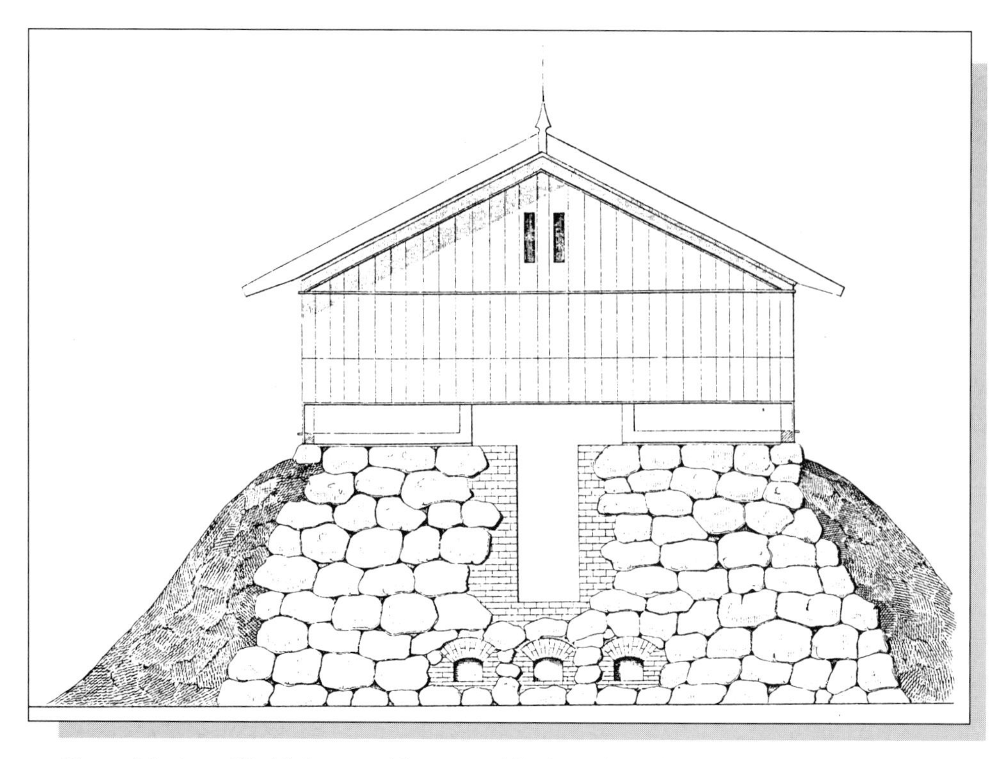

Figure 8.9: A small brick factory with an open kiln from the middle of the 19th century in Scandinavia. Source: Broch 1848.

kiln. It is interesting to note that development of the kiln and the baking oven have run parallel.

The open charcoal kiln is the earliest type, used in smaller brick works as late as the early twentieth century. It consists of two permanent, parallel kiln walls in brick. At the bottom of the walls or between them at the ends there are a series of openings for feeding the fuel. Clay blocks to be fired are stacked up according to a very exact system. The top layer is a solid layer of ready-fired bricks with some openings for the smoke. They are then covered with earth. The firing takes about two days of intensive burning. The bricks are left in the kiln to cool down slowly over a period of several days before the earth and the bricks are removed.

A brick factory should have two or three kilns to guarantee continuous production. Firing in an open charcoal kiln is not very economical with regard to energy consumption. If production is local, the compensation for this is that transport energy is drastically reduced.

A small, unusual and totally new version of the open charcoal kiln has recently been developed in the Middle East. The kiln is in fact a whole house, which is fired. The clay blocks are stacked up into walls and vaults in their air-dried state.

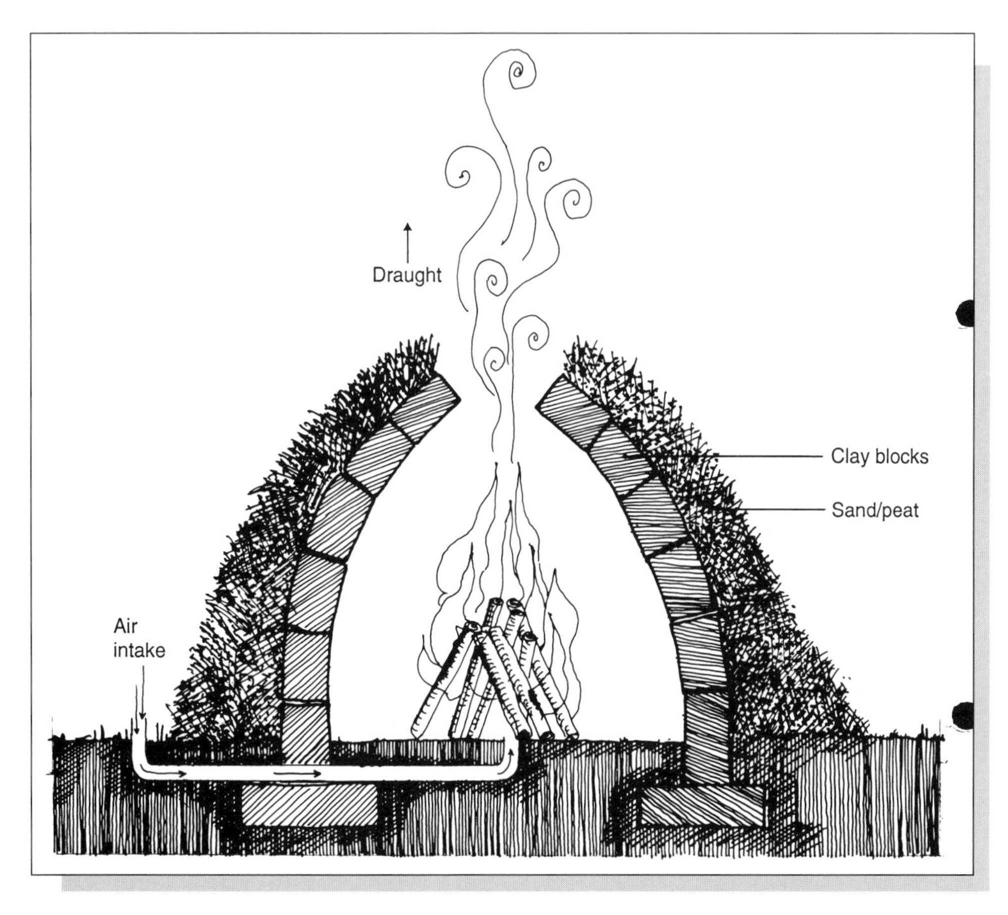

Figure 8.10: Firing clay blocks that in themselves form the walls of the kiln.
Source: Khalili 1983

There is a hole in the roof and an air duct in the ground to feed the fire. A thick layer of earth is placed over the whole building and a huge bonfire is then lit inside the building. A door or hole in the roof is required so that the fire can be loaded with wood. After a couple of days, firing is complete. The building then needs a couple of days to cool down. The earth is removed, the windows are knocked out and any cracks in the walls are filled.

The Hoffman kiln, unlike the charcoal kiln that has to be cooled after each firing, can be kept in continuous use. The firing zone can be simply moved from chamber to chamber. Each chamber is firing for a set period before the heat moves onto the next chamber. A complete rotation takes about three weeks. The bricks are fired with sawdust or fine coal-dust sprinkled down through small openings in the roof of the chambers. In modern brickworks where circular kilns are still used, it is more usual to use oil as a fuel.

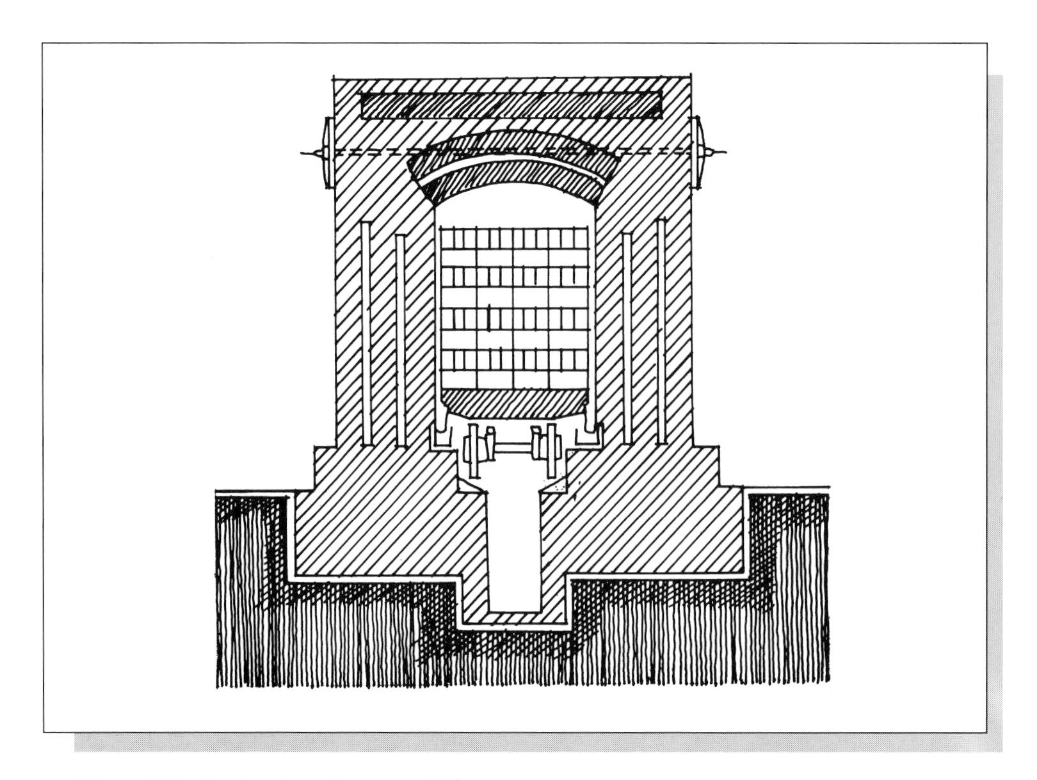

Figure 8.11: Section through a tunnel kiln.

The tunnel kiln came into use after the Second World War. The kiln can be up to 120 m long and is divided up into a warming-up zone, a firing zone and a cooling zone. The unburned clay bricks are placed on a truck which moves slowly through the kiln. The energy source can be coal, gas, oil or electricity.

In the brick industry there is a big difference in the energy consumption of different kilns. The open charcoal kiln uses approximately twice as much energy as the circular kiln, while the circular kiln uses slightly less energy than the tunnel kiln. Energy consumption during firing in the circular kiln and the tunnel kiln varies a great deal depending upon the product being fired, and falls considerably with lower firing temperatures, to about 60 per cent for medium fired products.

Sorting

There is an uneven distribution of heat in an open charcoal kiln. The bricks at the outside are usually less well fired than those in the middle. There is some shrinkage in the circular kiln, but much less than that occurring in the open kiln. Tunnel kilns give the most even heat distribution and shrinkage is minimal, even if the outermost bricks have a tendency to sinter.

Figure 8.12: Examples of English patterns for tiles from around AD 1200.

Manufacture of ceramic tiles

In the third dynasty in Egypt, small glazed tiles in light blue, green and black were used to decorate the steps of the Saqqara pyramid. Nowadays ceramic tiles are widely used in both public buildings and dwellings. Their increased use in housing is largely a result of the development of the private bathroom with associated ceramic plumbing fixtures.

Quarry tiles and terracotta are produced from damp pressed clay in the same way as bricks, using the same raw materials. It is normal to fire the clay until it sinters, at up to 1000°C.

Vitrified ceramic tiles and faience are fired from dry pressed clay, often with ground kaolin, a white clay used in the porcelain industry. Finely-ground waste glass can be added to increase the volume of the mix. The product is fired until vitrified, and the resulting tile is much more exact and smooth than products made from damp pressed clay.

All tiles can be glazed. There are three forms of glazing: cooking salt glaze, lead glaze and earth glaze. Earth glaze is mainly a lime glaze, which can also have pigments added in the form of metal oxides or salts. Many of these are environmental poisons, and there are very strict rules as to how these materials are disposed of as waste products. Salt glaze is pure sodium chloride (NaCl) which is sprinkled on during firing and reacts with clay to produce a silicate glass. This process needs high temperatures and requires a particularly high-quality clay. Lead glaze and earth glaze are applied to the ready-fired products, which are then fired again.

Tiles that are coloured all the way through are usually vitrified and the added pigments are the same as those used in glazes. Pigments used in glazes (see Table 8.5) can be mixed to achieve other colours.

Production of light expanded clay

All clays can be expanded, though some expand more easily than others. The ideal clay is very fine, with a low lime and high iron content. Smelting must not occur before the clay has expanded – this mainly depends upon the minerals in the clay.

Clay used for the production of expanded clay pellets needs to air for about a year before being used. It is then ground, mixed with water and made into pellets. Medium-quality clay can have chemicals added, mostly ammonia sulphite in a proportion of 3 per cent volume of the dry clay, and sodium phosphate in a proportion of 0.1 per cent. The lower the iron content in the clay, the lower the use of energy in the kiln.

Expansion can occur in a vitrifying kiln where sawdust, oil or coal can be mixed with the clay and then fired. Alternatively, the more efficient rotating kiln

Table 8.5: Examples of pigments used for glazing ceramics

Colour	*Alternative pigments*	*Percentage*
Yellow	Ferric oxide	1–2
	Uranium oxide (rare)	4–10
	Sodium aranate (rare)	5–15
	Potassium aranate (rare)	5–15
	Chrome chloride	0.5–1
	Antimony trioxide	10–20
	Vanadium oxide	2–10
Red	Cadmium oxide	1–4
	Chrome oxide	1–2
	Manganese carbonate	2–4
Green	Copper carbonate	1–3
	Chrome oxide	1–3
Blue	Cobalt carbonate	1–3
	Nickel oxide	2–4

can be fired with coal dust, oil vapour or gas, natural gas or bio-gas. The rotating kiln usually consists of a metal cylinder with a diameter of 2–3 m and a length of 12–60 m. But there are also smaller, molbile models (see Figure 8.13). The temperature in the kiln is about 1150°C and the firing time from clay pellets to expanded clay pellets is approximately seven minutes.

For the manufacture of a light clay, thermal block Zytan moulds are filled with light expanded clay, then gases are blown through this mass at temperatures of

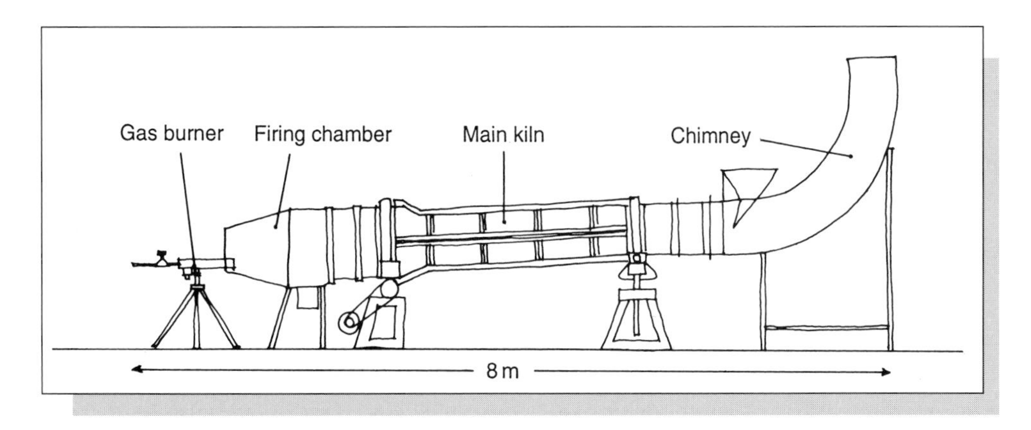

Figure 8.13: Section through a Pakistani mobile rotating kiln for the production of expanded clay. The kiln is about 5 m long with inside diameter of 500 mm. The rate of production is about 125 kg of expanded clay per hour. Source: Asfag 1972

about 1000°C (Brien, 1978). The light clinker expands even more. The spaces within the mould are filled and the material becomes a solid block. Once the moulds have cooled down, the result is a homogeneous and highly-insulating thermal block which can be used immediately. The density of the blocks can vary from 200 kg/m^3 to 1200 kg/m^3 depending on the firing temperature. All blocks are load-bearing, but have different bearing capacities. Holes can be sawn and drilled into these blocks, just as in other light clay blocks. At the moment, these blocks are not produced commercially.

Fired clay products and reduced energy consumption

The energy consumption in the manufacture of fired clay products is very high and thereby also energy polluting.

The brick industry uses large amounts of oil-fired energy to dry the unfired bricks. The required temperature is relatively low, which means that solar energy could be used as an energy source.

The consumption of energy in the kilns can be reduced considerably by the use of bricks with different firing temperatures in building. Many bricklayers will remember the routine of using low- and medium-fired bricks in internal partition walls and well-fired bricks outside. Today, only vitrified and well-fired bricks are available, and these are used inside and out. The use of energy increases by about 0.2 MJ/kg for very 100°C increase in the firing temperature: the brick industry could reduce its total energy consumption by approximately 20 per cent by going back to old methods. This system could go a step further by introducing unfired earth bricks in internal or rendered non-load-bearing walls. There is no technical barrier to the use of this technique, even in large buildings. Unfired brick also has exceptionally good moisture-regulating qualities.

Because of the high temperatures needed for firing clay the use of heat exchangers would be a potential source of energy-saving. One problem that has arisen is the fast erosion of ducts and equipment, mainly because of aggressive sulphur gases. By adding lime, the sulphur can be released in the kiln.

Energy consumption is also related to transport needs. Fired clay products are heavy, and industries producing them are relatively centralized. It is therefore worth considering whether it is ecologically correct to use brick in an area with no local brick factory. This is especially relevant for areas that cannot be reached by water.

Simple technology and the relatively widespread availability of clay gives brick and clay tile production many potential advantages for local manufacture.

Also in the case of light expanded clay products, it should be possible to have competitive manufacturing works at local or regional level, especially in the case of a mobile manufacturing plant.

Recycling must also be considered, since the energy consumption in manufacture is so high. The durability of fired clay products is very high, and the energy needed to remove and clean up old material only represents 0.5 per cent of the energy required for the manufacture of bricks and tiles. However, the re-use of bricks is only possible if a weak- or medium-strength mortar has been used. Products such as roof tiles which have no mortared joints, have a very high re-usability potential. Bricks can also be ground to pozzolana powder, if they were originally fired at temperatures no higher than 600°C.

Light expanded clay that is free from mortar, e.g. in capillary beds or in insulation underneath a building, can be easily re-used in the same way if it has been protected from roots, sand and earth.

References

ASFAG H *et al*, *Pilot plant expanded clay aggregates*, Engineering News No. 17, Lahore 1972

BRIEN K *et al*, *Zytan - a new building material*, Bahia 1978

BROCH T, *Larebog: bygningskunsten*, Christiania 1848

KHALILI N, *Racing alone*, San Francisco 1983

LÅG J, *Berggrunn, jord og jordsmonn*, Oslo 1979

PARRY J P M, *The brick industry: Energy conservation and scale of operation*, *Appr. Techn.* Vol. 2 1975

SPENCE R J S, *Small scale building materials production in India*, unpublished, Cambridge 1974

9 Fossil oils

The most useful type of oil is oil extracted from the Earth. Oil can also be extracted from coal or from oliferous slate or clay. Natural gas is a form of gaseous oil and has approximately the same properties.

Refined oil is the starting point for many products used in the building industry. Tar and asphalt by-products of oil can be used directly, mostly for making roofs, joints etc. watertight. Other refined products provide raw materials for a whole spectrum of products: solvents for painting, glue, waxes, oils, and also plastics. Plastic has developed greatly over the past 40 years. By 1971 an average apartment contained about 1 ton of plastic. A modern Swedish apartment contains approximately 3 tons of plastic in everything from the covering for electric cables to floor coverings and window frames. The building industry uses 25 per cent of all plastic produced.

Distillates from coal tar, natural oil and natural gas are formed by hydrocarbons. These are chemical compounds containing only carbon and hydrogen.

The explanation of how oil has been formed has changed somewhat over the centuries. Oil was once considered to come from the corpses of those who died during the great flood described in the Bible; theories later claimed that it came from rain from outer space. Today, most researchers agree that the oil within the Earth is formed from animal and plant remains that have sunk in shallow stretches of sea in prehistoric times, and have later been exposed to certain pressure and temperature conditions.

It is estimated that 6000 years ago oil from the Earth was used for building in the form of asphalt. Noah used the material to make his ark watertight, and the Babylonians jointed their clay block houses with bitumen from asphalt lakes. Wider use of oil did not really start until the nineteenth century, when the industry began with the huge exploitation of reserves on the American continent. The main use of oil was as a fuel, and later for waterproofing. It was not until the twentieth century that is was first used for the commercial production of plastics.

Table 9.1: Basic materials from oil and gas

Material	*Areas of use*
Bitumen	Vapour barrier, damp-proofing, mastic
Asphalt	Mastic, vapour barrier, damp-proofing
Organic solvents	Paint thinner, glue, mastic, impregnation
Plastics	Sheeting, window frames, wallpaper, cladding, flooring, thermal insulation, electric insulation, pipes, door and window furniture
Other chemicals	Additives in concrete and plastics, organic pigments, impregnation, additives and binders in pain and glue, constituents for the production of plastics

Oil resources are very limited. This is particularly the case for oil from the Earth, where the supply is estimated to last 40 to 50 years at the present rate of exploitation (British Petroleum, 1993). Oil is extracted by pumping if from subterranean reservoirs to the surface. It is then transported to refineries where the crude oil is distilled into different fractions, which are further refined at plants producing paints, plastics or other materials. Extraction, refining and production of the final material all cause industrial pollution. Every time an oil tanker unloads, tons of hydrocarbons are released into the air. If an oil blow-out occurs on land or at sea, oil and chemical tankers can go aground, leaving coastal areas in ecological ruin for decades. The catastrophic potential of oil can be used as a political weapon, as in the Gulf War when the oil wells of Kuwait were set on fire. The oil industry is similar in character to the atomic power industry.

The refining of oil to plastics and other basic materials requires a great deal of energy – as much as in the metal industries. The greenhouse gas carbon dioxide and acidic sulphur dioxide are released during processing. Many of the pollutants from the production process are highly poisonous, including hydrocarbons from oil-based products or chlorine and heavy metals required for processing. This does not affect the natural environment alone. Cancer and chemically-induced nervous problems are more frequent amongst workers in these industries than in the general population. Children born with deformities are more commonly registered in areas near to plastics factories than elsewhere.

Oil-based products, when used in building, can release transitory organic compounds either as direct emissions or as a result of a chemical reaction with other materials, e.g. concrete. Many of these pollutants irritate the mucous membranes and can produce traditional symptoms of a bad indoor climate such as irritation in the eyes, nose and throat, unusual tiredness, headache, giddiness, sickness and increased frequency of respiratory illnesses. Other more serious emissions have also been registered; these can cause allergy, cancer or embryonic malformation. There has also been a marked increase in deaths due to smoke inhalation from fires during the last few years – one reason for this is the increased use of plastic in buildings (Curwell, 1990).

Tar, solvents and other oil-based chemicals and products that contain them have a strong risk factor as waste products – these substances are highly poisonous and have to be placed at special disposal depots. The dumping of waste plastic can lead to the release of poisonous substances such as heavy metals into the environment. Plastic materials in themselves are not usually poisonous, but pose a problem mainly because of their volume, as they break down very slowly in the natural environment.

Old asphalt can be recycled quite effectively by mixing it with new asphalt. Recycling is also possible for a few plastics. All plastics, however, contain additives and impurities which lead to a lower quality plastic after recycling (down-cycling). Even if the primary energy consumption through down-cycling is only 10 per cent of the cost of manufacture of new plastic, the high energy costs of transport still have to be taken into account, as the plastics industry is highly centralized. Plastic materials can be recycled at least four or five times before they finally have to be abandoned as waste.

Most of the waste products from solvents, oil-based chemicals and plastics can be transformed into energy. With few exceptions, the materials must be burned in furnaces with special facilities for cleaning the emissions. Even so, there is a

Table 9.2: Primary energy consumption for different oil products

Product	*MJ/kg*
Bitumen	10–11
Asphalt	3
Solvents	14–36
Other chemicals:	
Urea	14
Formaldehyde	14
Phenol	18
Ethylene (gas)	27
Acetone	13
Ethanol	15
Plastics:	
Polyvinyl chloride	56–84
Polypropylene	71
Polystyrene	75
Polyethylene	67
Polyester	22
Phenolic plastic	22
Acrylic plastic	56
Polyurethane	98–110

Note: All the products except for gas and asphalt weigh about 1 kg/l.

chance that very poisonous pollutants such as dioxin and heavy metals will be released.

The basic materials

Bitumen and tar

Bitumen is obtained by distilling oil at 200–300°C. It is a strong waterproofing substance used to impregnate materials such as paper, sheets and jointing mastics, or applied directly to a surface. The products usually have organic solvents, or are in a suspension of water and finely ground clay. By adding powdered stone, sand or gravel different varieties of asphalt are produced which can be used for road surfaces, damp-proofing on foundations or independent roof covering on a flat roof. Asphalt also occurs naturally, for example in Trinidad, where it is called Trinidad asphalt.

Coal tar can be extracted from coal by condensation. This substance was once used widely in the building industry, but is now almost completely replaced by bitumen.

The chemical composition of tar and bitumen differ greatly. Tar is composed of almost 50 per cent polycyclical aromatic hydrocarbon (PAH) compounds which are almost non-existent in bitumen. Both materials can include early stages of dioxin and are a potentially dangerous source of organic compound seepage. Materials that contain tar or bitumen, need to be safely disposed of (Strunge, 1990).

Solvents and other chemicals

Light distillates can be used directly as solvents or as a chemical base for other products. The monomers, which constitute essential components of plastics, are important. Solvents are substances that break down other materials without changing them chemically, and usually evaporate from a finished product (as in paint that has dried).

The following substances are products directly and indirectly used in the building industry.

Aliphatic and aromatic hydrocarbons

Amongst the aliphatic hydrocarbons are paraffins, naphthenes and n-hexane, while the aromatics include substances such as xylene, toluene, trimethylbenzene, ethyl benzene and styrene. These substances can be used directly as solvents. Naphtha is also the most important raw material for the production of plastics from intermediary substances such as propylene, ethylene and acetylene. Styrene, benzene, toluene and xylene are also necessary chemicals for the plastics industry and the latter two are used in the production of organic pigments. Benzene is the initial source of creosote, which is mixed with coal tar to make the impregnating poison carbolineum.

Chlorinated hydrocarbons

These are formed when hydrocarbons react with hydrochloric acid. They include important solvents such as trichloroethane, trichloromethane, trichloroethene, dichloroethane and dichloromethane. These substances are used mainly in varnishes, paints and paint removers. Dichloroethane is also an important solvent for synthetic rubber and is often used in mastics with a bituminous base. Polychlorobiphenyls (PCBs) have been used widely in the past as fire retardants in electrical cables and as softeners in mastics, but are no longer used because of their high toxicity. The chloroparaffins are very common in plastic products as flame retardants and secondary softeners, in PVC floor coverings, as softeners and binding agents in putty and mastics and as fire retardants in synthetic rubber.

Chlorofluorocarbons

Chlorofluorocarbons are produced by replacing hydrogen with fluorine in chlorine compounds and are used for foaming plastic-based mastics and insulation materials. Chlorofluorocarbons that contain bromine are used as fire retardants in a range of plastics. These substances are very stable in the lower part of the Earth's atmosphere. When they reach the stratosphere the sunlight is strong

Table 9.3: The relative effect of the different chlorofluorocarbons on the ozone layer

Chlorofluorocarbon	*Destruction factor*
CFC 11 — (Trichlorofluoromethane)	1
CFC 12 — (Dichlorodifluoromethane)	1
CFC 22 — (Chlorodifluoromethane)	0.05
CFC 113 — (Trichlorofluoromethane)	0.8
CFC 114 — (Dichlorotetrafluoroethane)	1
CFC 115 — (Chloropentafluoroethane)	0.6
CFC 132b — (Dichlorodifluoroethane)	Less than 0.02
CFC 134a — (Tetrafluoroethane)	0
CFC 142b — (Chlorodifluoroethane)	0.05
Halon 2111 — (Bromochlorodifluoromethane)	3
Halon 1301 — (Bromotrifluoromethane)	10
Halon 2402 — (Dibromotetrafluoroethane)	6.2

enough to break down their molecular structure, releasing chlorine atoms which react with natural ozone and break down the ozone layer (see Table 9.3).

Alcohols and aldehydes

The alcohols that are mostly used as solvents, especially in varnishes, are ethanol, propanol, isopropanol, buthanol, isobuthanol and methanol. Phenol is an important ingredient in different building glues. Through further oxidation of alcohol, formaldehyde (an important glue substance when mixed with phenol and urea) is formed.

Ether alcohols and ketones

Important ether alcohols are glycol ethers such as methyl and ethyl glycol. They are used as solvents and plasticizers in varnishes. Methylketone and methyl-isobutylketone are the ketones used as solvents in chloroprene glue.

Amines

Amines are produced from hydrocarbons which react with ammonia. Amines are most common as additives in plastics, e.g. silicon and polyester, mainly as a hardener or anti-oxidizer. Amines are the starting point for the production of isocyanate, which is the main constituent of polyurethane. Amines are also used in the production of certain organic paint pigments.

Alkenes (or olefines)

This is the group name for hydrocarbons with double combinations. Amongst the most important are ethylene and propylene, which are produced from naphtha and function as monomers in the production of polyethylene and polypropylene. Vinyl chloride is produced by chlorinating ethylene and it is the main constituent of PVC plastics.

Esters

Esters are formed when hydrocarbons react with acetic acid. Butyl acetate, ethyl acetate and methyl acetate are commonly used as solvents in glue, while polyvinyl acetate (PVAC), is an important binding agent in certain water-based glues and paints. The acrylates are esters of acrylic acid, which is oxidized from propylene, and is used as a binding agent in paints and the production of plastics such as polymethyl metacrylate ('Plexiglas').

Phthallic acid esters

These esters are produced when phthallic acid reacts with alcohols. They are used mainly as plasticizers in a range of plastics and can constitute as much as 50 per cent of a plastic. The most important types are diochtylphtalate (DOP) and diethylhexylphtalate (DEHP).

Table 9.4: The environmental effects of solvents used in the building industry

Solvent	*Environmental effects*
Aromates:	
Xylene	Irritates mucous membranes; can damage the heart, liver, kidneys and nervous system
Toluene	Irritates mucous membranes; can damage the nervous system
Benzene	Carcinogenic; mutagenic
Styrene	Mutagenic; irritates mucous membranes
Aliphatic substances: Paraffin Naphthene n-hexame	Generally irritate skin and inhalation routes; can act as promotor for carcinogenic substances
Chlorinated hydrocarbons: Dichloroethane Trichloroethane Trichloroethylene	Generally highly poisonous to the majority of organs; irritate mucous membranes; can damage the liver and kidneys; carcinogenic; mutagenic; narcotic
Alcohols[1]**:** Ethanol Propanol Methanol Isopropanol Butanol	Generally irritate mucous membranes; large doses can damage the foetus
Esters: Butyl acetate Ethyl acetate Methyl acetate	Generally irritate mucous membranes; medium strong nerve poisons; mutagenic
Ether alcohols: Methyl glycol Ethyl glycol	Generally weak nerve poisons; can slightly damage the foetus
Ketones: Methyl keton Methyl isobutyl ketone Acetone	Generally weak nerve poisons; can slightly damage the foetus
Terpenes[2]**:** Limonen Turpentine	Slighty allergenic; slightly irritates mucous membranes; slightly acting as promotor for carcinogenic substances

(1) Can be produced by plants.
(2) Produced by plants

Plastics in building

During the last 20 years distillates from oil and natural gas, mainly naphtha, have become almost the only raw material used in the plastics industry.

Table 9.5: Oil based chemicals with high environmental risk

Oil based chemical	*Areas of use*	*Environmental effects*
Formaldehyde	Glue in chipboard and plywood	Carcinogenic; allergenic; irritates air inhalation routes; poisonous to water organisms
Phenol	Glue in laminated timber	Carcinogenic; mutagenic; poisonous to water organisms
Chloroprene	Synthetic rubber, glue	Carcinogenic; damages liver and kidneys; irritates inhalation routes
Butadiene	Synthetic rubber (SBR)	Probably carcinogenic
Vinyl chloride	Polyvinyl chloride (PVC)	Persistent carcinogenic; can cause damage to liver, lungs, skin and joints; irritates inhalation routes; poisonous to water organisms
Ethylene (ethene)	Polyethylene	Probably carcinogenic
Propylene (propene)	Polyethylene	Probably carcinogenic
Phthalates	Softeners in plastics	Persistent; irritates the mucous membranes; allergenic; probably carcinogenic; environmental oestrogen: damages reproductive organs
Amines	Silicone, polyurethane, epoxy	Irritate inhalation routes; allergenic; possibly mutagenic; very acidifying in water
Epichlorohydrin	Epoxy	Carcinogenic; highly poisonous to water organisms
Acrylonitrile	Synthetic rubber	Carcinogenic; highly poisonous to water organisms
Acrylic acid	Acrylic plastics and paints	Poisonous to water organisms
Styrene	Polystyrene, polyester, synthetic rubber (SBR)	Irritates air inhalation routes; damages the reproductive organs
Isocyanate (TDI, MDI, etc.)	Polyurethane, glue	Strongly allergenic; difficult to break down; irritates skin and mucous membranes
Alkyl phenol toxilates	Pigement paste, alkyd varnish	Environmental oestrogen; damages reproductive organs

Previously oil from coal and partly natural materials such as cellulose, animal and vegetable proteins were used.

The definition of a plastic is: a substance that contains natural or synthetic high molecular organic material which can be liquefied and thereby cast in specific

Table 9.6: The use of plastics in a typical dwelling

Use	*kg*	%
Flooring	800	30
Glue, mastics	700	26
Pipework	425	16
Paint, filler	275	10
Wallpaper, sheeting (e.g. vapour barrier)	200	8
Thermal insulation	100	4
Electrical installation	100	4
Cover strips, skirtings, etc.	50	2
Total	2650	100

moulds. The 'building blocks' are called monomers, the completed plastic is called a polymer and the reaction is polymerization. During production processes substances such as chlorine, hydrochloric acid, fluorine, nitrogen, oxygen and sulphur are used, as well as oil-based chemicals. Almost all plastics have a rich variety of additives including plasticizers, pigments, stabilizers against solar radiation, preservatives and perfumes.

Plastics are divided into two categories: thermoplastics and thermosetting plastics (see Table 9.7). Thermoplastics leave the factory complete, but can be worked to a certain extent with pressure and warmth, and can even be cut. Common thermoplastics in the building industry are polyvinyl chloride, polypropylene, polyethylene and polystyrene. Thermosetting plastics differ from thermoplastics in that they are not finished products – the product is completed by smaller companies or at the building site where hardeners are added using two component plastics, amongst them polyester, epoxy and polyurethane. The synthetic rubbers are a sub-group of thermosetting plastics with almost permanent elasticity. The basic thermoplastics can be foamed up, extruded, moulded, rolled out to thin foil, etc.

Polyvinyl chloride was the first plastic. Polymerization was discovered by accident by the French chemist Henri Regnault in 1838. PVC was first produced commercially 100 years later. In 1865 celluloid (a mixture of cellulose nitrate and camphor) was patented. Bakelite plastic was the first really successful plastic. It comprised mainly synthetic phenol formaldehyde resins and was patented in 1909. Other milestones in plastics include the first production of polystyrene in Germany in 1930, polyethylene and acrylates in 1933, polyester in 1942 and silicones in 1944.

Pollution related to the most important building plastics

Depending on their type, plastics give off environmentally damaging substances during production and use, and when they are deposited or dumped. Primary

Table 9.7: The use of plastics in the building industry

Type of plastic	*General areas of use*
Thermoplastics:	
Polyethylene (PE):	
hard	Drainpipes, water pipes, interior furnishings and detailing
soft	Sheeting (vapour barrier, in foundation work, false ceilings), cable insulation
Polyisobutylene (PIB)	Roofing felt
Polypropylene (PP)	Sheeting, boards, pipes, carpets (needle-punched carpet), electric fittings, electric switches, cable insulation
Polyamide (PA)	Pipes, fibre, carpets (needle-punched carpet), electric fittings, electric switches, cable insulation; tape
Polyacetal (POM)	Pipes, boards, electric fittings
Polytetrafluorethylene (PTFE)	Thermally-insulated technical equipment, electrical equipment, gaskets
Polyphenyloxide (PPO)	Thermally insulated technical equipment
Polycarbonate (PC)	Greenhouse glass, roof lights
Polymethyl methacrylates (PMMA)	Rooflights, boards, flooring, bath tubs, paint
Methyl metacrylate (AMMA)	Paint
Polyvinyl chloride (PVC)	Sheeting, boards, sections/profiles, window frames, pipes, cable, artificial leather, flooring, wallpapers, gutters, sealing strips
Polystyrene (PS, XPS, EPS)	Sheeting, thermal insulation (foamed), electrical insulation, light fittings
Acrylonitrile butadiene styrene (ABS)	Pipes, door handles, electric fittings, electrical switches
Polyvinyl acetate (PVAC)	Paint, adhesives
Ethylene vinyl acetate sampolymer (EVA)	Paint, adhesives
Cellulose acetate (CA)(1)	Tape, sheeting
Polyacryl nitrile (PAN)	Carpets, reinforcement in concrete
Thermosetting plastics:	
Butadiene styrene rubber (SBR)	Flooring, sealing strips
Butadiene acrylonitrile rubber (NBR)	Hoses, cables, sealants
Chloroprene rubber (CR)	Sealing strips
Ethylene propylene rubber (EPDM)	Sealing strips
Butyl rubber (IIR)	Sealing strips
Silicone rubber (SR)(2)	Electrical insulation, sealants
Polysulphide rubber (T)	Sealants
Casein plastic (CS)(3)	Door handles
Phenol formaldehyde (PF) (Bakelite)(4)	Handles, black and brown electrical fittings, thermal insulation (foamed), laminates, adhesives for plywood and chipboard
Urea formaldehyde (UF)	Light-coloured and white electrical fittings, socket outlets, switches, adhesive for plywood and chipboard, toilet seats, thermal insulation (foamed)
Melamine formaldehyde (MF)	Electrical fittings, laminates, adhesives

Epoxide resins (EP)	Filler, adhesives, paint, floor finishes, clear finishes, moulding of electrical components
Polyurethane (PUR)	Thermal insulation (foamed), adhesives, clear finishes, floor finishes, moulding of electrical components, paint, sealants
Unsaturated polyester (UP) (reinforced with fibreglass)	Roof lights, window frames, gutters, adhesives, clear finishes, floor finishes, rooflight domes, tanks, bath tubs, boards, paint

Notes:

(1) Based on cellulose
(2) Based on silicon dioxide, but polymerization requires the help of hydrocarbons
(3) Based on milk casein with the additional help of formaldehyde
(4) Bakelite is the trade name for phenolic materials manufactured by Bakelite Xylonite Ltd.

Table 9.8: The use of additives in plastic products

Area of use	*Additive /type of plastic (abbreviation, see Table 9.7)*
Anti-oxidants and ultraviolet stabilizers (0.02–1.8% by weight)	Phenols/various • phosphorous compounds/various • hydroxyphenyl benzotriazoles/various • soya oil/PVC • lead compounds/PVC • organic tin compounds/PVC • organic nickel compounds/PVC • barium–cadmium compounds/PVC • calcium–zinc compounds/PVC
Lubricants	Stearates, paraffin oils, paraffin waxes, amide waxes/various
Colour pigments (0.5–1% by weight)	Zn, Cu, Cr, Ni, Nd, Pb (as shown in Table 18.2)/various
Fire retardants (up to 10% by weight)	Chlorine compounds/PE, PP and PVC • bromine compounds/various • phosphorus compounds/PVC, PPO • phosphates/ABS, PE and PP • boron compounds/various • tin oxide/various • zinc oxide/various • zinc borate/various • aluminium trioxide and trihydrate/various • antimony silicates/various
Smoke reducer (approx. 2.5–10% by weight)	Aluminium trihydrate/various • antimony trioxide metals/various • molybdenum oxide/PVC
Anti-static agents (up to 4% by weight)	Ammonia compounds of alkanes/various • alkyl sulphonates, sulphates and phosphates/various • polyethylene glycol, esters and ethers/various • fatty acids-esters/various • ethanolamides/various • mono- and diglycerides/various • ethoxylated fatty amides/various
Softeners (up to 50% by weight)	Phthalatesters/various • aliphatic esters from dicarbon acid/various • esters from phosphonic acid/various • esters and phenols from alkylsulphonic acid/various • esters from citric acid/various • trimellitate/various • chlorofied paraffins/various • polyesters/various
Fillers (up to 50% by weight)	Challac, zinc oxide, wood, flour, stone flour, talcum, kaolin/various
Foaming agents	Pentane/PS, PF, PUR • trichlorofluoromethane/PS, UF, PUR, PF • dichlorodifluoromethane/PS • oxygen/UF • water/PF

energy consumption for all plastics is high and they are also energy polluting. Extraction and refining of crude oil also has a considerable impact on the environment. The different plastics have the following properties.

Polyethylene (PE)

Polyethylene is produced 99.5 per cent from polyethylene, which is polymerized from ethylene (ethene) and to which 0.5 per cent antioxidant, light-stabilizer and pigment is added. The antioxidant is usually a phenol compound and the ultraviolet stabilizer consists of amines or carbon black. Other additives are also used in larger or smaller proportions. Exposure to ethylene (ethene) may occur in the workplace. The finished product probably does not emit anything. As waste it is difficult to decompose, but it can be burned without giving off dangerous fumes.

Polypropylene (PP)

This is produced through polymerization of propylene. Ultraviolet stabilizers, anti-oxidants and colouring are usually added. Phenol compounds are used as antioxidants and amines as ultaviolet stabilizers, to a total of about 0.5 per cent. Other additives are used in variable proportions.

Exposure to propylene during its manufacture can be damaging. There are no dangerous emissions from the finished product. As waste it is difficult to decompose.

Polystyrene (PS)

Polystyrene is produced by the polymerization of styrene to two different products: foamed-up expanded polystyrene (EPS) and extruded polystyrene (XPS). The end product for both is insulation, but the latter is also vapour-proof. EPS comprises 98 per cent styrene; in XPS only 91 per cent is used. Additives include an antioxidant, an ultraviolet stabilizer and even a fire retardant. Phenol propionate in a proportion of 0.1 per cent is usually the antioxidant, amines are the ultraviolet stabilizer and the flame retardant is organic bromine compounds with or without antimony salts, up to one per cent in EPS and two per cent in XPS. An inhibitor can also be included in the product to prevent spontaneous polymerization; this is usually hydrochinon in a proportion of about 3 per cent. EPS is then foamed up with pentane and XPS with chlorofluorocarbons.

During production emissions of benzene, ethyl benzene, styrene, pentane and chlorofluorocarbons are quite likely. In production plants the effects of benzene, ethylene and styrene have been registered.

The finished product can have some unstable residues of monomers of styrene (less than 0.05 per cent) which may be released into the air, depending upon how the material has been installed in the building. XPS can also release smaller amounts of chlorofluorocarbons. As a waste product it can be environmentally damaging through the leakage of certain additives. It is also difficult to decompose.

Table 9.9: Other registered pollution from plastics

Type of plastic	*Pollution*
Polyester (UP)	Styrene (P)(H), dichloromethane (P)
Epoxy (EP)	Phenol (P), epichlorohydrin (P), amines (H)
Polyamide (PA)	Benzene (H), ammonia (H)
Polymethylmethacrylate (PMMA)	Acetonitrile (P), acrylonitrile (P)
Ureaformaldehyde (UF)	Formaldehyde (P)(H)
Melamineformaldehyde (MF)	Phenol (P), formaldehyde (P)
Polysulphide (T)	Toluene (P)(H), chloroparaffin (P)(H)
Silicone (Si)	Xylene (P)(H)
Styrene rubber (SBR)	Styrene (P)(H), xylene (P)(H), butadiene (P), hexane (P)(H), toluene (P)(H)
Isoprene rubber	Xylene (P)(H), nitrosamines (P)
Ethylene propylene rubber (EPDM)	Benzene (P), hexane (P), nitrosamines (P)
Chloroprene rubber (CR)	Chloroprene (P)(H), nitrosamines (P)
Polycarbonate (PC)	Possible bisphenol-A (H)

(P), in production; (H), in the house

Polyurethane (PUR)

Polyurethane is produced in a reaction between different polyethers (4 per cent) and isocyanates (40 per cent), using organic tin compounds as the catalyst. Antioxidants and flame retardants are also used. Phenol propionate is the usual antioxidant, and the flame retardant is an organic bromine compound. Chlorofluorocarbons, pentane gas or carbon dioxide, in a proportion of 10–15 per cent, are used to foam up the plastic.

Materials released during production are chlorinated hydrocarbons, phenol, formaldehyde and ammonia, possibly even organic tin compounds and chlorofluorocarbons. Workers are exposed, amongst other things, to isocyanates.

Small emissions of unreacted isocyanates and amines can seep from the finished product and within the building, along with a smaller seepage of chlorofluorocarbons, if they were used for foaming-up. Environmentally-damaging substances can be washed out of the waste product. Polyurethane has a long decomposition period.

Polyvinyl chloride (PVC)

PVC is produced by a polymerization of vinyl chloride, which in turn is produced from 51 per cent chlorine and 43 per cent ethylene via ethylene chloride. Many additives are also used, in some cases up to 50 per cent plasticizers, 0.02 per cent antioxidants and ultraviolet stabilizers, a maximum of 10 per cent flame retardants, 2.5–10 per cent smoke reducers, a maximum of 4 per cent anti-static agents, pigment 0.5–1 per cent and a maximum of 50 per cent fillers. Constituents that are critical for the environment are substances such as plasticizers containing phtalaths, ultraviolet stabilizers containing cadmium, lead or tin (in the case

Table 9.10: Plastics and fire

Type of plastic	*Gas emitted when burnt*
Polyvinyl chloride (PVC)	CO, CO_2, CH_4, HCl, Ba, Cd
Unsaturated polyester (UP)	CO, CO_2, benzene, styrene, formaldehyde
Polyurethane (PUR)	CO, CO_2, benzo nitrile, acetonitrile, ammonia, prussic acid, NO_x
Polystyrene (PS)	CO, CO_2, benzene, styrene, formaldehyde
Chloroprene rubber (CR)	HCl, dioxines
Butadiene styrene rubber (SBR)	SO_x, NO_x

Note: When using halogenic fire retardants and chlorinated paraffins, dioxines can be formed

of windows) and flame retardants with chloroparaffins and antimony trioxide. In PVC gutters, cables and pipes lead is often used as the ultraviolet stabilizer.

There are likely to be emissions from production plants of chlorine gas, ethylene, dioxin, vinyl chloride, the solvent dichloretane, mercury and other damaging substances. Certain larger plastics works have emissions of tons of phthalate into the air every year. During production, workers can be exposed to organic acidic anhydrides.

Emissions of phthalates or organic acidic anhydrides (when heated) can occur from the completed product and within the building, together with a series of other volatile substances such as aliphatic and aromatic hydrocarbons, phenols, aldehydes and ketanes, though only in small amounts. Left-over monomers from vinyl chloride may also be released (approximately 10 mg/kg PVC). There is also greater microbiological growth in plastic with phthalates, which probably functions as a source of carbon and nitrogen.

As a waste product, PVC contains environmentally dangerous substances that can seep out, e.g. when heavy metals have been used as pigments or cadmium as an ultraviolet stabilizer. PVC is considered to be the largest source of chlorine in waste products. When burnt it can form concentrated hydrochloric acid and dioxin. PVC waste can form hydrogen chloride when exposed to solar radiation. It decomposes slowly.

Durability of plastic products

Many external factors can break down plastics: ultraviolet and visible light, heat, cold, mechanical stress, wind, snow, hail, ice, acids, ozone and other air pollutants, water and other liquids, micro-organisms, animals and plants. The lifespan of a plastic depends on its type, its position and the local climate.

Plastic products are used in floors, roofs and walls in such a way that it is difficult and expensive to repair or replace them. They should have a functional lifespan equivalent to other materials in the building – at least 50 years. It is unlikely that any of today's plastics can satisfy such conditions.

Table 9.11: The anticipated lifespan of certain plastics

Type of plastic	*Assumed lifespan (years)*
PMMA	Less than 40
PIB	11–less than 40
PVC	8–less than 30(1)
PE	2–15(1)
UP	5–less than 35
EPDM	Less than 30
PUR	7–10
CR	2–less than 40
IIR	2–less than 35
T	22–less than 50
Si	14–less than 50
ABS	15
MF	6–10
PF	16–18
NBR	10
EVA	3
PA	11–less than 30
PP	3–less than 10
SBR	8–10
PTFE	25–less than 50

Notes:
The evaluation includes both external and internal use and built-in situations. Positioning within water or earth is not included. The most protected locations achieve the best results.
(1) Does not apply to buried cold water pipes in thicker plastic, which lasts longer, especially PE.
(Sources: Grunau, 1980; Holmström, 1984)

The vast majority of plastic products currently on the market have been around for less than 15 years, so there is very little feedback on their lifespan. Other products have been on the market for a longer period, but amongst the polymer technicians it is well known that today's components are very different from those that were used in products of 20 years ago. The design of products has changed so much recently that it is difficult to find examples giving a picture of the lifespan of articles and products made today.

The assumed lifespan of a plastic is based on so-called accelerated ageing tests. The material is exposed to heavy, concentrated stresses and strains over a short period. Dr K. Berger from the plastics manufacturers Ciba Geigy AG states that present forms of accelerated ageing tests have a 'low level of accuracy at all levels' (Holmström, 1984).

The picture is not made easier by the fact that the plastics are often full of additives. PVC is considered a plastic with very good durability, but it has been known to undergo very rapid breakdown. In Sweden, 10-year-old plastic skirtings crumbled, not because of the PVC but because of an added acrylonitrile butadiene styrene (ABS)-plastic which should have increased the strength and durability. All plastics oxidize easily.

Polyethylene sheeting, which was in use as a moisture barrier until 1975, had an effective lifespan of 10 years. This is far too low considering that the sheeting is usually inaccessible within the fabric of a building, and often supposed to prevent condensation within the walls. Polyethylene has recently included additives which should make it more stable.

Sealing strips of ethylene propylene rubber (EPDM) are often used between the elements in prefabricated buildings of timber and concrete. Research has shown that certain makes have lost elasticity after only one year, which means that the joint is open and the material no longer functions.

Recycling

Even if plastics have a relatively short functional lifespan, it takes a long time for them to decompose in the natural environment. On tips, plastic waste is a problem in terms of volume as well as pollution because of the additives which seep into the soil and ground water, these problems can be reduced by recycling plastic.

Recycling through re-use is not really practicable. Recycling through melting down is possible. Thermoplastics, and even a few thermosetting plastics, can be recycled in this way. Amongst them are polyvinyl chloride, polyethylene and polypropylene. Recycling is also possible, in theory, for purified polyurethane products, but is not happening very much at present. Synthetic rubbers can be crumbled for use as a filler.

The maximum potential of future plastic recycling is estimated at 20–30 per cent in the form of down-cycling only. Almost all plastics are impure because of their additives, which makes reclamation of the original materials technically difficult. The uses for recycled plastic vary from park benches, sound barriers and flowerpots to huge timber-like prefabricated building-units for construction. The latter are now in production in Great Britain, Sweden and the USA, based on melted polystyrene waste with 4 per cent talcum powder and 11 per cent other additives. Polystyrene can also be ground and added to concrete to increase its insulation value.

References

British Petroleum Corporate Communications Services, *BP Statistical Review of World Energy*, London 1993

Curwell S, *et al*, *Building and Health*, RIBA, London 1990

Grunau A B, *Lebenswartung von Baustoffen*, Vieweg, Braunschweig/Wiesbaden 1980

Holmström A, *Åldring av plast och gummumaterial i byggnadstillämpningen*, Byggforskningsrådet rapp. 191:84, Stockholm 1984

Strunge *et al*, *Nedsiving af byggeaffald*, Miljøstyrelsen, Copenhagen 1990

10 *Plants*

'The forest gives generously the products of its life and protects us all.'
Pao Li Dung

Until the introduction of steel construction at the beginning of the industrial revolution, timber was the only material with which man could build a complete structural framework. Timber unites qualities such as lightness, strength and elasticity. Compared with its weight, it is 50 per cent stronger than steel. It is more hygienic than other similar materials – the growth of bacteria on kitchen benches of timber is much lower than that on benches of plastic or stainless steel. Timber also has good thermal conductivity. These qualities, mean that, in relation to most modern European building standards, timber can be used in up to 95 per cent of the components of a small building. This includes everything from roof covering to furniture, thermal insulation and framework.

Other plants can be used in building, though their use as a structural material is the exception rather than the rule. Examples exist along the rivers of eastern Iraq, where bunches of papyrus have been tied together to carry walls and ceilings, a building technique that is 5000 years old.

There are many non-structural uses for plants from living, climbing plants, which act as a barrier against wind and weather to linseed oil from the flax plant, used in the production of linoleum and various types of paint. Wood tar and colophony can be extracted from wood for use in the painting industry, the glue lignin, vinegar and fats in the form of pine oil for the production of green soap. Copal is extracted from many different tropical woods and is used as a varnish. Natural caoutchouc from the rubber tree can be used in its crude form as a water-repellent surface treatment or as the starting point for the production of plastics, e.g. chlorocaoutchouc, formed from a reaction with chlorine.

Table 10.1: Basic plant materials which need little processing

Material	*Areas of use*
Softwood and hardwood	Structures, cladding, floors, roof covering, windows, doors, plugs, wood fibre, tar, wood vinegar, cellulose, adhesives, alcohol, terpenes
Climbing plants	Wall cladding, improving internal climate and micro-climate outside
Roots	Starch
Straw and grass	Roof covering, wall cladding, cellulose
Grass turf	Roof covering, minor structures
Peat turf	Fibres, thermal insulation, cellulose, alcohol
Lichen	Pigment
Moss	Fibres, thermal insulation
Citrus fruits	Oils, terpenes
Plants containing silica	Pozzolana

Table 10.2: Basic plant materials which need a large amount of processing

Material	*Areas of use*
Cellulose	Wallpapers, paper in plastic laminates, ingredient in plastics
Oils	Paint, green soap, linoleum, solvents
Alcohol	Solvents
Terpenes	Solvents
Plant fibres	Thermal insulation, concrete reinforcement, building boards, sealants, carpeting, wallpaper, canvas, linoleum
Pozzolana	Ingredient in pozzolana cements
Vinegar	Impregnation, alcohol, acetic acid for the production of plastics
Wood tar	Impregnation, surface treatment
Starch	Adhesives, paint

All plants contain carbohydrates in the form of sugar, starch and cellulose. These are the most important nutritional and accumulative substances in the organisms. Sugar is formed in the green parts of the plant by carbon dioxide from air and water subjected to sunlight.

$$6CO_2 + 6H_2O + 2822\ kJ = C_6H_{12}O_6 + 6O_2 \quad (1)$$

During this reaction oxygen is released. The plant later transforms the sugar to starch and cellulose. Cellulose builds up the cells and the starch is stored. When the plant dies, it degrades back to carbon dioxide, water and ash. Oxygen is a necessary ingredient for this process. If there is very little or no oxygen, the plant becomes peat, which after millions of years may become coal and oil.

Flax, a plant of diversity

Flax is one of the oldest cultivated plants. The seeds can be pressed to produce oil for use in painting and for producing linoleum. Its fibres can be woven into very valuable textiles, pressed into strips for sealing joints around doors and windows, braided into light insulation matting or compressed into building boards.

Flax fibres are twice as strong as polyester fibres – they are considered the strongest of natural fibres, about 50–75 per cent stronger than cotton (Andersson, 1986). It is even stronger when wet. It is naturally resistant to most insects. It is relatively fire-proof and can be used as insulation in fire doors. If it does ignite, it smoulders and does not emit poisonous gases.

Plants are renewable resources that can be cultivated and harvested at regular intervals. With sensible methods of cultivation, they are a constant source of raw materials.

Pollution problems that have arisen in this area during recent years are a worrying development. In the Czech Republic and Poland more than half of the forests are dead or dying. Investigations into forest deaths in the USA show that productivity of American pine has declined by 30–50 per cent between 1955 and 1985. There are similar situations in Siberia (Brown, 1990) and in Scandinavia. Coniferous trees have suffered more from pollution than deciduous trees; other species of plants are also affected. The most damaging pollutants are considered to be ozone, sulphur and oxides of nitrogen, and the main producers are heavy industries and cars. The picture is made more complex because certain forms of pollution actually stimulate growth in the forest for a short period. This is especially relevant to nitrogen oxides – growth stops when the forest becomes saturated, and the apparent vitality ceases.

The importance of trees and plants to the global climatic situation has begun to be realized. They break down carbon dioxide (the dominant greenhouse gas) into oxygen. From this perspective it is amazing that the rain forests are threatened not only by pollution but also by clearing for development. This happens also in larger areas of Australia, Russia and the USA, where timber is felled without the necessary replanting. Siberia is in a very critical situation, forests of larch trees are being cleared in order to solve the country's economic crisis.

Timber from the tropics

The first shipments of tropical timber came to Europe via Venice during the fifteenth and sixteenth centuries. This timber was mainly extracted from the rain forests, which covered about 14 per cent of the Earth's surface at the beginning of the twentieth century. This has now been reduced to 6–7 per cent. The consequence of this is likely to be an increased greenhouse effect, more frequent flooding, the extinction of rare animal and plant species and an increase in the areas of desert.

Tropical timber is used for window and door frames, interior panelling, floors and furniture as solid timber and veneers. Some timbers, e.g. azobè, iroko and bankiria, have

Table 10.3: Primary energy required for basic plant materials

Material	*MJ/kg*
Split logs:	
air dried	0.5
artificially dried	1.9
Planed timber:	
air dried	1.0
artificially dried	3.8
Sawdust/wood shavings	0.6
Straw bales	0.2
Cardboard and paper	9.3

qualities useful to ecological building. They have a strong resistance to rot and can therefore be used in very exposed situations without impregnation. Despite this, using rain forest timber should be avoided altogether, except where the timber is managed under sustainable and well organized forestry.

The production of organic, plant-based building materials is mainly local or regional. Energy consumption for industrial processing and transport are relatively low as well as pollution occurs at the cultivating, harvesting and refining stages. This favourable environmental profile will be reflected in the building, in the form of a positive indoor climate. When the building starts deteriorating, the organic materials will return simply and quite quickly back into the natural environment. Some of the materials can be recycled for re-use or as a source of energy. Building materials based on plants act as a store for carbon, thus reducing the greenhouse effect. One kilogram of dry timber contains about 50 per cent carbon, which in turn binds 1.8 kg of carbon dioxide). In an average-sized timber dwelling, which contains about 20 tons of timber, there are 36 tons of carbon dioxide effectively bound in. The products must be durable and preferably recyclable. Carbon is bound within the timber until it rots or is burned.

Table 10.4: Potential pollution by basic plant materials

Material	*Potential pollution by processing*
Cellulose	Lye of organic chemicals, e.g. organic chlorine
Solvents	Alcohol, terpenes
Tar	Aromatic and aliphatic hydrocarbons
Plant fibres	Dust

Whilst most organic materials have this healthy environmental profile, there are a few exceptions. Cultivating some plants can involve the use of insecticides, fungicides, hormone additives and artificial fertilizers, which lead to environmental problems such as increased erosion, poisoned ground water and the damage or destruction of local ecological systems. This type of cultivation can produce defects in the product, e.g. enlarged and mouldy cell growth in timber. The finished products can also be impregnated or given a surface treatment, which pollutes the indoor climate. Such products need special dumping grounds when they become waste, in turn reducing the chance of recycling either as a new product or fuel. Gene manipulation has been suggested as a solution to these problems. By adding genes of a more resistant plant, it is possible to reduce the amount of insecticide sprayed on a crop during cultivation. This gives the modified species an 'unfair' advantage over other species in the ecosystem, however, and may lead to the collapse of the whole system. This solution is at present too dangerous to accept as a long-term environmental strategy.

Generally it can be said that it is desirable to increase the use of organic materials in the building industry. Only a small percentage of the potential organic building materials available are used today. Timber is the most common structural element in building. The use of more varied species will stimulate different methods of application and a richness and diversity of species within forestry and agriculture. This is beneficial to both the farmer and to nature.

Living plants

Plants that can be used in buildings in their living state include grass or turf, climbing plants and hedges. Many indoor plants bind dust and absorb gaseous pollution, which makes them especially useful in towns and heavily polluted indoor climates. Besides carbon dioxide, many other gases that can be absorbed or broken down by plants, e.g. benzene, formaldehyde, tetrachloroethylene and carbon monoxide. Green plants produce oxygen.

Turf

Turf roofs represent the oldest-known form of roof covering in the northernmost parts of Europe, and are still in use. In towns and cities in central Europe there is a renaissance of the turf roof and roof gardens. Turf has also been used as insulation in walls. In Iceland, the construction of pure turf walls with structural properties was widespread right up to the twentieth century.

Ordinary grass turf is used for building. It should preferably be taken from old mounds or fields to ensure that it is well bound with grass roots. If the grass is relatively newly planted (three to four years old), the root system will be undeveloped, so the turf may break up when removed. Turf should not be taken from a marsh.

Modern turf roofs often start as loose earth that is then sown with grass seed. The recommended grasses are 70 per cent sheep's-fescue (*Festuca ovina*), 10 per cent timothy grass (*Phleum pratense*) and 20 per cent creeping bent grass (*Agrostis stolonifera*). In dry areas, generous amounts of house leek (*Semper vivum*) and rose root (*Sedum roseum*), which are very resistant during dry periods, should be added. The sedum can be sown when the *Semper vivum* is planted, because it will spread through the root system. *Semper vivum* contains a lot of sap and therefore has a certain degree of fire resistance. When turf roofs were common in towns, laws ensured the use of *Semper vivum* on the roof.

Redcurrants, gooseberries and blackcurrants thrive in roof gardens and on flat roofs with a deep layer of earth. Trees planted on roofs should have a very shallow root system, e.g. birch.

Climbing plants and hedges

Climbing plants and hedges are not used very much in building despite their interesting characteristics. They can reduce the effect of wind, increase warmth and sound insulation, and protect wall materials.

There are two main types of climbing plant: those that climb without support, and those that need support.

Self-climbers

Self-climbers need no help to climb up and cover a wall. They climb by means of small shoots that have small roots or sticky tentacles. The smallest unevenness on the wall gives them the opportunity to fasten. Over a period of time an even green screen will form, requiring a minimum of care. These types of plants are best suited for high, inaccessible façades.

The most important climber in the northern European climate is ivy. It grows slowly, but can spread out to a height of 30 m, and is evergreen.

Trellis climbers

Trellis climbers are dependent upon some form of support to be able to climb a wall. There are three types:

- Twining plants need to twist around something to climb. They do not grow well on horizontal planes. Wisteria, honeysuckle and hops are the most common examples.

Table 10.5: Climbing plants

Plant	*Maximum height (m)*	*Growth conditions*
Self climbers:		
Ivy (*Hedera helix*)	30	Shade
Climbing hydrangea (*Hydrangea anomala* and *H. petiolaris*)	4–8	Sun and shade
Virgina creeper 'five-leaved ivy' (*Parthenocissus quiquefolia*)	20	Sun and shade
Virginia creeper 'Lowii' (*Parthenocissus tricuspidata*)	20	Sun and shade
Trellis climber:		
Chinese wisteria (*Wisteria sinensis*)	6–10	Strong sun
Honeysuckle (*Lonicera pericymenum*)	10	Sun
Winter jasmine (*Jasminum nudiflorum*)	2–5	Strong sun
Blackberry, bramble (*Rubus fruticosus*)	2–3	Sun
Common Virginia creeper (*Parthenocissus vitacea*)	10	Sun and shade
Climbing rose (*Rosa canina*)	3–4	Strong sun
Common hop(1) (*Humulus lupulus*)	10	Sun and shade

Note:
(1) Plant withers in winter

- Self-supporting plants have a special growth which attaches to unevennesses on the wall or to a trellis of galvanized steel or timber. These plants grow strongly and need regular cutting and care. The Virginia creeper is the most common. The wall does not have to be particularly uneven for the plant to be able to fasten onto it – in some cases these plants can be classified as self-climbers.
- Some plants that need support do not fasten either to walls or to other objects. They grow upwards quickly and chaotically, and can form thick layers. Blackberry bushes are an example.

If there is no earth along the external walls of a building, most climbing plants can be hung from a balcony. Virginia creeper and blackberry are good hanging plants.

Hedge plants

Hedge plants can be planted against a wall and grow independently with a strong trunk, but do not fasten to the wall. They have to be trimmed regularly, with openings made for windows. Quite a few grow in the northern European climate, e.g. rose hip.

Timber

Trees are mainly composed of long cells stretched vertically, forming wood fibres. Across the trunk are pith divisions, forming rectangular cells. This structure gives timber elasticity and strength. Cells vary in form from timber to timber, but they all contain carbon, oxygen, hydrogen and nitrogen as their main chemical constituents. They also contain small amounts of minerals, which are left over in the ashes if the tree is burned. A healthy tree consists mainly of cellulose, lignin and other organic substances such as proteins, sugar, resin and water. In softwoods the main constituent is cellulose; in hardwoods it is primarily lignin.

A cross-section of a tree trunk divides into bark, bast, heartwood, sapwood and pith. The growth rings in a tree are visible because summer wood is darker than spring wood. The number of rings gives the age of the tree, and the width of the rings indicates the growth conditions and therefore the quality. In coniferous trees, narrow rings usually indicate better quality than wide rings. In deciduous trees wide rings indicate better quality timber.

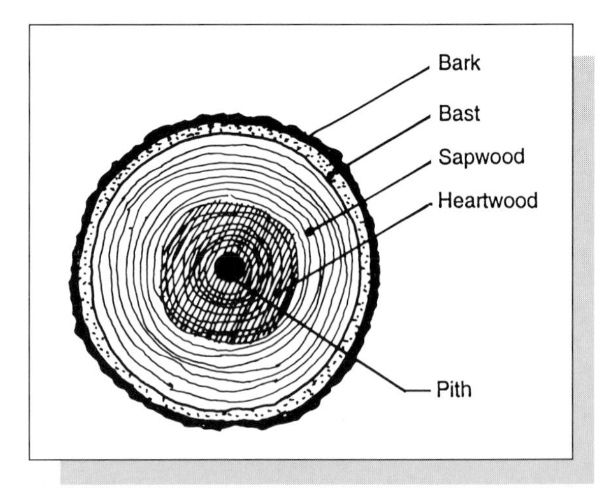

Figure 10.1: Cross-section of a tree trunk.

On the island of Madagascar there are 1000 species of tree. In northern Europe there are approximately 35 species, of which about two thirds can be used for construction. Despite this, usually only two coniferous trees are used (spruce and pine), and increasingly large areas supporting deciduous trees are taken over for the cultivation of spruce and pine forests. There is also a tendency to replace pine with spruce, as it produces less waste and is more practical to handle in the sawmill.

Many deciduous timbers have qualities which should encourage their more widespread use in building. In certain areas they are superior to spruce and pine, because of their higher resistance to moisture and greater strength. Ash, for example, is 60–70 per cent stronger than spruce.

Building using only accessible deciduous trees, and the use of materials according to their strength, could reduce the amount of structural timber needed by 25 per cent (Bunkholt, 1988).

In India, 300 different types of timber have been analysed to assess how useful they are in building. Factors such as weight, strength properties, durability and damage through shrinkage have been investigated. Timber varieties are then graded according to their properties. By doing this, a whole new range became available for use in building, including types previously classified as non-resources or firewood.

Forestry

Forestry is often managed as a monoculture of coniferous trees, mainly spruce. This is especially the case when producing timber for the cellulose industry.

Table 10.6: Use of timber in building

Timber	*Properties*[1]	*Areas of use*
Scots pine (*Pinus sylvestris*)	Soft, elastic, strong, durable, easy to cleave and work, denser and more resin than in spruce, difficult to glue and paint, can be pressure impregnated	Structures, floors, cladding, windows, doors, tar, roofing, foundations below ground level, plugs
Norway spruce (*Picea abies*)	Soft, elastic and medium hard wearing, sensitive to moisture, easy to glue and paint, difficult to pressure impregnate	Structures, roofing, cladding, laminated timber, fibreboard
European larch (*Larix decidua*)	Tough, strong and durable, good moisture resistance, easy to work, cannot be painted	Structures, floor plate, doors, windows, roofing
Common juniper (*Juniperus communis*)	Tough, firm and very durable, difficult to split but easy to work	Cladding plugs
English oak (*Quercus robur*)	Dense, heavy, hard, hard wearing, elastic and durable, tendency to twist, quite difficult to work moisture resistant	Structures, floors, windows, doors, thresholds, plugs, cladding, roofing
Aspen (*Populus tremula*)	Moisture resistant but strongest when dry, does not twist	Floors, plywood, suspended ceiling, smaller structures, cladding, piping for water and gutters, piles
White birch (*Betula pubescens*) Silver birch (*Betula pendula*)	Tough, strong, elastic, low resistance to moisture, twists easily, easy to work	Floors, stairs, internal panelling, veneer, chipboard, bark for damp proofing, smaller structures
Norway maple (*Acer platanoides*)	Hard, dense, tough, elastic, flexible, hard wearing, low resistance to moisture, easy to work	Floors, balustrades, stairs, plugs
Common ash (*Fraxinus excelsior*)	Hard, dense, tough, elastic, low resistance to moisture, easy to bend under steam	Floors, veneer, internal panelling, stairs, internal structural details
Common beech (*Fagus sylvatica*)	Hard, strong, medium resistance to moisture, twists easily, no smell, easy to work	Floors, balustrades, smaller structures, veneer, internal panelling, tar, vinegar
Wych elm (*Ulmus glabra*)	Strong, tough, elastic, durable, moisture resistant, not particularly easy to work	Floors, balustrades, piles, stairs, panelling, internal structural details
Lime (*Tilia cordata*)	Tough, medium strong, slightly elastic, easy to work	Smaller structures (used for log buildings in the Carpathians), internal panelling, veneer, fibre for woven wallpaper and rope
Common alder (*Alnus glutinosa*)	Not particularly durable in air, very durable under water, soft, light, brittle, twists easily, easy to work	Piles, gutters, veneer, internal cladding

Table 10.6: Use of timber in building – *continued*

Timber	*Properties*[1]	*Areas of use*
Common hazel (*Corylus avallana*)	Strong and elastic, not particularly durable	Wattle walling in timber framework
Grey alder (*Alnus incama*)	Not particularly durable, light and brittle, easy to work	Internal panelling, veneer
Wild cherry (*Prunnus avium*)	Stable, hard wearing	Floors
Plum (*Prunus domestica*)	Splits easily when dried	Veneer
Holly (*Helix aquifolium*)	Hard, homogeneous, hard wearing	Veneer
Apple (*Malus pumile*)	Hard, homogeneous, hard wearing, low resistance to moisture	Wooden screws, dowels, thresholds
White willow (*Salix alloa*)	Tough, elastic, easy to cleave	Veneer, wattle cladding on external walls
Rowan or mountain ash (*Sorbus aucuparia*)	Heavy, hard, tough, durable, hard wearing, easy to work	Wattle cladding on external walls

Note:
(1) Varies according to place of origin and the conditions of growth

These trees induce an acidic soil and reduce the pH level in water and rivers, and the forests are, ecologically speaking, deserts – local ecological systems do not function. This form of forestry leads to increased erosion of soil through comprehensive drainage systems which quickly channels rainfall into rivers and streams. In Scandinavia this form of forestry threatens more than 200 different species of plants and animals with extinction.

Forestry can be run on ecological principles. The secret lies in the natural regeneration of the forest. This requires sowing seeds of a multitude of local tree species, including deciduous trees that prevent acidity, and careful harvesting so that younger trees and other plants are preserved. There is clear evidence that timber from these mixed forests is of a higher quality than that cultivated in monocultures (Thörnquist, 1990). The bark from the trees is kept in the forest, which leaves nutrition on the forest floor, including nitrogen from the needles which avoids adding nitrogen in the form of artificial fertilizer.

People were once much more careful when choosing trees for felling. They chose mature trees: conifers more than 80-years-old and deciduous trees between 30 and 60-years-old. Hardwoods such as beech and oak need to be well over 100-years-old to be ready for felling. The definition of a mature pine is that pith and heartwood forms at least half of the cross-section of the trunk. In both spruce and pine the heartwood begins to form around the age of 30 to 40 years.

Figure 10.2: A traditional method of cultivating special qualities in timber.

The best quality conifers grow in lean soil. Heartwood timber shrinks less than other timber and is more durable, making it well-suited to the construction of doors, windows or external details. The demands of quality are lower for the production of cellulose, internal panelling etc.

In order to be economical with the use of heartwood timber, it used to be worked while the tree was still standing. This process, called self-impregnation, is known in most cultures from the British Isles to Japan. The most common method is to chop the top of the tree and remove a few strips of bark from the bottom to the top. Three or four of the highest branches are left to 'lift' the resin. After two to seven years the whole trunk is filled with resin. There is little growth during these years, but it produces a high timber quality. The method is especially effective on pine, which contains three times as much resin as spruce. Economically speaking, it is quite possible that the reduced growth is balanced by the higher strength and the reduced amount of impregnation needed, both of which are valuable assets.

Before timber for felling was categorized, people tried to find suitable features in timber for use as diagonal ties and bracing in post and lintel construction or framework construction. Crooked trees and round growths on the roots of trees proved particularly interesting. The tree could be worked with while it was still growing to achieve certain effects. English framework is, in many cases, based on bent timber. A 'bulge' occurs when a coniferous tree that was bent straightens itself up, the bulge occurs on the underside of the bend. Timber at this point is close knit and strong and has been used for exposed items such as thresholds.

There is no great value in hand picking timber with today's production techniques. Even the quality of timber is given little attention apart from the desire for straight trunks with few knots, and concern focuses upon volume. However, there are still possibilities for small, more specialized industries in this field.

In Sweden, research is now being undertaken to evaluate the possibilities of differentiating qualities of timber in modern forestry, in order to return to a situation where the best quality timber is used in the most exposed situations.

Timber damaged by air pollution is considered to be normal quality, as long as it is not mouldy in any way.

Felling

Both deciduous and coniferous trees intended for construction purposes should be felled in winter when the quantity of sap is at its lowest and the state of swelling, acidity, etc. are at their most favourable. Timber felled during spring is more readily susceptible to mould. Trees to be used in damp earth or in water should, however, be felled during the sap-period. Another advantage of the winter felling of ordinary construction timber is that the sawn timber dries out more slowly and is therefore less likely to split. Some felling traditions were related to the phases of the moon. Coniferous trees were felled at full moon, because the resins were well drawn out of the roots and into the trunk.

It has been assumed that the large amount of mould damage to newer Swedish timber buildings, especially in windows and the outside panelling, relates to the fact that the timber was felled during the summer – a usual occurence in Sweden during the 1960s (Thörnquist, 1990).

Storage

Although newly-felled timber should be treated as soon as possible, it is usually some time before this can be done. The timber should be stored in water, where there is hardly any oxygen. This reduces the risk of mould and insect damage. If it is stored in salt water, there is a risk of attack by marine borer.

Timber stored in water during the summer often becomes porous through the action of anaerobic bacteria which eat the contents of the cells and pore membranes. This can dramatically increase its resistance to rotting later, because the timber can easily cope with damp.

Splitting

The trunk is transported to the site where it is to be milled. Splitting should take place while it is still very damp. For log construction and certain other forms of building the log is used as it is, occasionally with its sides trimmed slightly flat with an axe. Pine has a longer life span if it is split in this way along two sides, because the hardness and amount of resin increases towards the centre of the trunk. Spruce should not be chopped along its sides, because the outside wood is stronger and heavier than the wood in the middle of the tree.

The oldest way of splitting a trunk is by cleaving through the core of the tree. The halves can be used as logs almost as they are, or they can be trimmed to a rectangular cross section. They can be further cleft along the radii, giving triangular profiled planks.

With the invention of the vertically-adjustable saw during the sixteenth century, splitting timber by saw became the dominant technique. The method was

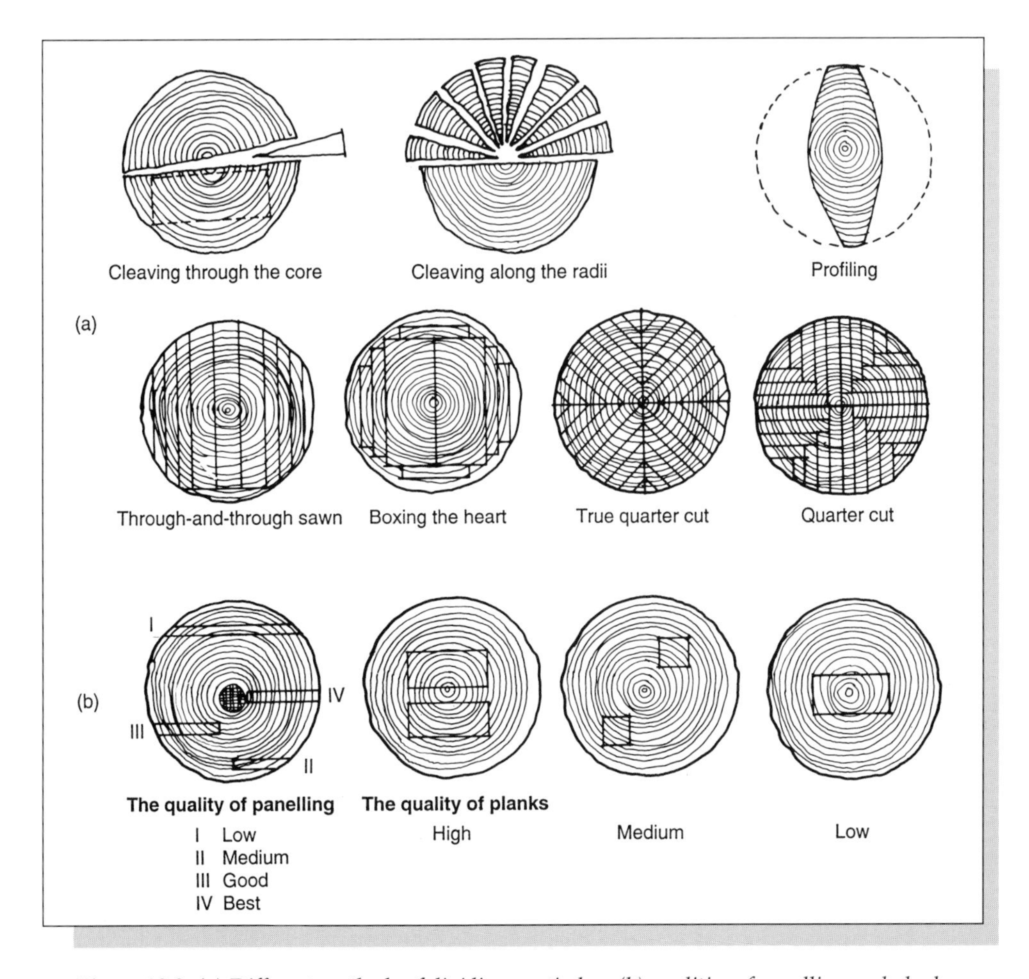

Figure 10.3: (a) Different methods of dividing up timber; (b) qualities of panelling and planks.

particularly effective for dividing logs into panelling. Today the even more efficient circular saw is usually used. For this method there has to be a rotational force, usually produced by electricity from the national grid. Rotational energy can also be produced directly by simple wind or water turbines. In this way the loss of energy through the transferrence of electricity is eliminated, and energy consumption can be halved. This source of energy is also free of pollution. Saw mills create a lot of dust in the working environment. Dust from oak and beech is carcinogenic.

There are different ways of sawing a log to produce planks: sawing through and through, boxing the heart, true quarter cutting and quarter cutting. Boxing the heart works well with the circular saw and is almost the only method used today.

The wedge is more sensitive with wood than the axe, and the axe is more sensitive than the saw. By using a wedge, the cells are kept whole when the wood is split; the saw cuts straight through the cell walls. This is critical to the timber's absorption of water, which governs the risk of attack by mould or insects. In spruce, which when whole has an impermeable membrane between the pores, this is particularly important. A carefully-divided spruce can be as durable as pine heartwood.

Timber from deciduous trees often has high inner tensions. To avoid this developing into twisting in the sawn timber, it is important to keep to smaller dimensions, preferably not above 50 mm.

Drying

Some researchers say that the drying routines for freshly sawn timber are much more important for its durability than the time of felling. Spring- and summer-felled timber should be dried as soon as possible (Raknes, 1987).

Timber shrinks 15 times more in its breadth than in its length when being dried, so when a newly-felled log dries it forms radial splits. By putting a wedge into one of these splits, further splitting can be controlled. In the same way, sawn timber has a tendency to bend outwards on the outer side when wet, and outwards on the inner side when it is dried. This is why the way in which a log has been sawn determines the degree of movement in a sawn plank.

In order to use newly sawn timber, 70–90 per cent of the original moisture in the trunk must be dried out, depending upon the end use. The sawn timber is stacked horizontally with plenty of air movement around it, and is dried under pressure. The stack can be placed outdoors or in special drying rooms. The outdoor method is more reliable for drying winter-felled trees during spring, as artificial drying produces some problems. Certain types of mould tolerate the temperatures used in this technique, and develop quickly on the surface of the wood during drying, emitting spores which can cause allergies. It has also been noted that the easily soluble sugars which usually evaporate during the slow drying process are still around in artificially dried timber, and become the perfect breeding ground for mould. It is also possible that the natural resins in the timber do not harden properly. This could be, for example, the reason why there often are considerable emissions of natural formaldehyde in buildings made purely of timber. Formaldehyde is an unwelcome substance to have in an indoor climate, and can cause irritation in the ear, nose and throat, allergies, etc. Another reason for drying timber outside is the lower energy consumption, which for an ordinary load of timber rises by 300 per cent when dried artificially.

Drying outside is best carried out in the spring. The number of months required for drying can be roughly estimated by multiplying the thickness of the timber in centimetres with 3.2 for spruce and 4.5 for pine. Normal planks take about three months, deciduous trees take longer.

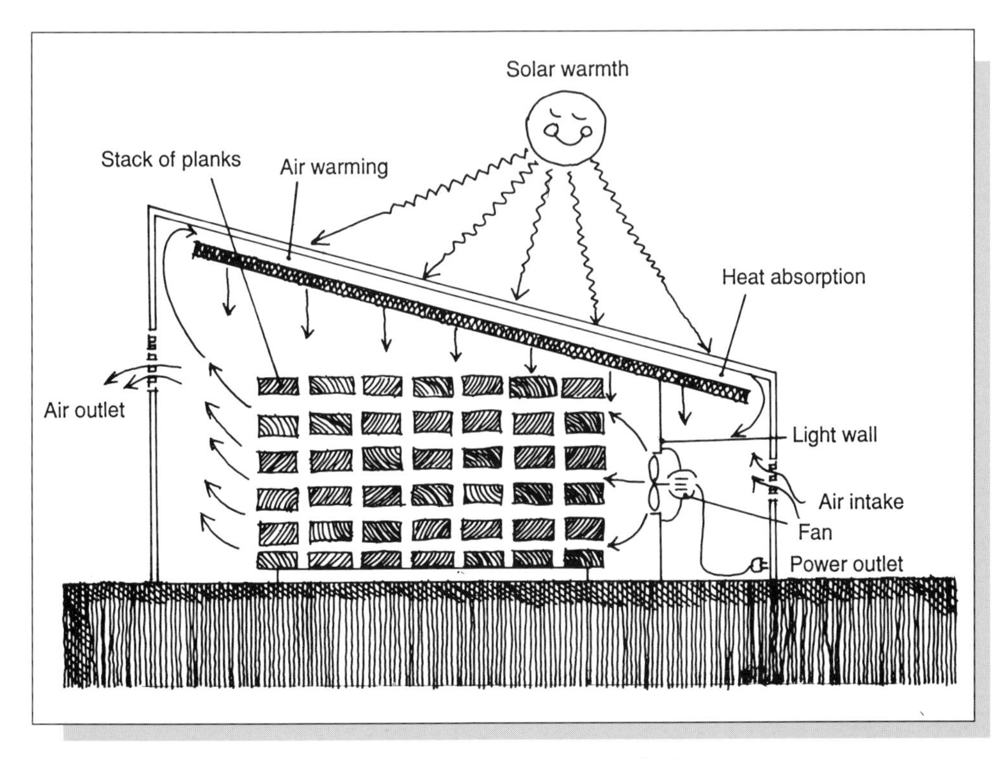

Figure 10.4: Principle for solar drying of timber. Source: Hall 1981

When building with logs it is best to fell, notch and use the timber while it is still moist. Logs with large dimensions have a long drying time – it can take years! A log building will therefore shrink between 5 and 10 cm on each floor. When the moisture content has decreased to 15–20 per cent, windows and panelling can be installed. Around 1900 this method of building fell out of favour because it was labour intensive and slow. If a framework construction is built of ready-dried timber, there is no noticeable shrinkage.

The durability of timber

All timber breaks down eventually. This can usually happen either through oxidization caused by oxygen in the air, or through reaction with micro-organisms which attack the proteins and therefore the sugars. These methods of deterioration usually work together. Timber that is submerged in water is more durable because of the lower amount of oxygen; in swamps timber can lie for thousands of years without deteriorating.

Timber keeps as long as it is not attacked by fire, insects or mould. The oldest-known timber building in existence is the Horiuji temple in Japan, which was built of cypress in AD 607. There are also completely intact timber beams in the

Table 10.7: Durability of timber in years in different situations

Timber	*Always dry*	*Sheltered outside*	*Unsheltered outside*	*In contact with earth*	*Underwater*
Pine	120–1000	90–120	40–85	7–8	500
Spruce	120–900	50–75	40–70	3–4	50–100
Larch	1800	90–150	40–90	9–10	More than 1500
Juniper	–	More than 100	100	–	–
Oak	300–800	100–200	50–120	15–20	More than 500
Aspen	–	Low	–	Low	High
Birch	500	3–40	3–40	Less than 5	20
Maple	–	–	–	Less than 5	Less than 20
Ash	300–800	30–100	15–60	Less than 5	Less than 20
Beech	300–800	5–100	10–60	5	More than 300
Elm	1500	80–180	6–100	5–10	More than 500
Silver fir	900	50	50	–	–
Willow	600	5–40	5–30	–	–
Poplar	500	3–40	3–40	Less than 5	–

1900–year-old ruins of Pompeii. An untreated timber surface can last for 150 years under favourable conditions. As a rule of thumb, heavier timber will last longer.

Timber is very resistant to aggressive pollution in the atmosphere – evidence of such damage occurring in timber has not been found.

Some factors are now beginning to threaten timber's good reputation. The extensive use of artificial fertilizer is probably reducing its durability, as the fast growth of cells produces wide annual rings and gives a spongier, more porous timber. Fast-growing species were introduced in the 1950s which have proved to yield lower quality timber. These conditions also led to a greater need to impregnation timber with chemicals.

Recycling

Timber is a recyclable material, and in the form of prefabricated components it can be re-used in many different situations. The re-use of logs, in part or as a whole, has been ubiquitous in most of Norway and Sweden. Both log construction and stave construction are building techniques where the components can be easily dismantled and re-erected without any waste. The Japanese have developed a whole series of techniques for timber joints without glue, the most well known being the so-called 'timber locks'. Most structures in the twentieth century have been based upon less flexible principles. Gluing and nailing have been the dominant methods of jointing. Modern timber-frame construction is at best firewood after demolition! Some chemicals, glues and surface treatments make timber unsuitable for use as fuel, and it has to be considered a problematic waste.

In Denmark, comprehensive timber recycling is now developing. Old oak beams are split up after the central core (malmen) has been removed for use in floor boards or windows. The renewal of timber windows has also become significant in the industry.

Old timber has the advantage that, since it is dead, it does not twist, and therefore provides good material for floors, for example. Nails and other metal details which may be part of the timber components can cause problems.

In Belgium and France old quality timber costs about 25–50 per cent of the price of new timber, while in Holland old timber costs up to 75 per cent of the price of new.

Grasses and other small plants

Plants mainly of the grass species produce straw which often has a high cellulose and air content, making it strong, durable and well-suited for use as insulation material.

Species such as rye, wheat and flax also contain natural glues and can be compressed into building sheets without additives. Elephant grass is a large grass plant first imported to Europe from Asia in 1953. It produces large quantities of grass straw well suited to building sheets. These plants can also be used as reinforcement for traditional earth structures and as roof covering. The cleaned plant fibres of flax, hemp and, stinging nettles can be woven into linen, carpets, wall coverings and rope.

Peat and moss have always been used to seal joints between materials and between different parts of buildings, e.g. as the sealant between the logs in log

Table 10.8: The use of plants in building

Plants	*Part used*	*Areas of use*	*Location*
Cultivated plants:			
Wheat (*Triticum*)	Stalk	Roofing, external cladding, building boards, thermal insulation, reinforcement in earth and concrete	Northern Europe
Rye (*Secale cereale*)	Stalk	Roofing, external cladding, building boards, thermal insulation, reinforcement in earth and concrete, woven wallpaper	Northern Europe
Flax (*Linum*)	Stalk	Roofing, external cladding, building boards, thermal insulation, reinforcement in earth and concrete, rope and woven wallpaper	Northern Europe
	Seed	Oil	

continued overleaf

Table 10.8: The use of plants in building – *continued*

Plants	*Part used*	*Areas of use*	*Location*
Oats (*Avena*)	Stalk	Roofing	Northern Europe
Barley (*Hordeum*)	Stalk	Roofing	Northern Europe
Hemp (*Cannabis sativa*)	Stalk	Building boards, concrete reinforcement, thermal insulation	Northern Europe
	Seed	Oil	
Jute (*Corchorus capsularis*)	Stalk	Sealing joints	Bangladesh
Elephant grass (*Miscanthis sinensis gigantheus*)	Stalk	Building boards, thermal insulation	Central Europe
Rice (*Oryza sativa*)	Stalk	Building boards	Asia
Sugar-cane (*Saccharum officinarum*)	Stalk	Building boards	South America
Cotton (*Gossypium*)	Stalk	Building boards	America, Africa
Coconut (*Cocos nucitera*)	Nutshell	Thermal insulation, sealing joints	The Tropics
Wild plants:			
Reed (*Phragmites communis*)	Stalk	Roofing, reinforcement in stucco work and render, insulation matting, concrete reinforcement	Northern Europe
Ribbon grass (*Phalaris arundinacea*)	Stalk	As reed	Northern Europe
Greater pond sedge (*Care riparia*)	Stalk	Roofing	Northern Europe
Cat-tail (*Typha*)	Seed	Thermal insulation	Northern Europe
Stinging nettle (*Urtica*)	Stalk	Thermal insulation, building boards, textiles	Northern Europe
Eelgrass (*Zostera marina*)	Leaves	Roofing, external cladding, thermal insulation, building boards	To the Artic regions
Marram grass (*Ammophila areniaria*)	Straw	Roofing	Northern Europe
Scotch heather (*Calluna vulgaris*)	The whole plant	Roofing, thermal insulation	Northern Europe
Common bracken (*Pteridium aquilinum*)	The whole plant	Roofing	Northern Europe
Moss (*Hylocomium splendens*) and (*Rhytriadiadelphus squarrosum*)	The whole plant	Sealing of joints, thermal insulation, building boards	Northern Europe
Peat-moss (*Sphagnum* spp.)	The whole plant	As moss	Northern Europe

Note: Many of the wild plants can be cultivated, as for example, nettles, reeds and cat-tail.

construction. The main constituent of peat is cellulose from decomposed plants. Dried peat can be used in building sheets and as thermal insulation. For use as thermal insulation, it has to be worked and must contain plenty of fibres. Sphagnum moss in peat contains small quantities of poisonous phenol compounds which impregnate the material.

Grasses and other small plants represent a very large potential resource. As far as cultivated plants are concerned, e.g. wheat, rye, oats and barley, the waste left over after the grain has been harvested can be used.

Plant resources are seldom used in today's building industry, probably because of their perceived ineffectiveness and because of the lack of efficiency in the handling of the raw material, the production of the final building material and the on-site handling.

Cultivating and harvesting

Most cultivated plant products are by-products from the production of grains. Intensive production of grain requires the extensive use of artificial fertilizers and pesticides.

Flax is immune to mould and insects and needs no pesticide treatment. Grain is harvested when it is ripe, usually during late summer. Cutting of wheat and rye for roof-covering must be carried out without breaking the stalk or opening it up. Many wild species grow in water, e.g. reed, ribbon grass and pond sedge. These plants live for many years, sprouting in spring, growing slowly through the summer and withering during winter. From 1000 m^2, 0.5–3 tons of material can be produced. Harvesting either by boat or from the ice occurs during the winter when the leaves have fallen.

Moss grows ten times as much in volume per unit area than forests. When harvesting moss, care must be taken not to destroy its system of pores. It is technically better and functionally easier if it is pulled up in pieces.

Harvesting peat is best done during the summer when the peat is at its driest. Summers with high rainfall can cause problems during harvesting as well as in the quality of the final product. There are machines which shave 3–5 cm off the surface of the peat. When large areas are harvested the local ecology of the area must be taken into account, particularly when harvesting moss and peat. Marshes have very sensitive ecological systems which include complex animal life. It is best to use peat resources which is likely to be wasted in cultivation for agricultural purposes, road-building, etc.

Preparation

Most of the smaller plants must have their leaves, seeds and flowers removed before direct use as roof-covering, thermal insulation, etc. Extraction of fibre

Table 10.9: The durability of exposed components made of plants

Type of plant	*Unfertilized/fertilized naturally (years)*	*Artificially fertilized (years)*
Reed	50–100	30
Straw from rye/wheat	20–35	10–12
Eelgrass	200–300	
Bracken	8–10	
Heather	More than 25	

from plants such as flax, stinging nettles and hemp is carried out by retting on the ground. The first stage of the cleaning process is left to mould, bacteria, sun and rain for two to three weeks until the fibres loosen from the stalk and can be harvested for the final cleaning process. Certain plant products are chemically impregnated to increase durability. Jute, produced in Bangladesh, is often impregnated with a copper solution for shipping to Europe. When producing building boards, it is quite usual to add glue even if many of the plants used contain natural glues which are melted out when the board is heated under pressure.

Building chemicals from plants

Plants can be the source of many building chemicals, which can be pressed out or distilled through warming in the absence of oxygen, a process known as dry-distillation. The main chemicals are as follows.

Wood vinegar

This is extracted from trees and can form a raw material base for methanol and acetic acid. It has a disinfectant effect on timber that is beginning to rot, and can form the basis of the production of synthetic substances. Other plants can form alcohol through fermentation. This can be used as a solvent for, amongst other things, natural resin paints and cellulose varnish.

Wood tar

Wood tar can be distilled to any consistency, from a thin transparent liquid to a thick black viscous liquid. One liquid, 'real' turpentine, is used as a solvent for paints.

The amount obtained from distillation depends upon the speed of the process and the type of wood used. Rapid distillation produces more gas and less liquid. Deciduous trees such as beech and birch produce the most wood vinegar, whereas coniferous trees contain more tar.

Wood tar consists mainly of hydrocarbons. Dry distillation from coniferous trees requires temperatures of 1000°C, and polyaromatic hydrocarbons such as benzo-a-pyrene are formed. Extraction of beech wood tar takes place at temperatures of around 250–500°C. The PAH content of this tar is low – about 0.1 per cent of the equivalent for carbolineum (chlorinated anthracene oil) which is extracted from coal. When extracting beech wood tar there is no emission of phenols – something that does occur with other timbers.

Lignin

After cellulose the main constituent of timber is lignin, whose function is to fix cellulose fibres and protect against mould. In the construction industry, lignin is sometimes used as a glue in wood fibre boards.

Cholofonium

This is a resin extracted from pine resins used in the paint industry and in the production of linoleum.

Drying oils

These are extracted from soya beans, linseed and hemp seeds and are used extensively in paint production.

Glycerols

Fatty substances in plants, known as glycerols, can be extracted from fatty acids by adding lye, and used in the production of soap.

Etheric oils

These are extracted from herbs such as rosemary and lavender and are often used as aromatic additions to paint products.

Starch

Starch can be extracted from potatoes and wheat and used as a glue or binder in paint.

Silicates

Siliceous plants contain large quantities of active silicates which react very strongly with lime, and the ash left over after burning the plant can be used as pozzolana in cements. Common horsetail (*Equisetum arvense*) is particularly rich in silica.

Potassium carbonate

Deciduous trees contain a particularly high amount of potassium carbonate, which is the main constituent of ash after timber has been burnt. Potassium carbonate (often called potash) is an important ingredient in the production of glass. Today, it is almost exclusively produced industrially from potassium chloride.

Cellulose

Cellulose can be produced from peat, straw and timber; the majority comes from timber. The main constituent of timber is cellulose $(C_6H_{10}O_5)_n$. Carbon makes up 44 per cent by weight, hydrogen 6.2 per cent and oxygen 49.4 per cent.

In the sulphite chemical process of the paper industry timber is ground and put under pressure with a solution of calcium hydrogen-sulphite, $Ca(HSO_3)_2$, releasing the lignin. The pure cellulose is washed again and may be bleached to a clean white pulp, rich in fibre. To produce paper glue and filler substances such as powdered heavy spar, kaolin or talcum are mixed in. Leaving out glue will produce more brittle, porous paper.

Viscose, rayon, cellulose acetate, celluloid, cellulose varnish, cellulose glue and cellulose paste are all produced from cellulose. For the production of viscose, cellulose from spruce timber is best. Other chemicals are often added in these processes, e.g. acetic acid and methanol (extracted from wood vinegar). Cellulose acetobutyrate (CAB) and cellulose propionate (CAP) are plastics made by adding a mixture of acetic acid and butyric acid to cellulose. These materials are as clear as glass and can be used to produce half-spherical roof lights.

The cellulose industry uses large quantities of water and creates high pollution levels. The cooling process leaves a high concentration of lye as a by-product. This contains different organic process chemicals, of which a few are recycled; the rest is released into rivers or lakes near the factory. These industries could reduce effluent to a minimum, if not completely, given the appropriate technology.

If the cellulose is bleached with chlorine, the pollution increases drastically. Organic chlorine substances can accumulate in the nutrient chain and act as poisons. Alternatives are bleaching paper with oxygen or hydrogen peroxide, but ideally all bleaching should be stopped.

References

ANDERSSON A, *Lin kommer igjen*, Fåra 1986
BUNKHOLT A, *Utnyttelse av lauvtrevirke til produksjon av skurlast og høvellast*, NLH, Ås 1988
BROWN L R (ed.), *State of the world*, Washington 1990
HALL G S *et al*, *The art of timber drying with solar kilns*, Hannover 1981
RAKNES E, *Liming av tre*, Universitetsforlaget, Oslo 1987
THÖRNQUIST T, *Trä och kvalitet*, Byggsforskningsrådet rapp 77:1990, Stockholm 1990

11 Materials of animal origin

Animals are mainly herbivores. Certain species, such as cows, can digest carbohydrate cellulose and change it into food. Man is mainly dependent on an intake of carbohydrates in the form of sugar and starch, but also needs protein, carbohydrate, vitamins, minerals, etc.

Humans and animals depend entirely on air for respiration. Oxygen enriches the blood and makes the body capable of burning food in an exothermic reaction releasing heat, approximately 80–150 W for an adult, depending upon their activity.

Figure 11.1: The woollen fibres of a sheep can be used as the main ingredients in paper, sealing strips and insulation. The bones, milk and blood can form the basis of materials for binders for glue and paint.

Table 11.1: Building materials from the animal kingdom

	Part	*Areas of use*
Coral	The whole coral	Building blocks, structures
Bees	Wax	Surface treatment of wood and hide
Fish	Oil/protein	Binder in paint and adhesives
Poultry	White of the egg	Binder in paint and adhesives
Hoofed animals	Wool (sheep/goat)	Textiles, wool-based sheeting, sealing around doors and windows, thermal insulation
	Hair (horse, pig, cow)	Reinforcement in render and earth floors
	Hides and skins	Internal cladding, floor covering, boiled protein used as binder in paint and adhesives
	Bone tissue	Boiled protein used as binder in paint and adhesives, ash used as pigment (ivory black)
	Blood	Protein substances used as binder in paint and adhesives, colour pigment
	Milk casein	Binder in paint, adhesives and fillers, base material for casein plastic
	Lactic acid	Impregnation

Products from the animal world have a limited use in modern building. Sheep's wool is useful for carpets, wallpapers, paper and more recently as thermal insulation and for sealing of joints. Wool of lower quality which would otherwise be wasted can be used for insulation and joint sealing. Beeswax has become a popular substance for treating timber surfaces. Protein substances extracted from milk, blood or tissues are still used as binding agents for paint and glue. Animal glue is the oldest known, and was used in ancient Egypt.

Traditional animal glue is produced by boiling skin and bone to a brown substance. When it is cleaned, gelatine, which is also used in the food industry, is obtained. Casein glue is made from milk casein, produced from curdled milk by adding rennet, and has a casein content of 11 per cent. Casein contains more than 20 different amino-acids and is a very complex chemical substance, but has no binding power in itself. Lime or another alkyd must be added to make the casein soluble in water. Casein plastic is produced from milk casein by heating the casein molecules with formaldehyde (HCHO) under pressure. This plastic, also called synthetic bone, is sometimes used for door handles.

When lactic sugar ferments, lactic acid ($C_2H_4OHCOOH$) is produced, which is a mild disinfectant.

In the south, a future can be envisaged in which organisms from the ocean such as coral (which depends upon warm water for quick growth) will be used in manufacturing building components. On tropical coasts this has considerable moral implications – as poisons may have to be used to hinder growth and produce the right dimensions.

The purple snail

The purple snail, *Purpur lapillus*, lives along most European coasts. It is so-named because it has a gland containing a coloured juice. The juice smells bad, but after painting with it in full sunlight, a purple colour appears after ten minutes which is clear, beautiful, durable and does not fade. A huge amount of snails are needed for the smallest amount of decoration. The development of this colour technique occurred in the eastern parts of the Mediterranean after the Phoenicians settled, about 5000 years ago. In Asia the purple painters had their own workshops at the royal courts, and purple became the colour of the rulers. The snail was worth more than silver and gold, but with the rise and fall of the Mediterranean empires almost the whole population of snails disappeared. Today the purple snail is no longer considered a resource, as the surviving snails are threatened by pollution from organic tin compounds used in some PVC products and the impregnation of timber.

The use of animal products has the same environmental impact as the use of plant products. They are renewable resources and the amount of energy used for production is relatively small; durability is usually good and the materials are easily decomposed. The level of pollution is low, but factories producing animal glue do smell if there is no appropriate cleansing equipment.

Protein substances can cause allergies in sensitive people. These substances can be released into the air when moistened, and internal use of paint, glue and fillers should be limited to dry places. It has been noted that casein mixed with materials containing cement, e.g. in fillers used to level floors, can develop irritating ammonia fumes.

12 Industrial by-products

Industrial processes often release by-products during the cleaning of materials, e.g. smoke, effluent, cooling water, etc. Materials such as slag and ash are also considered to be by-products. By-products have interesting qualities as raw materials:

- They are abundant without necessarily being in demand
- Chemically speaking, they are relatively pure
- It is usually difficult to dispose of them
- Other raw materials are saved by using them
- They are often formed of materials which produce an environmental problem such as pollution of air, earth or water.

The last point indicates some risks relating to by-products. Planned use of industrial by-products is a relatively new phenomenon in building, but this is in the process of changing and is mainly a consideration of substances that can be used as constituents in cement and concrete products. Widespread use of by-products which have properties similar to pozzolana, for example, will drastically reduce energy consumption within the cement industry, as well as saving other raw material resources. By planning industrial areas so that different industries support each other with their by-products, it should be possible to reduce transport costs in time and energy.

Industrial gypsum

It is necessary to differentiate between 'power station' gypsum, which is released in desulphurizing plants at power stations using coal, and phosphorous gypsum, from the production of artificial fertilizers.

Table 12.1: Industrial by-products and their uses in building

Material	*Industry*	*Areas of use*
Gypsum	Zinc works, oil- and coal-fired power station, brick factory, production of artificial fertilizer	Plasterboard, Portland cement
Sulphur	Oil- and gas-fired power station, refineries	Sulphur-based render, sulphur-based concrete, paper production
Silicate dust	Production of ferro-silica and silica	Reinforcement in concrete products, pozzolana
Blast furnace slag	Iron foundries	Pozzolana, thermal insulation (slag wool)
Fly ash	Coal-, oil- and gas- fired power stations	Pozzolana
Fossil meal	Oil refineries	Pozzolana, thermal insulating aggregate in render and concrete

Power station gypsum has similar technical properties to natural gypsum. Even the content of heavy metals and radioactivity is about the same as in the natural substance. Power station gypsum is therefore appropriate for both plasterboard and plaster and as a raw material for Portland cement.

Phosphorous gypsum has a higher likelihood of unwanted constituents because of the raw material used. Gypsum is also a by-product of other industries, e.g. in the production of phosphoric acid and titanium oxide, but contains large quantities of unwanted materials such as heavy metals.

Sulphur

Sulphur has been used for a long time in the building industry to set iron in concrete, e.g. for setting banisters in a staircase. At the end of the nineteenth century the first sulphur concrete blocks came onto the market.

Sulphur has a melting point of a little less than 120°C, and when melted binds well with many different materials. It can replace other materials used in casting, e.g. Portland cement. Sulphur concrete is waterproof and resistant to salts and acids. It should not be used with alkaline substances such as cement and lime. Sulphur can also be used in mortar and render, but because of its short setting time this can cause practical problems.

Sulphur dioxide is emitted in large quantities from industries where gas and oil are burned, but it is possible to clean up 80–90 per cent of these emissions. The temperature for working molten sulphur is around 135–150°C. There is probably little chance of the emission of hazardous doses of either hydrogen sulphide or sulphur dioxide at these temperatures, though even the slightest emission of the former gives a strong, unpleasant smell. The workplace should be well ventilated. Sulphur burns at 245°C, and large quantities of sulphur dioxide are emitted. Under normal circumstances there is very little risk of the material igniting.

Silicate dust

This is removed from smoke when ferro-silica and silica, used in steel alloys and the chemical industry, are produced. Silica dust, also called micro silica, is mainly composed of spherical glass particles. It does not react with lime and is a very good form of concrete reinforcement. It can, for example, replace asbestos. Silica dust is relatively new on the market, but is already used in products such as thermal light concrete blocks, concrete roof tiles and fibre cement tiles.

Blast furnace slag

This is produced in large quantities at works where iron ore is the main raw material. The slag is basically the remains of the ore, lime and coke from the furnaces. This is considered to be a usable pozzolana and can be used in Portland cement to bulk it out.

It is also possible to produce slagwool which can be used as thermal insulation in the same way as mineral and glasswool. The constituents of blast furnace slag increase the level of radioactive radon in a building, but this is negligible.

Fly-ash

Fly-ash reacts strongly with lime and is used as an ingredient in Portland cement and in the brick industry. It is a waste product from power stations that use fossil fuels. It contains small amounts of poisonous beryllium and easily soluble sulphates which can seep into and pollute a ground water system when they are dumped. Fly ash from waste-burning processes should not be used because it will probably contain heavy metals.

Fossil meal

Oil refineries that use oil from porous rock formations on the sea bed will produce fossil meal as a by-product. This can be used as thermal improvement for mortars and is also a good pozzolana.

Section 2: Further reading

AIXALA J N, *Small scale manufacture of Portland Cement*, Moscow 1968

ASHFAG H *et al*, *Pilot plant for expanded clay aggregates*, Engineering News no. 17, Lahore 1972

ASHURST J *et al*, *Stone in building. Its use and potential today*, London 1977

BHATANAGAR V M, *Building materials*, London 1981

CLIFTON J R *et al*, *Methods for Characterizing Adobe Building Materials*, NBS Technical Note 977, Washington 1978

CURWELL S *et al*, *Building and Health*, RIBA Publications, London 1990

DAVEY N, *A History of Building Materials*, London 1961

EMERY J J, *Canadian developments in the use of wastes and by-products*, CIM Bulletin Dec. 1979

HALL G S, *The art of timber drying with solar kilns*, Hannover 1981

HILL N *et al*, *Lime and other alternative cements*, Intermediate Technology Publications, London 1992

HØEG O A, *Planter og tradisjon*, Universitetsforlaget, Oslo 1974

HOLMGREN J, *Naturstenens anvendelse i husbyggingen i Scotland*, NGU no.78, Kristiania 1916

HOLMSTRÖM A, *Åldring av plast och gummimaterial i byggnadstillämpningen*, Byggforskningsrådet rapp. 191:84, Stockholm 1984

KEELING P S, *The geology and mineralogy of brick clays*, Brick Development Association 1963

LIDÈN H-E, *Middelalderen bygger i sten*, Universitetsforlaget, Oslo 1974

ORTEGA A, *Basic Technology: Sulphur as Building Material*, Minamar 31, London 1989

ORTEGA A, *Basic Technology: Mineral Accretion for Shelter. Seawater as Source for Building*, Minamar 32, London 1989

PROCKTER N J, *Climbing and screening plants*, Rushden 1983

RINGSHOLT T, *Development of building materials and low cost housing*, Building Research Worldwide Vol. 1a, 1980

RYBCZYNSKI W, *Building with leftovers*, Montreal 1973

SMITH R G, *Small scale production of gypsum plaster for building in the Cape Verde Islands*, Appr. Techn. Vol. 8 no. 4, 1982

SMITH R G, *Cementious Materials*, Appr. Techn. Vol. 11, no. 3, 1984

SPENCE R J S (ed.), *Lime and alternative cements*, London 1976

SWALLEN J R, *Grasses, their use in the building*, US Department of Housing and Urban Development, Washington 1972

TRYLAND Ø, *Kartlegging av miljøskadelige stoffer i plast og gummi*, SFT rapp. 91:16, Oslo 1991

UNITED NATIONS ECONOMIC AND SOCIAL COUNCIL, Timber Committee, *Industrial production and use of woodbased products in the building industry*, UN 1976

section 3

The construction of a sea-iron-flower

Building materials

13 Structural materials

A building structure usually consists of the following parts:

- *The foundation*, which is the part of the building that transfers the weight of the building and other loads to the ground, usually below ground level. In swamps and other areas with no load-bearing capacity the load must be spread onto piles going down to a solid base.
- *The wall structure*, which carries the floor, roof and wind loads. The walls can be replaced by free-standing columns.
- *The floor structure*, which carries the weight of the people in the building and other loads such as furniture and machinery.
- *The roof structure*, which bears the weight of the roof covering and possible snow loads.

These standard elements can be separated in theory, but in practice the different functions usually have no clear boundaries, as in the construction of a spherical building such as the Globe Sports arena in Stockholm. The different structural elements have a very intricate interaction in relation to the bracing of a building, for example, a particular wall structure can be dependent upon a specific floor structure for its structural integrity. Some structures also cover other building needs, such as thermal insulation, for example.

Structural materials have to fulfil many conditions. They are partly dependent upon the construction technique to be used, and their properties are defined in terms of bending strength, compressive strength, tensile strength and elasticity. These factors give an idea of the ability of the material to cope with different forces. How this happens depends upon the design and dimension of the structure.

A steel cable has its strength in its capacity to take up tensile forces, e.g. in a suspension bridge. A brick, however, almost entirely lacks any such stretching

Table 13.1: Materials and related structures

Material	*Foundations*	*Walls*	*Floors*	*Roof*
Steel	In general use	In general use	In general use	In general use
Aluminium		Limited use/at experimental stage	Limited use/at experimental stage	Limited use/at experimental stage
Concrete with air-curing binder		Not in use	Not in use	Not in use
Concrete with hydraulic cement	In general use	In general use	In general use	In general use
Stone	Limited use/at experimental stage	Limited use/at experimental stage	Not in use	Limited use/at experimental stage
Bricks, well-fired	Not in use	Limited use/at experimental stage	Not in use, except for special structural elements or as a vault	Not in use, except for special structural elements or as a vault
Bricks, low-fired		Limited use/at experimental stage	Not in use, except for special structural elements or as a vault	Not in use, except for special structural elements or as a vault
Stamped earth		Not in use	Not in use	Not in use
Plastic, formed from recycled material				
Softwood	Not in use, except for pine in extra foundations below the water table	In general use	In general use	In general use
Hardwood	Not in use, except for aspen, elm and alder in extra foundations below the water table	Not in use	Not in use	Not in use
Peat		Not in use		

properties and must be used in a building technique which is in static equilibrium due to its compressive strength. Structures that are in a state of static equilibrium tend to have a longer life span than those with different tensile loads, which in the long run are exposed to material fatigue.

The proportion of structural materials in a building vary from 70–90 per cent of the weight – a timber building has the lowest percentage, and brick and concrete have the highest percentage.

Structural materials usually provide very few negative environmental effects per unit of weight compared with other building materials. They are usually based on renewable resources such as timber, or on materials with rich resource reserves such as clay, lime or stone. The production is preferably local or regional. The amount of primary energy consumption including transport is approximately 30–40 per cent of the complete house. Pollution due to greenhouse gases carbon dioxide and acidifying sulphur dioxide will vary from 35–70 per cent. The level of environmental poisons will probably be much lower, and as waste products the majority of structural materials are not a problem. As these materials are relatively simple combinations of elements with large dimensions they are well suited for recycling, but the quantity of binders and the size of the units are decisive factors.

Despite their relatively good environmental profile, the choice of structural materials is a decisive factor in a building's environmental profile because of their large volume and weight.

Metal structures

Metal structures are relatively new in building history. Despite this, they have, together with concrete, become the most common structural systems in large modern buildings over the past 100 years. Even if metal melts and bends during a fire, it does not burn, and it is strong and durable in relation to the amount of material used, and it is 'industrial'.

Aluminium is used in light structures, but steel is without doubt the most important structural metal, and is used in foundations, wall, roof and floor structures.

The steel used in structural situations is most often unalloyed, pure steel recycled from scrap. High quality steel is alloyed with small amounts of aluminium and titanium. The resulting material is particularly strong, and means that the amount of material used can be reduced by up to 50 per cent.

Steel components are usually prefabricated as beams with different cross sections and as square hollow sections, round hollow sections or cables, put together to make different sorts of braced or unbraced framework structures. It is normal practice to weld the components together on site. Steel components can also be fixed together mechanically, with or without the use of bolts. This considerably increases the opportunities for recycling.

Metal components cause absolutely no emissions or dust problems within a building. They can, however, affect the indoor climate by picking up vagrant electrical currents from electrical installations and distributing them around the building. This can result in changes or increases in the electromagnetic fields in the building, which can affect health by increasing stress and depression. When dumping metals a certain level of seepage of metal ions to the soil and ground water must be assumed.

Both aluminium and steel components can be recycled by re-smelting. It has also proved profitable to re-use steel components in their original state. In Denmark, the market value of well-preserved steel components from demolition

Table 13.2: Concrete mixes, their properties, and areas of use

Type of concrete	*Materials and parts by volume in the mix*	*Properties*	*Areas of use*
Lime sandstone	Lime: 1 Quartz sand: 9	Durable, sensitive to moisture	Internal and external structures, cladding
Lime concrete	Lime: 1/1 Sand: 2/4 Aggregate: 4/6	Elastic, not very resistant to water and frost	Internal light structures, regulating of moisture
Lime pozzolana concrete	Lime/pozzolana: 3 Sand: 1 Aggregate: 2	Medium strength, elastic, frost and moisture resistant	Internal and external structures
Portland concrete	Cement: 2/1 Sand: 6/3 Aggregate: 5/3	Strong, durable, not particularly elastic, frost and moisture resistant	Internal and external structures, foundations
Portland-pozzolana concrete	Cement/pozzolana: 1 Sand: 3 Aggregate: 3	Strong durable, little to moderate elasticity, frost and moisture resistant	Internal and external structures, foundations
Gypsum concrete	Gypsum: 1 Sand: 1 Aggregate: 2	Not very resistant to water and frost	Light internal structures, moisture regulating
Sulphur concrete	Sulphur: 1 Sand/Aggregate: 3	Being researched	Internal and external structures, foundations

jobs has reached ten times the scrap value. Old railway lines have been used in the structure of office buildings in Sweden.

Concrete structures

Concrete is produced from cement, aggregate, water, and additives, when required. It is cast on site in shuttering, or as blocks or concrete elements. With few exceptions, the products are reinforced.

Concrete's important properties are compressive strength, fire resistance and a high heat capacity. Pure concrete structures are relatively rare in early building history, when cement was used mostly as a mortar to bind bricks or stones. Exceptions exist in the Roman Empire where the coffers in the ceiling vault of the Pantheon are cast in concrete using pumice as aggregate. In the 1930s, and again after the Second World War, the use of concrete in building became widespread. Today it is the leading building material for larger buildings in foundations, retaining walls, walls, roof and floor construction.

Concrete binders and, to a certain extent, reinforcement, have the most serious environmental consequences. It is important to try to choose the most appropriate

Table 13.3: Lightweight concretes, their properties and areas of use

Type of concrete	*Materials*	*Properties*	*Areas of use*
Aerated concrete	Cement, sand, lime, fine aggregate, aluminium powder	Relatively good thermal insulation, weak resistance to frost	Internal and external construction
Lightweight aggregate concrete	Cement, expanded clay or similar lightweight aggregate, sand	Relatively moderate thermal insulation, frost resistant	Internal and external construction, foundations
Punice concrete	Cement, punice, sand	Good thermal insulation	Internal and external construction
Concrete with wood chip	Cement, impregnated wood chip	Relatively low thermal insulation, not frost resistant	Internal construction
Woodwool cement	Cement, impregnated woodwool	Good thermal insulation	Light internal and external construction

Note:
All the different types of lightweight concrete are described in more detail in the next chapter. In many of the products, cement can be mixed with pozzolana, or be replaced with lime, gypsum or sulphur.

alternatives, at the same time reducing the proportion of these constituents. Some regions lack the required mineral aggregate, so the amount of this component must also be economised.

The composition of concrete

Binders

Air-curing binders and hydraulic cements can be used. Among air-curing binders, slaked lime and gypsum are the most important ingredients. Hydraulic cements include lime and pumice mixtures and Portland cement, with pumice additives if necessary. Sulphur is a binder in a group of its own because it cures when cooling, having passed through a melting down phase.

During building, contact with lime products can cause serious damage to the skin and eyes, so these products should be used with care. Portland cement contains chrome which can lead to a skin allergy, even though current products are usually neutralized, mostly with ferrous sulphate.

Melting sulphur for sulphur blocks is unlikely to produce dangerous levels of hydrogen sulphide or sulphur dioxide fumes.

Pure mineral binders usually have no effect on the indoor climate. Dust, however, can fall from untreated concrete surfaces. This can irritate the mucous membranes. Problems also occur when cement dust is left behind when building is completed, e.g. in ventilation ducts. If the cement is not completely hydrated, e.g. because of insufficient watering afterwards, it is capable of reacting with other materials such as fillers with organic additives and plastic coatings. As a waste product, Portland cement with fly-ash releases soluble sulphurs into the environment. Generally speaking, lime cements give the least environmental problems but they are slightly weaker than Portland cement.

In a final evaluation the environmental consequences of increased transport of both cement and aggregates must also be considered.

Aggregates

In ordinary concrete the aggregates are divided into three groups: sand, gravel and crushed stone. In lightweight concrete there are also many air-filled, thermally-insulating aggregates which are discussed in the following chapter.

Concrete can be increased in bulk by adding rubble. In walls with a thickness of 40 cm or more, up to 25 per cent stone, e.g. stones from a field, can be added. These stones or rocks must be properly cleaned before use.

In places with no sand, gravel or crushed stone, other types of building waste that do not attack lime can be used. Ground concrete, waste or crushed bricks give results as good as aggregates, as long as it is treated correctly. Crushed bricks from 1–40 mm can also be used, but the material must be good quality and has to be washed before use. Bricks made of fired clay cannot be used if they contain nitrate residue from artificial fertilizers, as this increases the decay rate of the concrete. The artificial fertilization of agricultural land started to take hold in the 1950s.

In many European countries, Portland cement-based concrete is recycled. The concrete is crushed to normal aggregate size and used in the casting of concrete slabs for foundations of small houses and parking blocks, where they can replace up to 20 per cent of the gravel. The wastage in demolition and crushing of old concrete is about 90 per cent, but with improved techniques and more experience it should be reduced to about 50 per cent (Lauritzen, 1991).

Little attention has been paid to the fact that different types of crushed stone make different demands on the concrete mix. The decisive factor is the tensile strength, and paradoxically a low tensile strength is more favourable. Crushed stone with a tensile strength of 200 kp/cm^2 needs much less cement than that with an ultimate strength of 500 kp/cm^2. Up to 10 per cent of the world's cement production could be saved if this was considered (Shadmon, 1983).

In some countries where deposits of gravel and sand are low, sand is sometimes removed from the beach zones and even from out at sea. This disturbs the shore and its sealife and can be damaging to existing ecological systems.

Different types of aggregate contain varying amounts of radioactive material. The levels are often low and usually have no effect on the indoor climate. Exceptions to the rule are pumice, some slates and industrial aggregates, which can affect the level of radioactivity quite strongly. It can also be affected by proximity to nuclear plants with (known or unknown) spillages.

Reinforcement

Steel is the most common material used to reinforce concrete. It is mainly recycled from scrap metal, but it is normal to add 10 per cent new steel to improve the strength. Steel reinforcement occurs in the form of bars or fibres that are 15 mm long. Fibres are usually mixed in in proportions up to 2 per cent of the volume of the concrete; the use of reinforcement bars takes up half as much volume as the fibres. The advantage with the use of fibres is that they are better at taking up the strains within the concrete and give a stronger concrete, which can reduce the thickness of a slab by 30 per cent. The distance between the expansion joints can also be increased considerably, therefore reducing the use of plastic joint mastics. Other fibres have been introduced more recently in the form of glass and carbon. Asbestos fibres were once used, but have been phased out because of their health damaging properties. Any products or components that may contain asbestos have to be identified and carefully removed from a site during demolition.

In smaller projects it is also possible to use fibres from plant material in a proportion of 2 per cent volume. No research has been carried out to find out which types are the most advantageous, but we can assume that long, strong fibres are well suited. They should be chemically neutral which is not always possible, but they should at least be cleaned of all active substances before being used (see 'Woodwool cement boards – production and use'). The most practical is hemp fibre (*Cannabis sativa*) which is very strong. Timber fibres are also used, and in the former Soviet states fibres from certain reed plants were tried, partly in industry and in schools up to three stories. There have been experiments with bamboo reinforcement in both the former Soviet Union and France in recent years with good results, even for larger buildings. *Sinarunddinarianitida* is a tolerant species of bamboo which can be cultivated in Northern Europe. *Thamnecolomus murielae* is also a possibility.

Additives

It is quite normal to put a whole range of additives into cement and concrete mixes (see Table 6.5). Additives are often organic and more or less volatile in ready concrete, and many of them can cause problems in the indoor climate.

Handling and demolishing concrete can cause a problem with dust from colouring pigments which contain heavy metals, including chrome, lead and cobalt. It is possible that the waste process allows seepage into the environment of added tensides, aromatic hydrocarbons, amines, borates, etc. Melamine-based plasticizers can develop poisonous gases during a fire.

Special concretes

Sulphur concrete

Sulphur concrete is most common in prefabricated blocks and elements which are cast by mixing smelted sulphur (120–150°C) with sand and pouring it quickly into a mould for cooling. This is a very simple process and the use of energy is low. Sulphur blocks are even waterproof as long as there are not many fibres in the mix. Sulphur concrete is visually attractive and virtually maintenance-free, without the 'ageing lines' which occur with Portland concrete. The development of a sufficiently sound sulphur concrete has not yet been achieved. For some reason the interest in this material disappeared after a very prolific period of use near the end of the nineteenth century, and the idea was first taken up again about 20 years ago by the Minimum Housing Group at McGill University in Canada, which has built a number of houses in sulphur concrete. Since then, experiments have been carried out in Germany and several other countries.

Figure 13.1: Building with sulphur blocks in both walls and vaults constructed in Rennes, France, in 1983, by the Institut National des Sciences. Source: Ortega 1989

One of the weaknesses of sulphur concrete is that it does not tolerate frequent changes of temperature, between freezing and thawing – small cracks appear in the block and it will start to decay. This can be remedied by adding materials such as talcum, clay, graphite and pyrites, in proportions up to 20 per cent by volume. Another problem to consider is fire risk, but it has proved difficult to set fire to a sand-mixed sulphur concrete, and if an accident should occur, the fire can be extinguished with water.

Lime sandstone

Lime sandstone is produced from a mixture of slaked and unslaked lime (5–8 per cent), mixed with 92–95 per cent quartz sand. The quartz sand is excavated from beaches or sandstone with a high quartz content. The stone is crushed to a grain size between 0.1 and 0.8 mm and mixed with pulverized lime. Water is added and the mixture is cast into blocks which harden for 10 hours in a kiln at 200–300°C. Lime sandstone is used structurally as brick, but is also used as stone lining. It cannot be recycled as new aggregate, but can be used as a stable mass.

The durability of concrete products

There are many examples of pure lime mortar keeping its functional properties for 2000 to 3000 years, but there are examples of Portland cement mortars that have crumbled within 10 years (Grunau, 1980). Some concrete buildings with Portland cement have stood undamaged for over 100 years.

Durability is clearly dependent on the quality of both workmanship and raw materials, as well as the proportions of the mix and the environment of the building. In recent years it has become evident that certain types of air pollution decompose concrete. Carbon dioxide and sulphur dioxide, both of which occur in high concentrations around industrial areas and towns, are particularly damaging.

It has been proved that carbon dioxide can carbonize up to 40 mm into concrete. The concrete loses its alkaline properties as a result and can be subject to corrosive attack. The next phase of breakdown usually occurs quite quickly, and involves the slow loss of the concrete. In the USA, one bridge per day is demolished as a result of such processes.

Much of today's concrete contains organic additives, and these types of concrete break down even more quickly. Mortars with artificial resins have been seen to decay within two to four years (Grunau, 1980).

The majority of Portland pozzolana concrete mixes have a much greater resistance to pollution than pure Portland concrete. There is no long-term experience of how lime sandstone and sulphur concrete last. The same can be said for lime concrete, which is seldom used in northern countries.

Concrete can be protected through constructional detailing. There are certain rules of thumb: avoid details that are continually exposed to rainwater. For example, in horizontal concrete surfaces exposed to soot and other pollution, the pollution is washed over the surface, intensifying decay of the concrete.

Recycling

The value of in-situ concrete in terms of recycling is low. It can, however, be crushed and ground to aggregate. The majority of it has to be sorted and used as fill. In theory, steel can be recycled from reinforcement, though this is a complex process using machines for crushing the concrete, electromagnets for separating, etc. Until 1950 smooth circular steel bars were used which were much easier to remove from concrete. Fibre reinforcement has no recycling potential.

Concrete units have considerably better recycling possibilities. By using mechanical fixings or mortar joints that make it possible to dismantle the units, the whole element can be re-used (see Figure 13.3).

The mortar used for constructions with concrete blocks is often Portland cement. This construction is very difficult to disassemble without destroying the blocks. Alternative are the different lime mortars, mainly based on hydraulic lime. In some cases, weaker mortar may require compensation in terms of reinforcement. Larger concrete units are usually fixed together by welding or bolting, which makes them easier to dismantle.

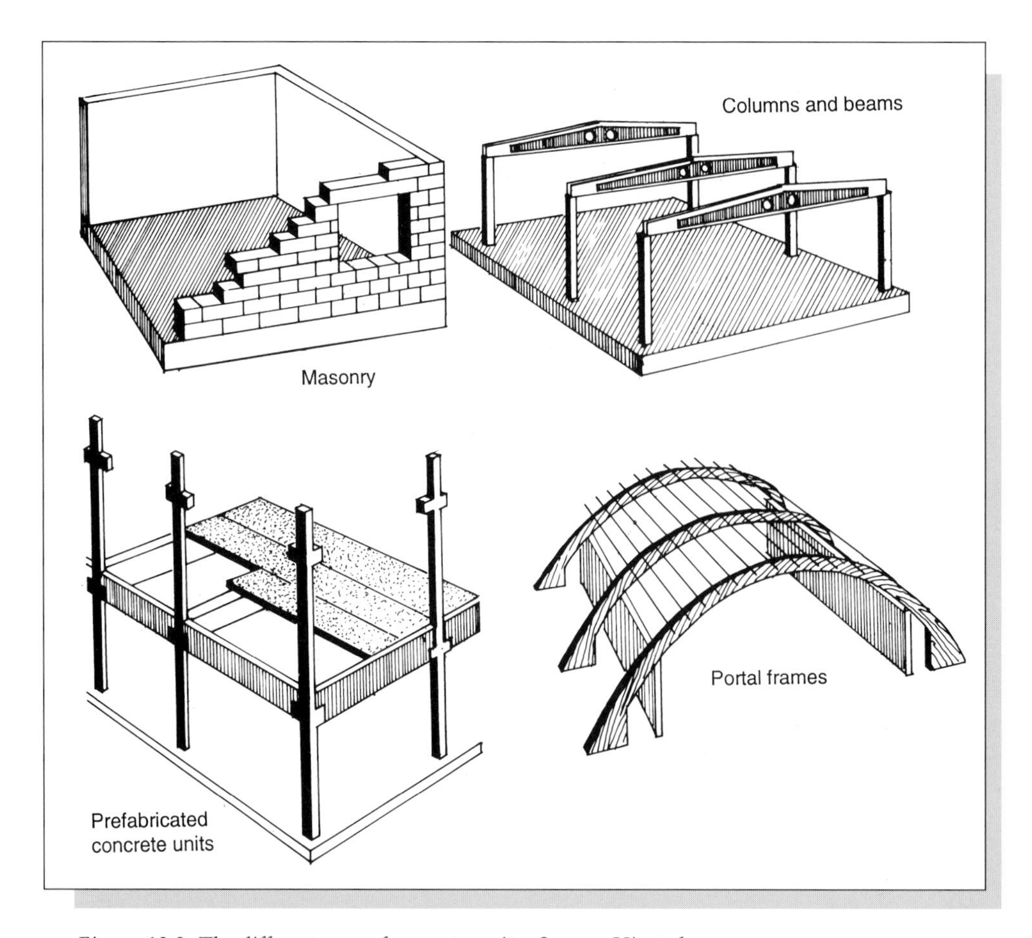

Figure 13.2: The different uses of concrete units. Source: Viestad

Holland already has a standard prefabricated system which can be taken down and rebuilt. In Denmark and Sweden there are many examples of industrial units and agricultural buildings built out of almost entirely recycled concrete units.

Figure 13.5 shows a Norwegian foundation system in concrete units. All the components are standardized and locked together internally with grooves or bolts. During demolition, the ties and pillars are lifted

Figure 13.3: Examples of blocks which do not need mortar. Their measurements are very exact, with a height difference of a maximum of ±1 mm. They are usually tongued and grooved. Their re-usability depends upon the strength of the render used on them. This method of building should reduce the amount of labour by about 30 per cent.

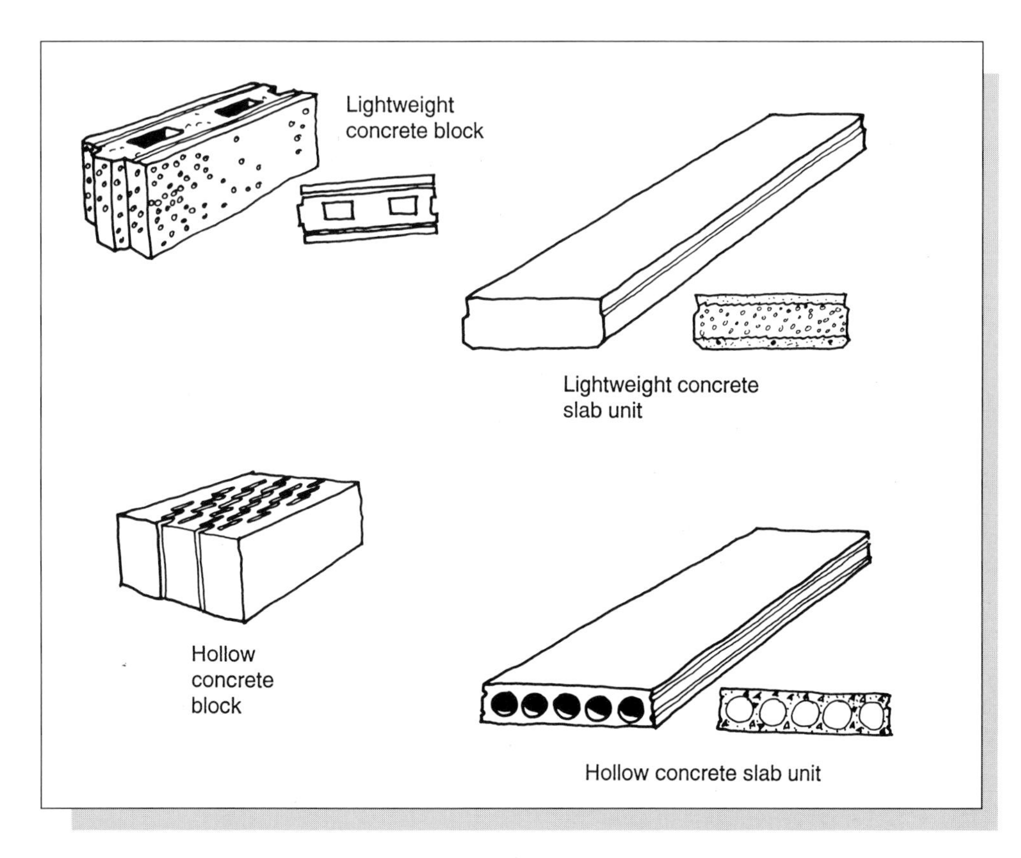

Figure 13.4: Standard concrete pre-cast units for walls and floors.

Figure 13.5: Norwegian foundation system of concrete units. Source: Gaia Lista, 1996

up, leaving only the bases of the pillars standing in the ground. The rest is quality-controlled on site and then transported direct to a new building site.

Sulphur concrete can be melted back to its original state, and aggregate can be removed through sieving and possibly be re-used.

Stone structures

The earliest remains of stone buildings in Northern Europe are of long communal buildings with low walls of stones taken from beaches and fields. They were probably jointed with clay. Walls of stone with lime mortar began to appear around AD 1000, with the use of stone cut from local quarries. The stone buildings of this period were almost without exception castles and churches. It was not until the twelfth and thirteenth centuries that cut stone was used for dwellings, and then it was used mainly for foundations and cellars. Foundation walls of granite were used until the 1920s, later in some places. During the Second World War stone became more widely used, but this was relatively short term.

Figure 13.6: The remains of a traditional dry-walled structure in Ireland. Source: Dag Roalkvam

Extraction and production of stone blocks has a low impact on nature and natural processes. Stone blocks use low technology plants and are well suited for decentralization. Energy consumption is low, as is pollution. Inside a building some types of stone can emit radon gas, though the quantity is seldom dangerous. The recycling potential is high, especially for well-cut stones that have been in a dry-stone wall.

Stones which lie loose in the soil in fields are easy to remove but are limited in their use. In larger buildings plenty of mortar is needed with this type of stone and it loses its ecological advantages. All the positive aspects of stone construction disappear if heavy construction materials are transported long distances. Stone, is and must be, a local building material.

Figure 13.7: A hydro-power station from the end of the 19th century, built of granite and concrete.

Structural elements

Solid stone or even flagstones can be used for structural stonework. There should be no trace of decay, splitting of layers or veins of clay. Sandstone and limestone can only be used above ground level; all other types of stone can be used both above and below ground level.

Free lying stones or stones from quarries can be used. Quarry stone can be divided into the following categories:

- *Normal quarry stone* which has been lightly worked;
- *Squared stone* which is produced in rectangular form and has rough surfaces
- *Cut stone* which is also rectangular, but the surfaces are smoothly cut.

The last two types are often called rough or fine-squared stone. If the dimensions of the stone are greater than 20 × 20 × 40 cm, it is too heavy to be lifted manually and must be placed by crane. Stone should dry for two months before being used.

Cutting granite, gneiss, sandstone and different slates releases quartz dust, which can cause serious lung damage.

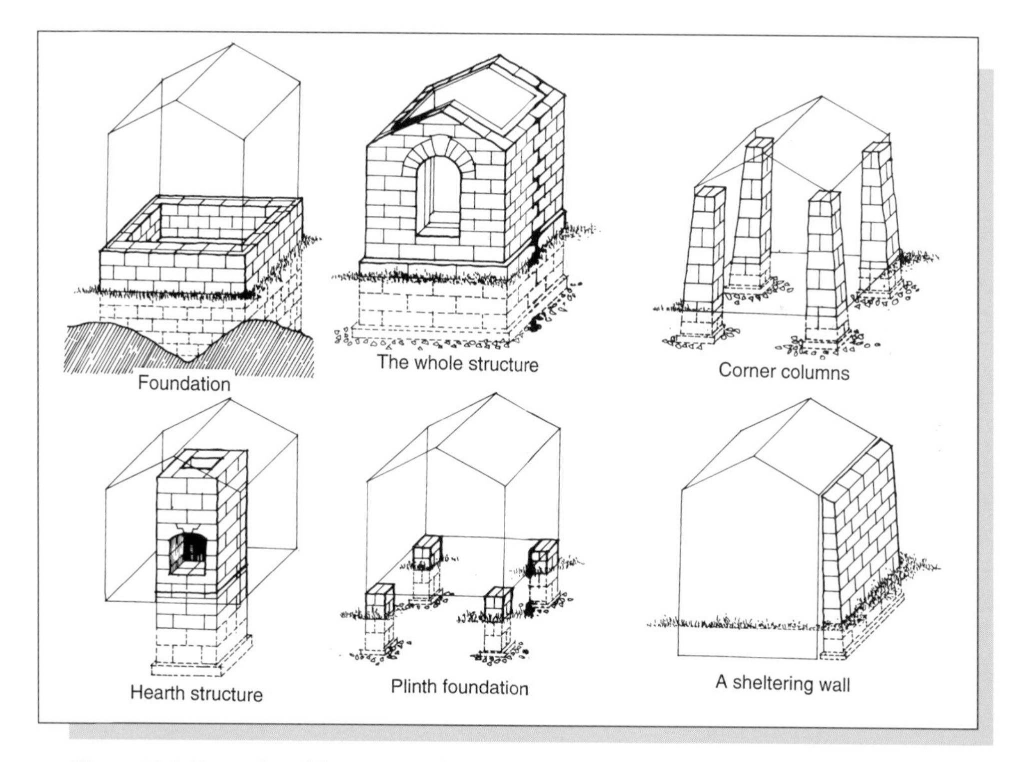

Figure 13.8: Examples of the structural use of stone.

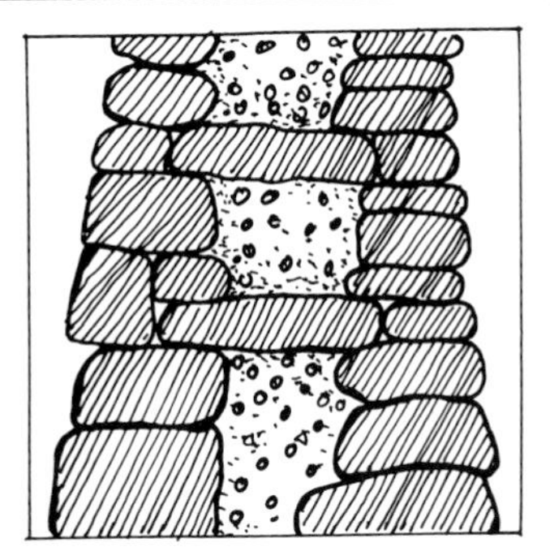

Cavity wall in field stones
The cavity is filled with small stones in mortar, clay, perlite, loose expanded clay or kieselguhr. On the outer leaf the stones lean outwards so that water runs off. On top of the wall there are large stones or a lime mortar. Good insulation and windproof as a sheltering wall.

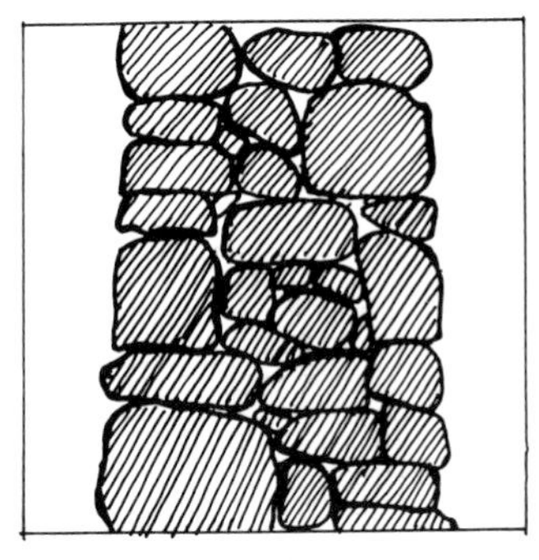

Solid wall in field stones
Can be rendered stable. Requires a lot of insulation as house wall. Best as foundation wall, or foundation to plinths.

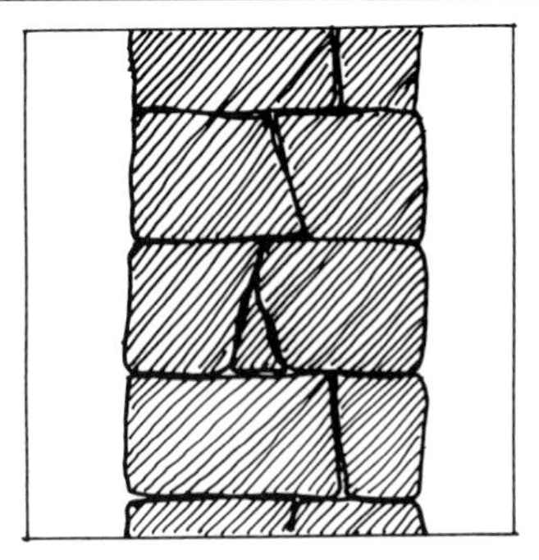

Solid wall in cut stone
Can be rendered, very stable. Requires a lot of insulation as house wall. Best as foundation wall or foundation to plinths.

Figure 13.9: Dry-stone walling techniques.

Masonry

When building with stone particular care needs to be taken with the corners of a wall. In many examples, a larger squared stone is placed on the corner, while the rest of the wall consists of smaller worked quarry stones or rubble.

Dry-stone walling

This technique demands great accuracy and contact between the stones. The stones have to be placed tightly against each other vertically and through the depth of the wall. Small flat angular stones can be put into the joints to fix the stones against each other. A quarter of the area should have bonders (or through stones) that go through the whole thickness of the wall between the inner and outer leaf.

Dry-stone walling is particularly appropriate for foundation walls as they have the function of stopping any capillary action from occurring – no water can be forced upwards in such a construction. This form of wall is not particularly windproof. One way of working is to have two parallel walls with earth or another fill between them. Better wind-proofing is achieved, but it has to be well drained to avoid expansion and splitting due to frost.

Walls bonded with mortar

Many different mortars can be used for masonry (see 'Mortars'). Generally, lime mortar and cement-lime mortar are the most suitable. The important properties are elasticity and low resistance to moisture penetration, because stone itself is so

Figure 13.10: Dry-stone walling of specially cut stone.

resistant to moisture penetration. This is especially important for igneous and metamorphic rock species, which can cause condensation problems on the external walls of a normal warm room, no matter which mortar is used. With the exception of marble, sedimentary rocks are best suited for this purpose.

For heated buildings stone is best used for foundations. The exceptions are limestone and sandstone which can be used for wall construction, but even sandstone is susceptible to frost. Both limestone and sandstone decay in the same way as concrete when exposed to aggressive air pollution.

Structural brickwork

Brick structures have been used for thousands of years in many cultures. In Europe it was not until the middle of the twentieth century that brick was

Figure 13.11: Openings in stone walls.

replaced by concrete as the main structural material, and since then it has often been used to clad concrete structures. In addition to being more durable than concrete, brick is easier to repair by replacement with new bricks.

Brick has a low tensile strength, which means that it is best used structurally in columns, walls and vaults of a smaller scale. Reinforcement and working with steel, concrete or timber, can expand its areas of use. Spans and the size of building units can increase and brick can be used in beams and floor slabs.

In normal brickwork, brick represents approximately 70 per cent of the volume – the rest is mortar. Brick is a heavy material completely manufactured at one factory, in contrast with concrete which has two components. Brick is normally used in large quantities, meaning that transport over large distances can have an environmental impact.

The production of brick seriously pollutes the environment and is very energy consuming, but bricks have a low maintenance level and are very durable, in the majority of cases outlasting all other materials in a building. Dieter Hoffmann-Athelm expresses this fact in his paradoxical critique of civilization: 'Brick is almost too durable to have any chance nowadays'. Bricks can withstand most chemical attacks except for the strongest acids. Drains made of the same material as bricks – fired clay – withstand acidic ground conditions; concrete pipes do not. It is therefore important that the design of brick structures considers the thorough planning of recycling. This would make brick a much more competitive and relevant ecological building material.

The polluted effluent from the brick industry can be relatively simply separated out or reduced by adding lime to the clay. The total energy consumption can be greatly reduced by differentiating the use of bricks in well-fired and low-fired products. Today only well-fired bricks are produced while low-fired alternatives could be used for most purposes in less weather-exposed parts of brick structures. This was common practice until around 1950.

Bricks fired at 200–400°C have kept for at least 4000 years without serious damage, mainly in warmer climates. In northern Europe the absorption of water would be so high that the bricks would run the risk of being split by frost during the winter if placed in exposed positions. A well-rendered brick wall, however, can cope with this problem, as demonstrated by northern Europe's rendered-brick buildings, many of which are built of low-fired bricks.

In a completed building, brick is considered a healthy material. The potential for problems can arise when radioactive by-products are used in the manufacture of the bricks, e.g. slag from blast furnaces. Otherwise brick has a positive effect on the indoor climate, especially bricks with many pores, which will regulate humidity. Conventional washing down brick walls with hydrochloric acid can cause problems in indoor climates.

Table 13.4: Structural uses of fired clay bricks

Types of bricks	*Firing temp (°C)*	*Properties*	*Areas of use*
Vitrified	1050–1300	Very hard and frost resistant	Exposed external walls, floors, lining of concrete walls, foundations
Well-fired	800–1050	Hard and frost resistant, slightly absorbent	External walls, lining of concrete
Medium-fired	500–800	Medium resistance to frost, very absorbent	Internal walls, inner leaf of cavity walls, rendered external walls, moisture-regulating layers
Low-fired	350–500	Not frost resistant, highly absorbent	Internal walls, inner leaf of cavity walls, well-rendered external walls, moisture regulating layers
Light-fired:			
Porous brick	Approx. 1000	Same as medium fired, plus moderate thermal insulation	Same as medium-fired, plus thermal insulation
Zytan	Approx. 1200 (twice)	Same as well fired, plus good thermal insulation	Same as well-fired, plus thermal insulation

Note:
Light fired clay products combine moderate structural properties with moderate to high thermal insulation properties, and are described in more detail in the next chapter.

Structural bricks and blocks

There are three main types of structural brick and block: solid bricks, perforated bricks, and blocks and light clay blocks. Blocks can also be composed of expanded clay pellets, fired together. Perforated bricks and blocks are the most common types. They use less clay, have a slightly better insulation value and are also lighter with a stronger structure, because the mortar binds them together more efficiently. The holes have to be small enough to prevent mortar filling them.

The size and form of bricks has varied widely, depend upon the culture and period of use. The Romans usually fired square or triangular bricks up to 60 cm in length with a thickness of 4 cm. They also produced semi-circular and ornamental bricks. The rectangular structural brick, with very few exceptions, has always been formed under the principle of its length being twice its breadth plus a mortar joint. The British Standard brick is 215 × 102.5 × 65 mm. The mortar joint is usually 10 mm.

On the continent the use of hollow blocks for floor slabs and beams is widespread. In hollow block beams the structure is held together by steel reinforcement in the concrete, while the slab units are only partly structural as they are held between beams of either hollow blocks or concrete.

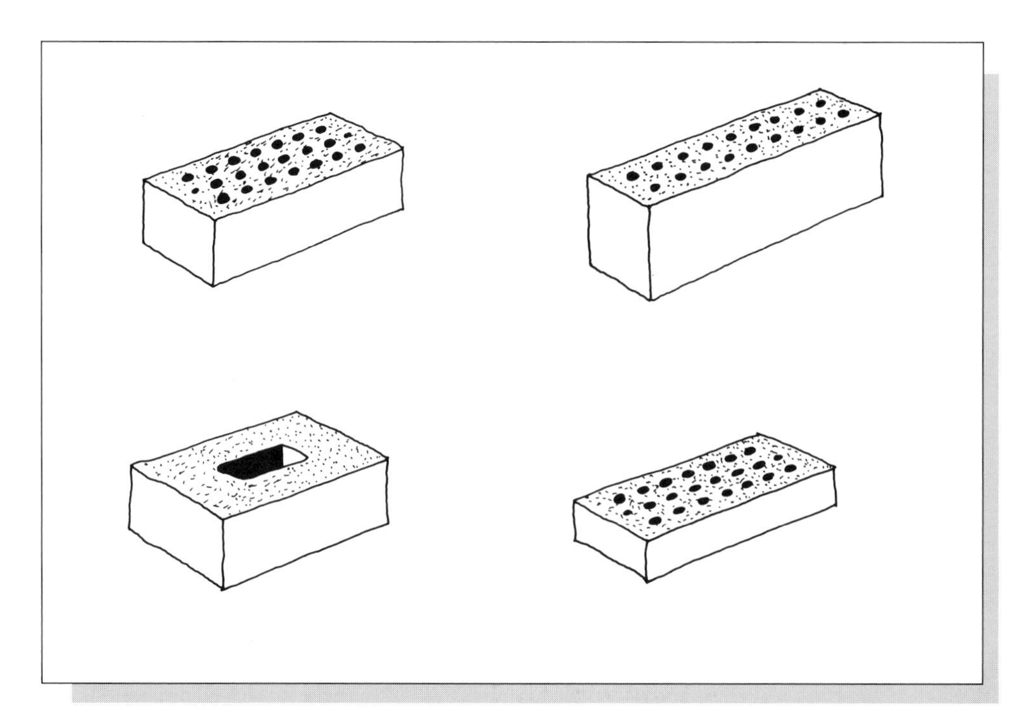

Figure 13.12: Examples of perforated bricks.

Figure 13.13: Examples of perforated blocks.

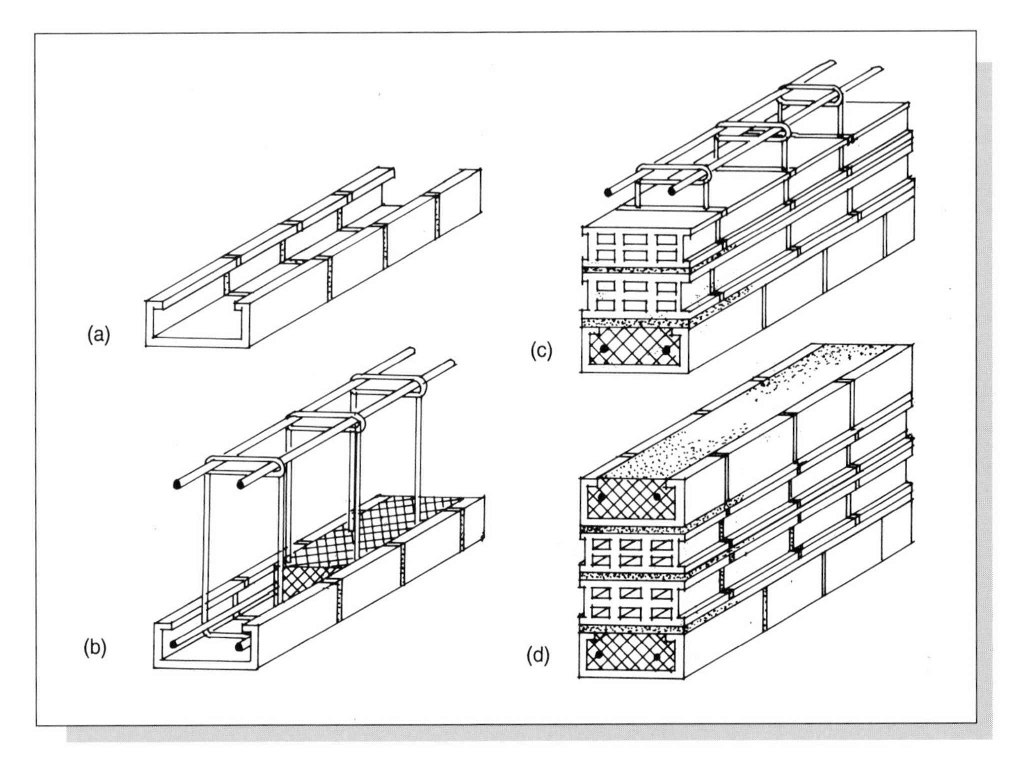

Figure 13.14: The process of steel reinforcement in hollow block beams.

Recycling

A brick can usually last many house generations. It needs to do this in order to justify its high primary energy consumption and highly polluting effluent during production.

To recycle brick, the mortar has to be weaker than the brick or the brick will break up before the mortar. Since 1935 strong mortars containing a large proportion of Portland cement have been used making walls from this period difficult to recycle. Lime mortars with a maximum of 35 per cent Portland cement make it possible to dismantle a wall. The necessary strength of brickwork is also achieved by using hydraulic lime mortar. When lime mortar is used, there is no need for expansion joints in the wall because of the high elasticity of the brickwork. Lime cement mortars should be used in districts with an aggressive climate, such as in towns or along the coast.

There is no technically efficient method for cleaning old bricks – it has to be done by hand and is relatively labour intensive. Recycled bricks are mainly usable in smaller structures such as party walls and external walls, where there is no heavy horizontal loading. In the pores of the brick, old mortar is chemically bound with the brick, making it more difficult to bind new mortar.

Recycled brick should be soaked before laying. If one side is covered in soot from a chimney, this must never face the outside as it would penetrate the render.

Bricks that cannot be dismantled can be ground and in certain cases used as an equivalent to pozzolana in cement. Larger pieces of brick can be used as aggregate in concrete. In Denmark, blocks are manufactured with beautiful pieces of brick used as aggregate.

Smaller brick structures

Brick structures above ground can be built as walls, columns, arches and vaults. Arches and vaults are used in roof construction, but they are labour intensive and require a good knowledge of the material. The arch is the most usual way of spanning an opening for windows or doors without having to use steel reinforcement. The following rules of thumb should be used when building a wall without reinforcement:

- The building should not be higher than two storeys
- The largest distance from centre to centre of the structural walls should not exceed 5.5 m; the distance between the bracing party walls should not be more than 4–5 m
- The main load-bearing walls should be at least 20 cm thick, i.e. two bricks wide. Alternatively they can be one brick thick with 30 × 30 cm piers
- Window and door openings should be above one another

Solid or cavity walls can be built. Solid walls are straightforward to build, and can be insulated either inside or outside, e.g. with woodwool slabs which can be plastered or rendered. If the woodwool is on the outside the brick's capacity to store heat when warmed is better utilized. Internal insulation causes colder brickwork and increases the risk of frost damage.

Cavity walls are normally two leaves of single brickwork with a distance between them of 50–75 mm. A hard fired brick that will withstand frost is necessary in the outer leaf to make use of the maintenance-free aspects. Extra- hard-fired bricks which are highly vitrified have a low capacity for water

Figure 13.15: A small Danish building entirely constructed in fired clay without using reinforcement.

Figure 13.16: Structural vault in brick.

absorption and should therefore be ventilated behind. If the outside surface is going to be rendered, bricks fired at lower temperatures should be used.

The inner leaf can be made of middle- or low-fired bricks. Such a differentiation of brick quality was completely normal until the 1950s, as energy-saving in production lowered costs. Today the hardest-fired bricks are used in all situations.

Low-fired and porous bricks must be soaked before laying so that they do not absorb all the moisture from the mortar, as with ceramic tiles on a similar surface. Low-fired brick binds well with clayey binders such as hydraulic lime, but less well with pure lime products (see Table 17.1).

The leaves are usually tied together with steel wall ties. The cavity is filled with insulation, preferably of mineral origin, such as perlite, loose light clinker, granulated glass and vermiculite. In areas where there is heavy driving rain it pays to render the inside of the outer leaf. Beams resting on the inner leaf are surrounded with impregnated building paper.

A vapour-tight render or paint should be avoided on the outside, as it will quickly result in frost damage. Good alternatives with open pores are hydraulic lime render and silicate paint.

Earth structures

Earth structures consist of either rammed earth carried out on site between shuttering, pisé, or earth blocks such as adobe. These are suitable for buildings of domestic scale. The material is fire-proof in itself even with plant fibres mixed in with it. Earth is also a good regulator of humidity. The oldest complete earth building that exists in Europe, dating from 1270, is in the town of Montbrison in central France. It now houses a library for moisture-sensitive books.

Earth buildings have many ecological precedents. Earth is a perfect material in terms of resources, pollution and indoor climate, and when the building is no longer needed, it reverts to its original material.

Earth has structural limitations as a building material as its compressive strength is low. This is compensated for by building thicker walls. The increase in the amount of material used does not really matter when the source of earth is near the site.

Earth does not have a particularly high thermal insulation value – slightly better than concrete, but more like brick. By adding different organic fibres the insulation value can be improved; dwellings cannot be built without extra insulation on the walls. Solid earth walls, possibly with fibre mixed in, are best for buildings with low internal temperatures or with external two-leaf walls containing a cavity. An exception to this is 'leichtlehm', or light clay (see p. 289).

Earth can only be used locally, as transporting it for building or rammed earth blocks over distances is uneconomical and ecologically unsound.

Figure 13.17: A six storey earth building erected in Weilburg (Germany) in 1827.

Suitable types of earth

For pisé construction earth must be dry enough for the shuttering to be lifted directly after ramming without damaging the wall. Shrinkage needs to be as little as possible to avoid small cracks. A well-graded earth with about 12 per cent clay is the best type, although even an earth mixture with up to 30 per cent clay is usable, but will be harder to form. If a mixture is less than 12 per cent clay, fine silt can be added. These types of earth need more preparation before ramming. Sand can be mixed with earth that has too much clay, and clay can be added to earth that has too little. This can be a very labour intensive and uneconomical task.

For adobe blocks a much more fatty earth with up to 40 per cent clay (or even more in blocks mixed with straw) can be used.

Stabilizing aggregate and other additives

In certain situations it may be necessary to add stabilizers. These usually have three functions:

- *To bind the earth particles together strongly.* These are substances such as lime, Portland cement, pozzolana cement and natural fibres. These strengtheners are necessary for buildings more than two storeys high, whatever the quality of the earth.

- *To reduce water penetration.* Lime, Portland cement, pozzolana cement and waterglass are examples. In areas where there is a great deal of driving rain it is advisable to have one of these additives in the earth mix as well as external cladding on the wall. In some case whey, casein, bull's blood, molasses and bitumen have been used for the same reason.
- *To avoid shrinkage.* This is mainly achieved by natural fibres, even though cement and lime also reduce shrinkage.

Lime and cement

Lime is the stabilizer for argillaceous (clayey) earth. Both slaked and unslaked lime can be used. The lime reacts with the clay as a binder. Lime can be used with silt containing a lot of clay, sand or gravel and is usually mixed by sieving into the proportion of 6–14 per cent by weight.

Portland cement is the stabilizer for earth rich in sand or containing very little clay. The proportion of cement to earth is 4–10 per cent by weight. This can also be used in foundation walls. The humus in the earth can attack the cement, so this construction technique is assumed to have low durability.

Pozzolanic cement can be used in both types of earth, either lacking or containing a lot of clay. It has about the same properties as Portland cement, but has to be added in slightly larger quantities.

All lime and cement additions reduce or remove the possibility of recycling the earth after demolition or decay.

Natural fibres

Natural fibres are best used in earth containing a lot of clay to increase thermal insulation and reduce shrinkage. A mixture of 4 per cent by volume of natural fibre will have a very positive effect on shrinkage and strength. The normal proportions in the mixture are 10–20 per cent by volume. Larger amounts than this will reduce its strength. In non-structural walls which are primarily for thermal insulation, it is normal to increase the fibre content to 80 per cent, but this wall will not hold nails.

Straw chopped into lengths of about 10 cm, preferably from oats or barley, is normally used. Pine needles are also good binders; alternatively stalks from corn, flax, dried roots, animal hair, twigs, sawdust, dried leaves and moss can be used.

If large amounts of organic material are used, mould can begin growing only a few days after erecting the wall. This is especially the case when blocks bound with a thin loose mixture of clay are used. These walls must dry out properly and cannot be covered until the moisture content has reduced to 18 per cent.

Expanded mineral products

Products such as exfoliated vermiculite or expanded perlite can be used as aggregate. There is no chance of mould, and higher thermal insulation is achieved. However, mineral aggregates require much more energy to extract and produce than natural fibres.

Waterglass

An earth structure can be waterproofed by brushing a solution of 5 per cent waterglass over the surface of the wall. The solution can also be used for dipping earth blocks before mounting them.

Methods of construction

All the different construction techniques require protection from strong sunshine and heavy rain. The easiest way is to hang a tarpaulin over the building. It is also advantageous to build during the early summer, so that the walls are dry enough to be rendered during the autumn.

Foundation materials for earth buildings are stone, lightweight expanded clay blocks, normal concrete or earth mixed with Portland cement. These should be built to at least 40 cm above ground level, and must be at least as wide as the earth wall, usually about 40 cm.

Stone and concrete walls can absorb a great deal of moisture from the ground through capillary action. Whatever happens, this moisture must not reach the earth structure, as this is even more sensitive to moisture than timber construction. Damp-proofing can be carried out with asphalt.

Pisé (earth ramming technique)

Earth suitable for ramming contains primarily sand, fine gravel and a small amount of clay which acts as a binder. Through ramming, these components are bound together. After the building process, the wall will be cured by substances in the air and eventually be almost as hard as chalk or sandstone. Shuttering and further equipment is required for ramming.

Figure 13.18: Recently renovated 200-year-old earth building in pisé construction in Perthshire. Source: Howard Liddell

Shuttering and ramming equipment

Shuttering must be easy to handle and solid. There are many patents.

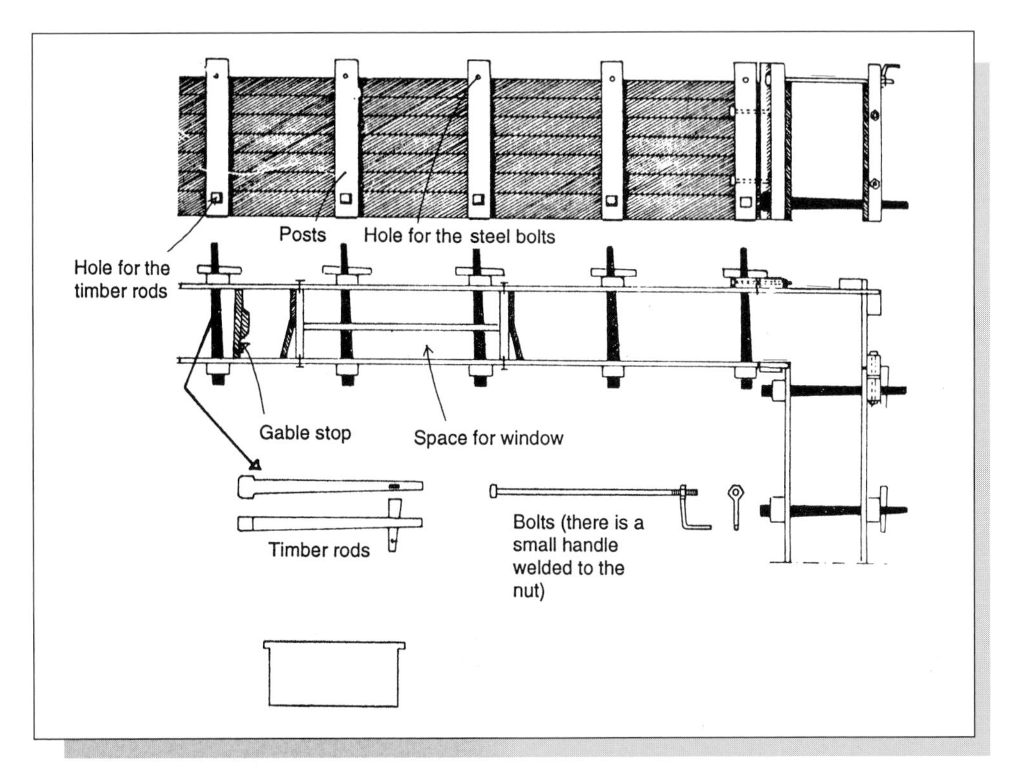

Figure 13.19: Swedish model for shuttering. Source: Lindberg 1950

Figure 13.19 shows a Swedish model which is easily self-built. It consists of two vertical panels fixed together by long bolts and wooden rods. The panels are made of 30 mm thick planks of spruce or pine. The length of the shuttering should be between 2–4 m depending upon the dimensions and the form of the building. The panels are 80 cm high and braced by 7 × 12 cm posts screwed to the boarding. The screws are 64 cm apart.

The spacing of the posts depends upon the thickness of the wall, usually 40 cm. On the bottom they are held together by timber rods, while the upper part are held together by steel bolts 18 mm in diameter. The rods are made of hardwood such as beech, ash or maple and are conical. The dimensions at the top of the rod are 6 × 6 cm and at the bottom 4.5 × 4.5 cm. The holes in the posts should be slightly larger so that the rods are loose. The gable ends of the shuttering have a conical post fixed with nails. To prevent the shuttering falling inwards, a couple of separating boards are needed inside the shuttering.

In order to form openings for doors and windows, loose vertical shuttering is placed inside and nailed through the shuttering panels. These can then be easily removed. It is quite possible to mount shuttering after each other as long as they are well fixed.

The ramming can be done either manually or by machine. When ramming by hand, three rammers with different forms are needed (see Figure 13.21). The handle is heavy hardwood and the rammer is made of iron. The weight of a rammer should be around 6–7 kg.

Ramming by machine is much more effective. This can be done using a compressed air hammer with a square steel head of 12 × 12 cm. The compressor's power should be around 5 hp per hammer. The job must be done by an operator who can steer the machine; it is heavy work. A robot-rammer which can follow the line of the shuttering is being developed in Germany.

Ramming is best carried out by a working team of two or three people. The wall shuttering is mounted on the foundation walls as in Figure 13.22 with gable ends and separating boards.

Figure 13.20: Ramming earth with a compressed air machine. Source: Gaia Lista 1991

When ramming by machine layers of 13–14 cm can be built. This is approximately two thirds of the volume of the original loose earth. When ramming by hand a layer thickness of not more than 8 cm is advisable. Clearly the two methods cannot be used together. It is important to ram at the edge of the shuttering when machine ramming – starting in the middle may cause stones and lumps to be pushed out to the edge and loosened. The ramming should make the earth as hard as rock – it should 'sing out' – and a pick should not make any marks when the surface is hit.

When the first layer is ready, the next layer is begin, and so on until the shuttering is full. The rods are then pulled out and moved up the shuttering. With each move it is necessary to check that the shuttering is vertical. The conical post on the gable end of the shuttering acts as a 'locking key' to increase the stability of the wall.

In the corners reinforcement of twigs or barbed wire are used, and after the first layer, holes are cut for the floor beams, which will be placed directly on the

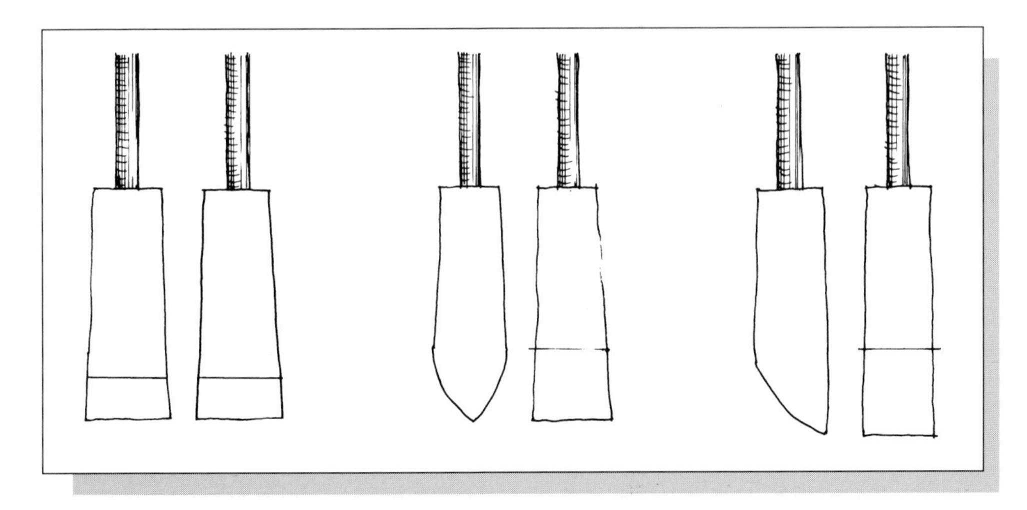

Figure 13.21: Different forms for the manual rammers.

damp proof course on the foundation wall. As the ramming progresses, openings for windows and doors are added, with timber or concrete reinforcing beams rammed in over them. Timber does not rot in normal dried earth walls. All timber that is rammed into the walls has to be dipped in water first. Timber blocks that are rammed into the wall for fixings should be conical, with the thickest end in the middle of the wall, so that it does not loosen. To hold the

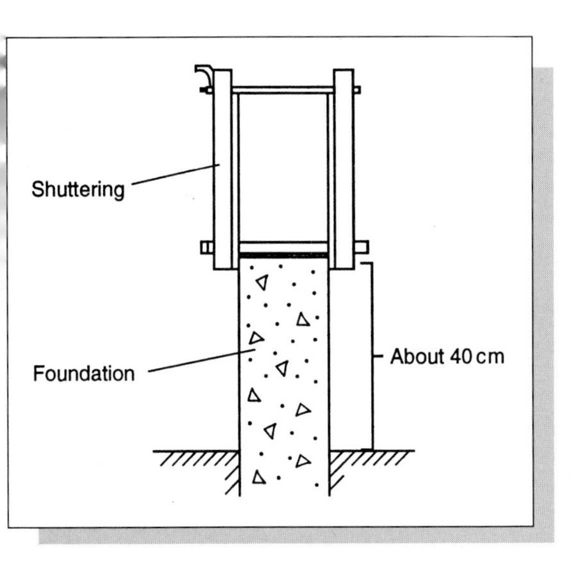

Figure 13.22: Putting up shuttering.

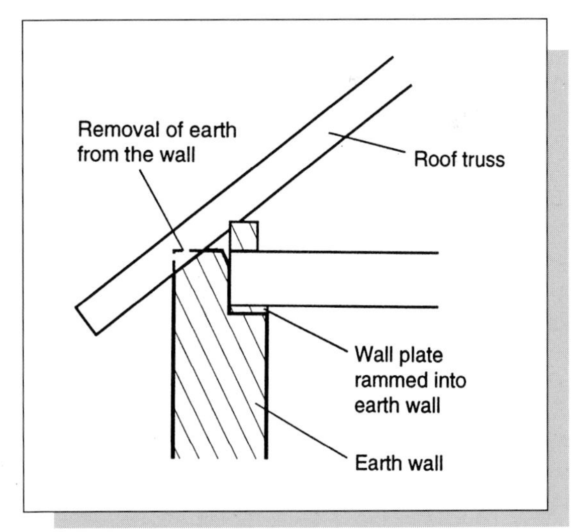

Figure 13.23: Ramming in the wallplate to carry the floor and roof structures.

floor joists further up the wall a timber plate the whole length of the wall must be rammed in (see Figure 13.23).

When the ramming is finished the roof is put on. A large overhang will protect the wall from rain, which is very important early in the life of the building.

Surface treatment
When the walls are complete the holes made by the rods are filled with crushed brick mixed with lime mortar, or expanded clay pellets, which give better thermal insulation. The outside and inside walls can be rendered with hydraulic lime or lime cement render. The inside can also be rendered with a normal lime mortar. Walls exposed to extreme weather conditions should be protected by timber panelling fixed to battens nailed directly onto the earth wall. The nails usually fasten to the earth wall without any problem. Internal walls can also be

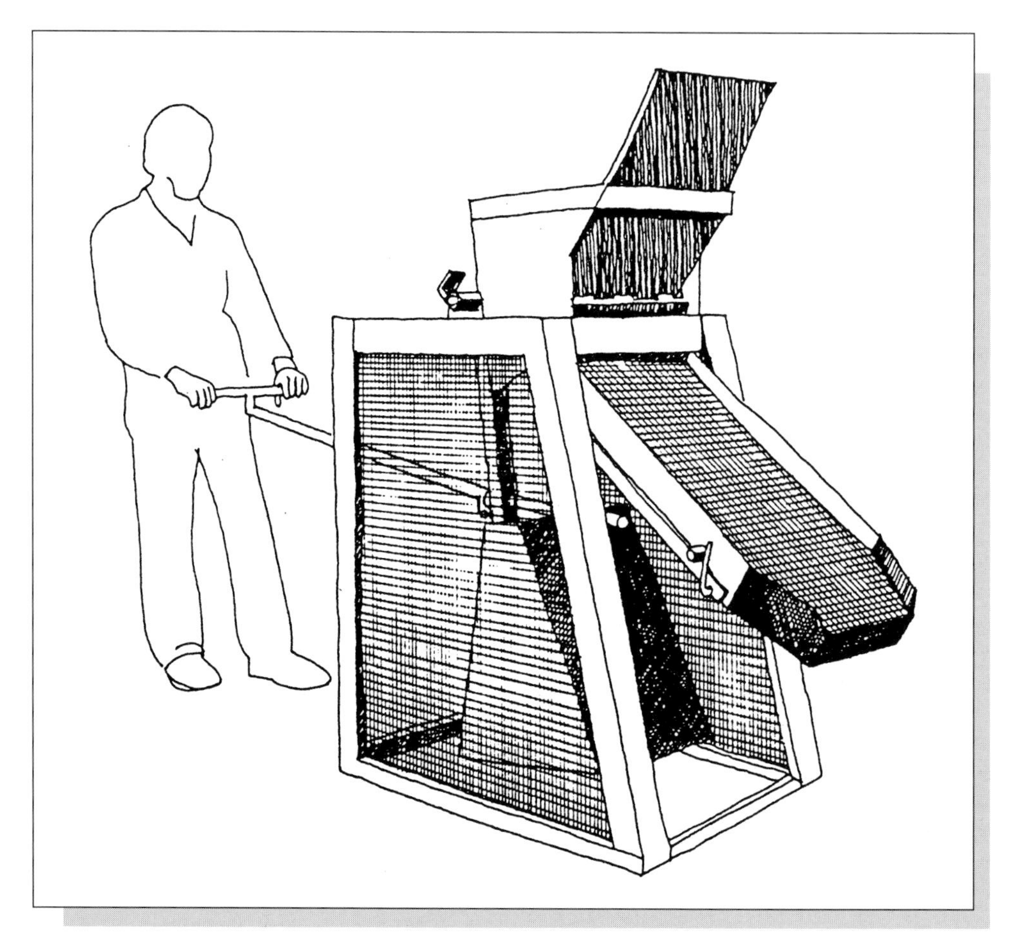

Figure 13.24: Manual clay crusher.

covered with panelling or wallpaper, or painted with mineral or casein paints. The surface of the walls must not be treated with a vapour-proof barrier, as this would quickly lead to moisture gathering inside the wall, thus allowing frost damage.

Adobe (earth blocks)

The advantage of building with blocks rather than pisé is that the building period is less dependent upon the time of year. The blocks can be made at any time, providing there is no frost, and can be stored until needed for building. Block-laying should be carried out during spring or early summer so that the joints can dry out before applying the surface treatment. As already mentioned, there must be a higher percentage of clay in earth for blocks. There should be no particles larger than 15 mm in the mix. Hard lumps of clay can be crushed in special crushers (see Figure 13.24).

Figure 13.25: Building with earth blocks. Source: Gaia Lista 1991

A certain amount of chopped straw is added to stop cracking due to shrinkage, and a little water, to make the earth more pliable before use.

Moulds

Loose moulds of wood or metal, and even mechanical block moulds, are available. The size of moulds can vary, but 'monolithic' blocks are 75 × 320 × 50 cm and mini-blocks are the same size as bricks. Larger blocks would require an impractically long drying time in some climates. Loose wooden moulds can be nailed together quite easily. Commercial block moulds have capacities that vary from 300 to 3000 blocks per day. These are easy to transport and are used manually or driven by a motor.

Pressing the blocks

The earth mix is rammed into the mould so that the corners are well filled, and excess earth is then scraped off with a board. After a few hours the blocks are ready to be removed from the mould, and after three days they are stacked so the air can circulate around them. During this period the blocks must be protected from rain, if they do not contain added cement. After two weeks the blocks are dried well enough for building.

Laying earth blocks

The mortar used is usually the same earth that the blocks are made of, mixed with water and even some lime. Portland cement should not be used, as it can split the stones during shrinkage. Blocks are laid in normal coursing after dipping in a waterglass solution to saturate them. Barbed wire, chicken wire or plant fibres are recommended in every third course as reinforcement.

It is also possible to construct ceiling vaults from earth blocks. Exposed earth roofs are not well suited to climates in which moisture and frost can quickly break down the structure.

Surface treatments are the same as those used for the pisé technique.

Other earth building techniques

Adobe and pisé are the most widespread of earth-building techniques, but other techniques also have interesting aspects. The most important alternative techniques are wet-formed earth walls, earth loaves, extended earth tubes and the 'sandbag' technique.

Wet-formed walls

As with earth blocks, earth used for wet-formed walls is relatively rich in clay. The earth and cut straw is mixed in a hole in the ground in the proportion of 50 kg straw to 1 m^3 of earth. The more clay the earth contains, the more straw is needed. The ready mixed earth and straw is then thrown up with a pitchfork into the shuttering of the wall and rammed down by foot. Between adding each course of about 50 cm the wall is left to dry out for about two days.

When the wall has reached full height, the vertical is checked and excess earth removed with a trowel, so that the wall has an even thickness. A clay mix or gruel is poured

Figure 13.26: The manor house of Skinnarebøl in south east Norway from the early 19th century is built in the wet-formed wall technique.

over the whole wall and it stands under cover until it is dry, from three months to a year. The shrinkage is quite considerable, about 1 cm per metre, so it could be disastrous to render a wall before it is totally dry. Because of the long and frequent intervals in the process, this building technique is seldom used nowadays, even if there are many historic examples which prove that it is a solid and well-tried method.

Earth loaves

This technique is a very simple earth building method brought to Europe by a missionary who learned it in East Africa. The German school of agriculture at Dünne further developed the method during the 1920s, and since 1949 about 350 buildings have been constructed in Germany using this technique. 'Loaves' are formed from well-mixed earth containing a high percentage of clay. These clay loaves measure about 12 × 12 × 25 cm.

The walls are built by laying the loaves on top of each other as in normal bricklaying, as soon as they have been kneaded, at a rate of four courses each day. They are reinforced with twigs every third course and every course in the

Figure 13.27: The earth loaf technique.

corners. After four to six weeks drying time the wall is strong enough to take the roof. The roof is often put up provisionally before hand to protect the walls against rain during the drying period. The earth loaf technique can of course be used for internal walls, with or without a load-bearing function.

Extended earth tubes
This method has been recently developed by the Technical High School in Kassel, Germany, and is a development of the earth loaf technique. In this case there is not as much clay in the mix, as shrinkage would cause a problem, but the amount of clay must be enough to give the mix a certain elasticity.

The earth is put in an extruding machine used for bricks (see Figure 8.7), compressed, and then extruded in tubes of 8–16 cm in diameter. The capacity of the machine is 1.5 m of tube per minute, and the length is unlimited. The material is so well compressed from the beginning that it can be combined and built without waiting for the lower layer to dry out. With a mobile extruding machine a house can be built in a few days in the same way that a vase of clay is made with long clay 'sausages'. Mortar is not necessary, but the walls must be rendered afterwards. This technique is still at an early stage of research.

The 'Sandbag' technique
Visually this building technique is similar to extruded earth tubes. The earth has to be as free of clay as possible, i.e. pure sand, which has no binding properties. The 'binder' is jute sacks which are 2.6 m long and about 0.5 wide. The sand-filled sacks are piled up as walls within a light timber framework. The sand can also be mixed with hydraulic lime mortar or cement, and the sacks dipped in water before being piled up, so the mix becomes hard enough to make the sacks superfluous. It is also possible to add some aggregate to increase the insulation value.

The efficiency of earth building

Constructing a wall of earth needs about 2 per cent of the energy used to build a similar wall in concrete. The building process for an earth wall is more labour- than capital-intensive. The material is almost free, but the amount of labour is very large. According to an investigation by the Norwegian Building Research Institute in 1952 the following proportioning of labour was found (Bjerrum, 1952) – the net time including only ramming and building up the wall, the gross time including the surface treatment:

Method	*Work hours/m² net*	*Work hours/m² gross*
Machine ramming	3.5	5.0
Ramming by hand	5.5	7.0
Blocks by machine	3.5	5.0
Blocks by hand	5.5	7.0

The equivalent for a fully completed concrete wall with surface treatment is 3.3 hours/m^2 whereas a brick wall takes 3 hours/m^2, but the figures only take into account the amount of work carried out on the building site. In the case of concrete and brick a large amount of work has been done before the materials actually arrive at the building site. The difference between these methods would be drastically reduced if these aspects were also considered, but there is little of a complete assessment of the different methods.

According to Gernot Minke of the Technical High School in Kassel, research and development of partly-mechanized earth building techniques is going to make this technique much more efficient in the near future. Working with the extruded earth method, an 80 m^2 house, both inner and outer walls, can be built in three days using four builders who know the techniques. A conventional earth house of the same size would take 14 days to build.

Earth buildings and indoor climate

A completed earth house has a high-quality indoor climate. Earth is a very good regulator of moisture compared to many other materials. The walls are relatively porous and can quickly absorb or release moisture into the room. The relative humidity of the inside air will usually be around 40–45 per cent. An investigation has been conducted in Germany amongst people living in stone, brick, concrete and earth buildings. Those in earth buildings were, without exception, satisfied with the indoor climate of their homes. This satisfaction was seldom found amongst the people in the other house types. These, perhaps subjective feelings, have only been partly scientifically proved. It is not only earth's property of moisture control that should be taken into account, but also other factors such as its absorption of gas and odours, its warmth capacity, its acoustic properties of reducing noise levels and even certain other psychological aspects.

Plastic structures

Plastic is seldom used as a structural material. The large amount of unspecified plastic waste which now exists in the Western world is a possible raw material for simple structural elements. Polystyrene waste can be cast into solid beams and columns if supporting substances are added in proportions of 10–15 per cent. The structural properties are approximately the same as timber, and components can be sawn and nailed. The concept is interesting and still being developed in England and Sweden. There is little evidence to assess its durability and workability with other products. Polystyrene and a large proportion of additives could possibly have unfavourable effects on the indoor climate, and pollution could occur when the products become waste materials.

Timber structures

Timber has been the main structural material for the nomad's tent and the farmer's house and fencing in all corners of the world, especially in the case of roof construction, in which its lightweight and structural properties have made it more attractive than any other alternative.

High-quality timber is stronger than steel when the relative weight is taken into account, and the environmental aspects are considerably better. Timber structures have been limited to small buildings because of fire risk, but now there are many developments in the use of timber in larger buildings. The reasons for this are the improved possibilities for technical fire protection and the revised view of timber's own properties in relation to fire, which are better than previously thought. In timber of a certain size, the outer carbonized layer stops further burning of the inner core of the timber.

History

The first mention of buildings constructed completely from timber in European history is in Tacitus. Tacitus writes about Germanian houses in his *Histories* in AD 98, characterizing them as something 'not pleasing to the eye'. The houses had either palisade walls with columns fixed into the earth or clay-clad wattle walls. They had thatched straw roofs. Excavations from a Stone Age village in Schwaben, Germany, showed that houses like these have been built over a period of at least 4000 years. Excavations of a Bronze Age village on an island in a Polish lake uncovered houses built of horizontal planks slotted between grooved posts. The palisade wall went through many improvements on the Continent and received a bottom plate, amongst other things.

Remains of log timber buildings from about 1200–800 BC have been found in the village of Buch outside Berlin. Even in China and Japan there are traces of this technique from an early period, but most likely from a completely separate tradition to that of Europe.

In areas where there is a milder climate, such as the British Isles, the coasts of the continent and Scandinavia, an alternative structural technique developed alongside log construction – the stave technique. This technique is best exemplified in all its magnificence by stave churches, and creates enormous airy timber structures from specially-grown timber, held together by wooden plugs.

The rendered wattle wall really started to develop when masonry walls were enforced by law. After a series of town fires during the seventeenth century, rendered wattle walls were almost the only alternative to brick and stone. At the end of the eighteenth century massive vertical load-bearing timbers were introduced as an alternative to log construction in Scandinavia. This technique was developed because builders wanted to be able to set up external panelling directly after the structure was ready, rather than having to wait for the building to settle, as is necessary in log construction. This structural technique disappeared around 1930. Log construction also started to disappear around this time, and by 1950 it had almost totally disappeared. It has enjoyed a sort of renaissance in the holiday cabin industry. In Scandinavia over the last 200 years the stave technique has been used mainly in outhouses. Immigrants in the USA, however, had access to timber of large dimensions, and further developed the stave technique for use in large storage buildings, barns etc., during the eighteenth and nineteenth centuries.

To a certain extent modern post-lintel construction can be seen as a further development of the stave technique. In Europe today, the main form of structural technique is the timber frame building, and this has gone through many improvements and different forms. There are also new methods in the structural timber industry: space frames and laminated timber beams have opened many new possibilities. Through looking at the history of building in other cultures shell construction has also been developed in Western culture.

Structural elements in timber

Materials in solid timber occur in different sizes, either as round logs or rectangular sections. There is an obvious limitation depending upon the size of the tree that is used, and this varies between different types of tree. Generally, the smaller the size of the element, the more effective the use of the timber available. The use of small timber sections from certain deciduous trees is important, as they are not particularly large trees. To resolve the problem of the limitations of some components, timber jointing can be used.

It is necessary to differentiate between timber jointing for increasing the length or increasing the breadth or cross-section. Jointing for increasing the length can be achieved with timber plugs, bolts, nails or glue. It is normal to use spliced joints for sills, logs, columns or similar components where compressive strength is more important than the tensile strength. Certain spliced joints, such as the glued finger joint, have a good tensile strength.

Increasing the breadth can be achieved by using solid connections or I-beams. Solid connections consist quite simply of the addition of smaller sized timbers to each other. The fixing elements are bolts, nails or glue. Bolted joints are often complemented by steel or timber dowels to stop any lateral movement between the pieces of timber, as in Figures 13.29 and 13.30. Dowels and toothing were

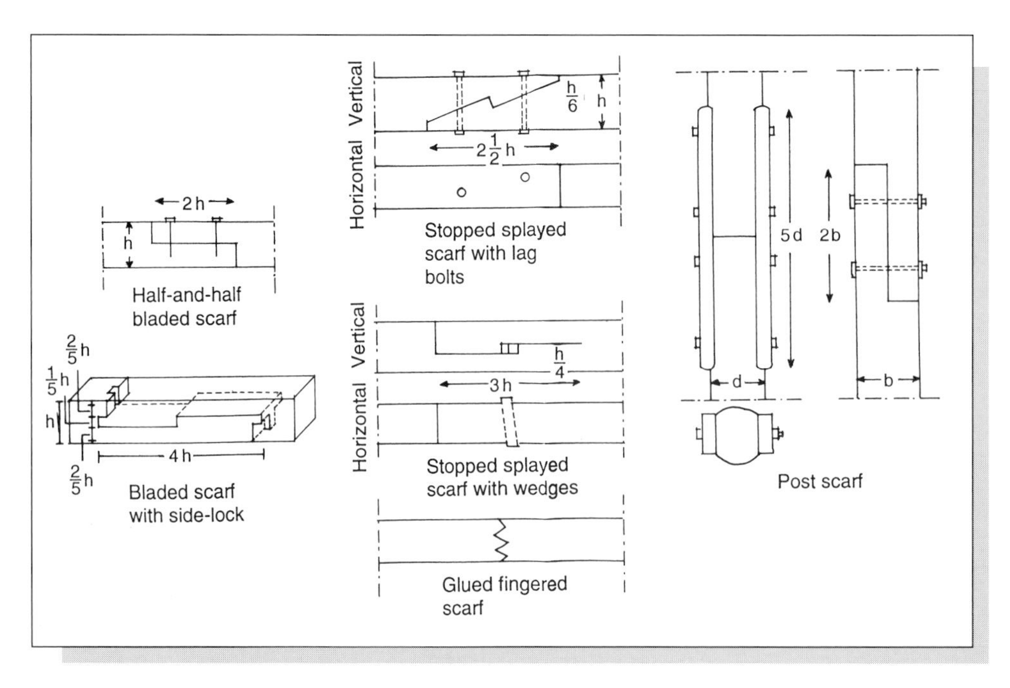

Figure 13.28: Timber joints for increasing the length.

used until the 1920s. Solid laminated timber joints have been in use since the turn of the century, and nowadays usually consist of 15–45 mm-wide spruce plank.

I-beams consist of an upper and lower flange with a web in between. The web can be formed of solid timber, steel, veneer, chipboard or fibreboard. The first two are usually fixed by plugging, bolting, nailing with nails or nail plates, while the others are glued. Depending upon how the I-beams are made and shaped they can also be roof trusses, which are used a great deal in prefabricated houses. I-beams are a very economical use of material in relation to their strength, and can be used in roof, floor and wall construction.

Figure 13.29: A roof joint bolted together, not glued, stiffened by dowels. Source: Gaia Lista, 1987

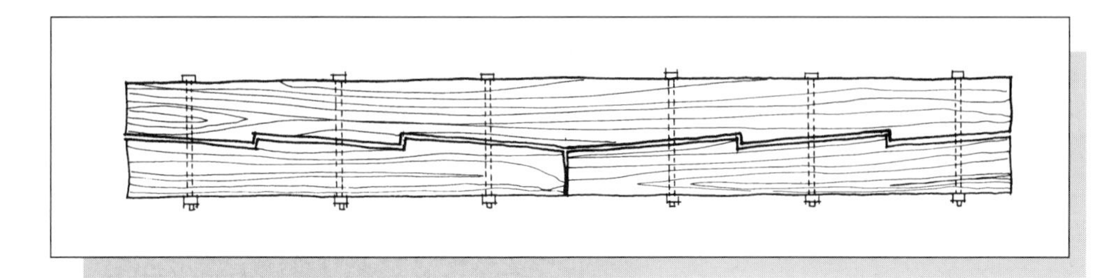

Figure 13.30: Toothed beam joint put together in three pieces.

The energy consumed in the production of laminated timber is considerably higher than for ordinary timber structures, especially if the laminates need warming before they are glued together. Even timber components which have metal bolts, nail plates etc., have a higher consumption of energy during production than pure timber construction. Structural elements that are bolted together can,

Figure 13.31: A lattice I-beam in a bakery. All joints are fixed by bolting; no glue is used. Source: Gaia Lista, 1990

Figure 13.32: Production of timber lattice beam on site. Source: Gaia Lista, 1990

however, be easily dismantled and have a high recyclable value, which can compensate for its energy consumption. Larger nailed and glued products offer a more difficult problem when recycling. In structures where dismantling and reassembly are anticipated, very high quality timber should be used. Glued products need to be assessed for their environmental qualities (see 'Adhesives and fillers').

Impregnated timber is as environmentally unsound during production and use as it is in its waste phase. It contains poisons derived from oil products or metal compounds such as arsenic, chrome or copper (see 'Impregnating agents' p. 429).

The use of timber in building

Timber is a many faceted structural material and can be used in foundations, wall and roof structures.

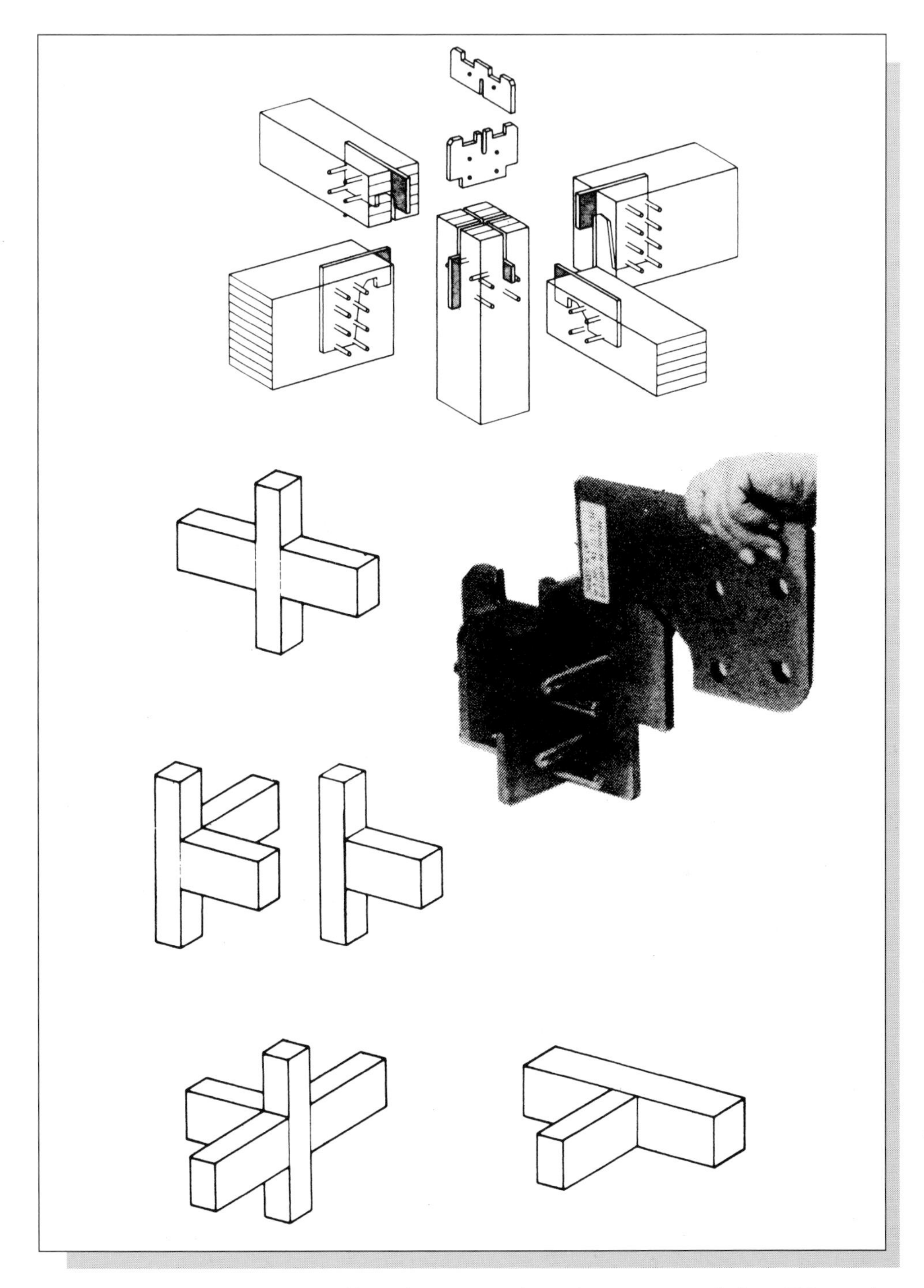

Figure 13.33: Modern demountable timber joints with metal components and plugs. This type is called Janebo. There are also stencils for the placing of holes and slits in the timber.

Foundations

The most important construction methods for foundations are raft and pile foundations. Their main areas of use are as bases for foundation walls and to stabilize weaker ground conditions.

Timbers have varying properties in relation to damp. Some timbers, such as maple and ash, decompose very quickly in both earth and water; spruce is similar. Many types of timber can survive longer in damp or low-oxygen environments than in normal country conditions. Pine, alder, elm and oak can last over 500 years in this sort of environment; larch can survive for 1500 years. As soon as the relative moisture content in timber drops below 30–35 per cent, rot sets in, and durability falls drastically. Certain types of timber are better than others even in these conditions. Oak can survive between 15 and 20 years, while larch and resin-filled pine can probably last seven to 10 years.

A key condition for a permanent timber foundation is an even, rich dampness. The timber should be completely concealed in earth and lie below the ground water level. Exposed logs can be impregnated, even though this is not particularly good environmentally, as it

Figure 13.34: A structure designed for re-use. The structure is made of prefabricated standard monomaterial components, timber and concrete, which can easily be dismantled and re-used. Source: Gaia Lista, 1995

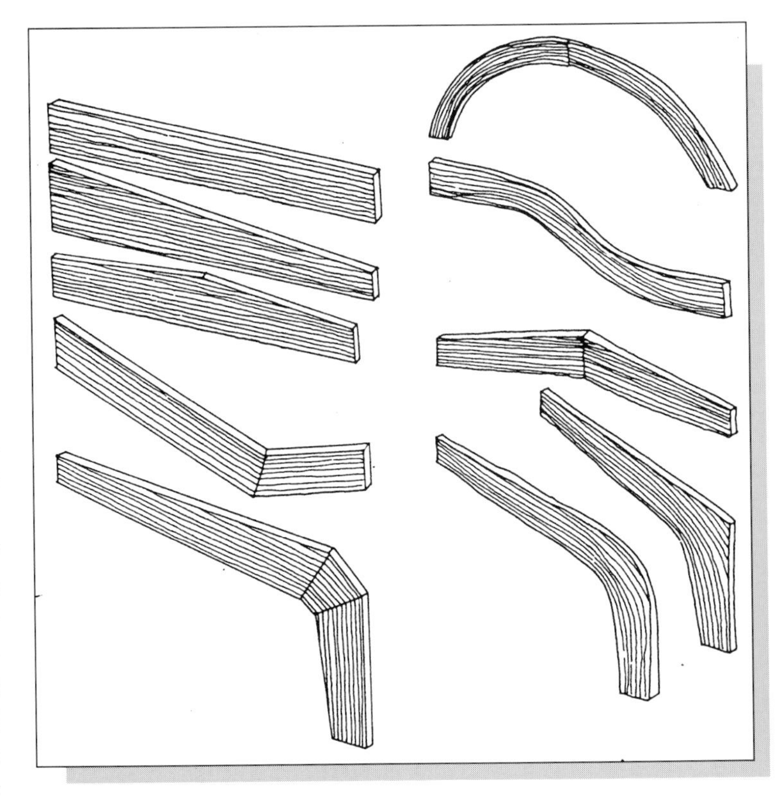

Figure 13.35: Structural possibilities for laminated timber.

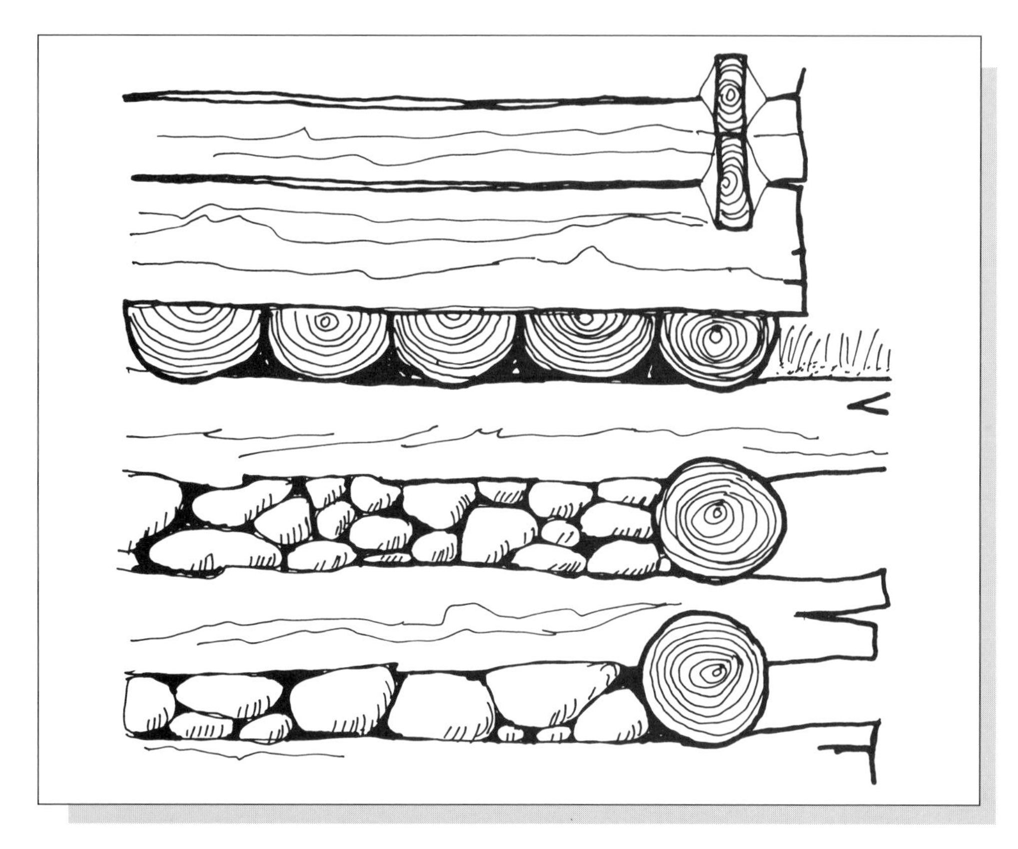

Figure 13.36: Bulwark method of foundation work. Source: Drange, 1980

causes pollution of the surrounding earth and water. Surrounding the timber with clay also helps.

Timber-based methods of foundation work

Bulwark

This technique has been used since the Middle Ages, especially when building along the edge of beaches and by farms. It is basically a structure of logs laid to form a square and, cut into each other at the corners, 2–3 m on each side. This form is then filled with stones to stabilize it. Bulwark has an elasticity in its construction which allows it to move, and it can therefore cope with waves better than stone or concrete. If the right solid timber is used, a bulwark can keep its functional properties for hundreds of years.

Raft and pile foundations

Many large coastal towns are built on raft or pile foundations. If the foundations are continually damp, then the durability is good. Excavations have discovered pile foundations of alder and aspen from the Middle Ages which are still in perfect condition, with even the bark of the tree preserved (Lidèn, 1974). Through the increase of tunnelling and drainage

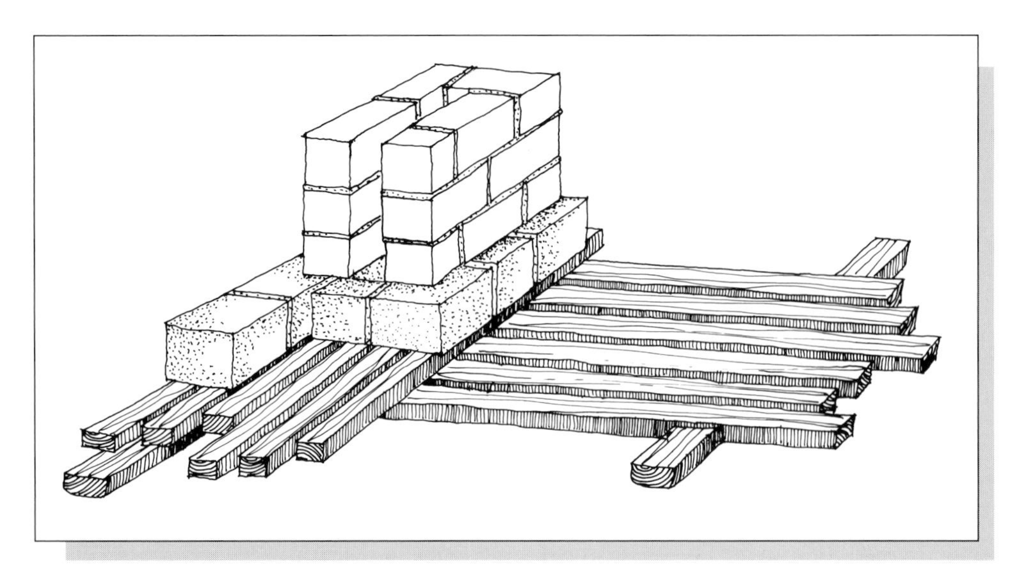

Figure 13.37: Raft foundation. Source: Bugge 1918

systems the level of the groundwater has been lowered, and because of this, fungus attack on the foundations will occur, causing a settling of the buildings.

The simplest form of raft foundation is a layer of logs laid directly onto the ground tied to logs laid across them. Masonry columns or perimeter walls are built on this foundation, and around the edges layers of clay are packed in. Raft foundations were probably in common use around the seventeenth century and quite normal up to about 1910, when they were slowly replaced by wide, reinforced concrete slabs.

In pile foundations the raft is replaced by vertical logs, which are rammed down into the ground. It is usual to lay three or four horizontal logs onto the piles to distribute the weight evenly, before building the walls. The weight of the building and the bearing capacity of the earth decide how close the piles need to be to each other. Foundations for smaller buildings usually have thinner piles, from the thickness of an arm down to the thickness of a finger. To distribute the load, a filled bed of round stones may be used.

In sandy earth lacking soil the piles above ground level can be taken to a bottom plate. This can provide a simplified solution in certain cases, but even with good impregnation and high-quality timber it is doubtful that the foundation will hold longer than 75 years.

Structural walls

Timber buildings are usually associated with load-bearing timber walls. It is necessary to differentiate between light and heavy structures. The most important aspect of lightweight building is the framework, which is economic in the use of materials and takes advantage of the tensile and compressive strengths of timber.

The log building technique is the most widespread technique of the heavy structures. This method uses a lot of timber and is statically based on the compressive strength of timber.

The Norwegian Building Research Institute has recently completed a research project on the environmental efficiency of different types of building, from the construction phase through a 50-year life span. As far as resource consequences and pollution effects are concerned, the log building technique came out best, despite the intensive use of timber. As the time span was only 50 years the possibilities of recycling the building materials was not taken into account, although this is an integral part of this technique, as is the high durability of such a structure. Log houses of more than 1000 years old exist in both Japan and Russia.

Types of structural walls

Log construction

In this method, logs are stacked directly over each other and notched together in the corners. These buildings are usually rectangular, but can have up to 10 sides. (A 10-sided log built barn exists at Fiskberg in Burträsk, Sweden.)

A solid timber wall has good acoustic properties and fire resistance. The thermal insulation is also good. For 700 to 800 years it has been considered the warmest alternative.

Pine has been the timber most used in log construction. It has been left open and exposed to all weathers, so it has been well tested for hardiness. In log construction with external panelling, spruce can also used. Larch makes a solid and durable log building and is very much in use in Russia. For outhouses birch, aspen and lime can be used. Lime is a large tree, common in the Carpathians (in the eastern part of Romania, where it is used for the log construction of dwellings. In particularly damp areas, exceptionally durable timber such as oak must be used for the bottom plate.

There are many ways of forming the logs and their joints, depending upon which timber is used (see Figure 13.38). Pine should have its surface worked by profiling, while spruce needs only the removal of the bark to keep its strength. Accessible technology and rationality have played a crucial role in the development of techniques. Type (a) in Fig. 13.38 belongs to the nineteenth century style of building and was well suited to the new machinery of the period – sawmills. The disadvantage was that it was difficult to make them airtight, and they were not as strong a joint as hand-worked logs. Types (b) and (c) from Finland and Canada come reports that the log-built house is on its way back, and in Canada and the USA between 50 000 and 60 000 log dwellings are built every year.

Vertical load-bearing panelling

This was developed in order to place a solid timber wall in a house without having to wait for settling, unlike log construction. The timber shrinkage along its length is minimal. Outer walls can then be panelled directly and windows installed (see Figure 13.39).

Stovewood house and firewood shed
Stovewood houses came from the last century. They represent a recycling building tradition and were built of bits of plank and spill from the sawmills, using a mortar of pure clay mixed with water and sawdust or chaff. The wall was more stable laterally than log construction, but needed a couple of years to settle before wallpapering and panelling.

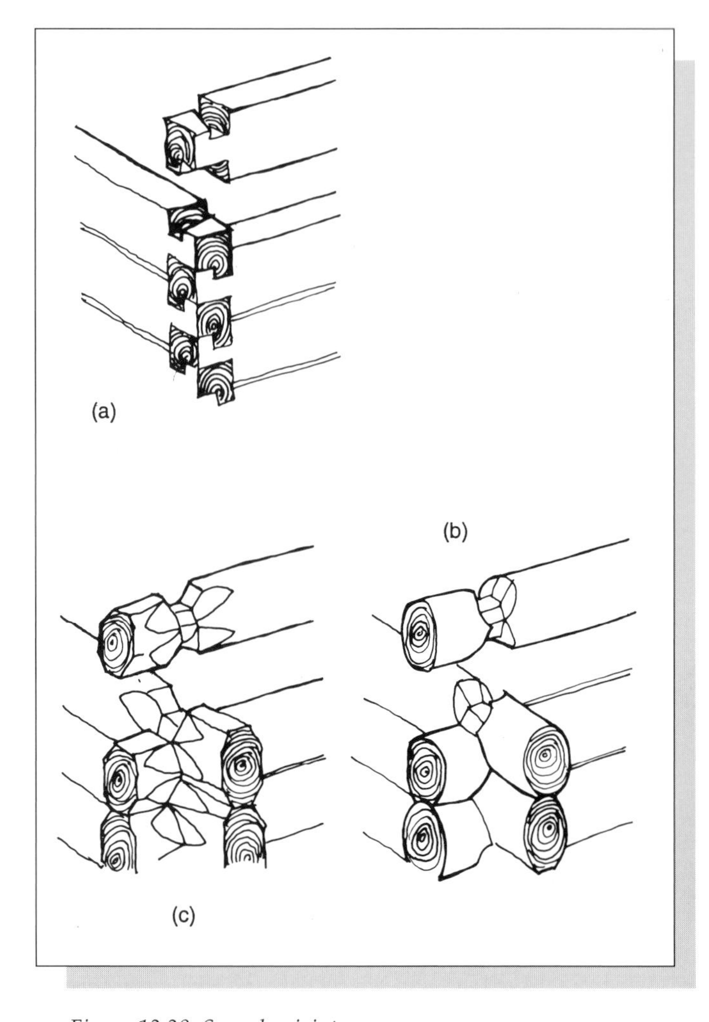

Figure 13.38: Some log joints.

Stave construction
This is a braced skeletal construction filled with vertical boards or plank tongued into a bottom and top plate. In modern post and lintel construction the space between is usually filled with boards and insulation which also braces the structure. The timber components are heavy and well-suited to recycling, providing that appropriate methods of fixing are used.

Structural framework
This consists of studs mounted between a top plate and a bottom plate and bracing. There have been many variations on this theme through time. The tendency has been toward small dimensions of timber components and more rational design. This has reduced the quality of the structure to a certain extent, particularly in relation to its strength. The distance between the studs can vary somewhat, from 300 mm to 1.2 m. Studwork was previously braced with diagonal lengths of timber, but nowadays it is more usually braced with sheets of fibre-, plaster- or chipboard.

The spaces in the wall are filled with different types of insulation. In earlier times they were filled with clay (in wattle walls), firewood, or bricks (known as half-timbered brick construction).

Structural framework uses timber very economically, but is seldom easy to recycle. The many and very strong fixings used make the material good only for energy recycling, i.e. burning. The timber used in frame construction has to have high-quality strength. It should not be too elastic or deform too much when exposed to moisture. The timbers best suited for this are fir, spruce, larch and oak. For smaller structures, birch, aspen, ash and lime can be used.

Figure 13.39: Vertical load-bearing panelling.

Figure 13.40: Traditional timber frame construction. Wooden plugs are used for fixing. Source: Gaia Lista, 1992

Timber frame construction is the dominant structural system in the timber building industry today.

Wattling

Wattle – poles interwined with twigs or branches – does not require large timber. This technique has been used up to the present day in Eastern Europe. It is usually combined with other structural techniques and is used mostly in the building of sheds, wind- or sun shelters, garages and outside kitchens, etc., in combination with free-standing houses, small industries and summer cottages. Many less attractive or less widely used trees can be employed, e.g. juniper, birch, ash, elm, lime, hazel, rowan and willow. The thicker pieces of wood should have their bark removed and the work should be carried out in spring when the wood is most pliable. (See also 'Wattle-walling'.)

Floor structures

Floor structures usually consist of solid timber joists, composite beams, laminated timber beams or a combination of these. As long as building standards are followed, most types of timber can be used in floor structures. High strength and rigidity during changes of moisture content must be guaranteed. Although softwood is mostly used, certain hardwoods can be used in small structures; they can save use of material, as they have a greater tensile strength than softwood.

Figure 13.41: Traditional way of filling spaces with brick in a timber framed building in Denmark.

A new form of heavy timber floor construction has been recently developed in Germany consisting of low quality planks nailed together to form 8-15 cm thick slabs. They can have a span of up to 12 meters and can also be used in walls and roofs. The surfaces can be sanded down and used as they are without any further finish or they can be covered with a screed on insulation board. Because of its solidity the structure has proven good properties. This technique has been used in Sweden for five storey housing units. The timber used can have the lowest quality, e.g. waste from saw mills or sitka-(norw.) spruce (*Picea sitchensis*). The construction method

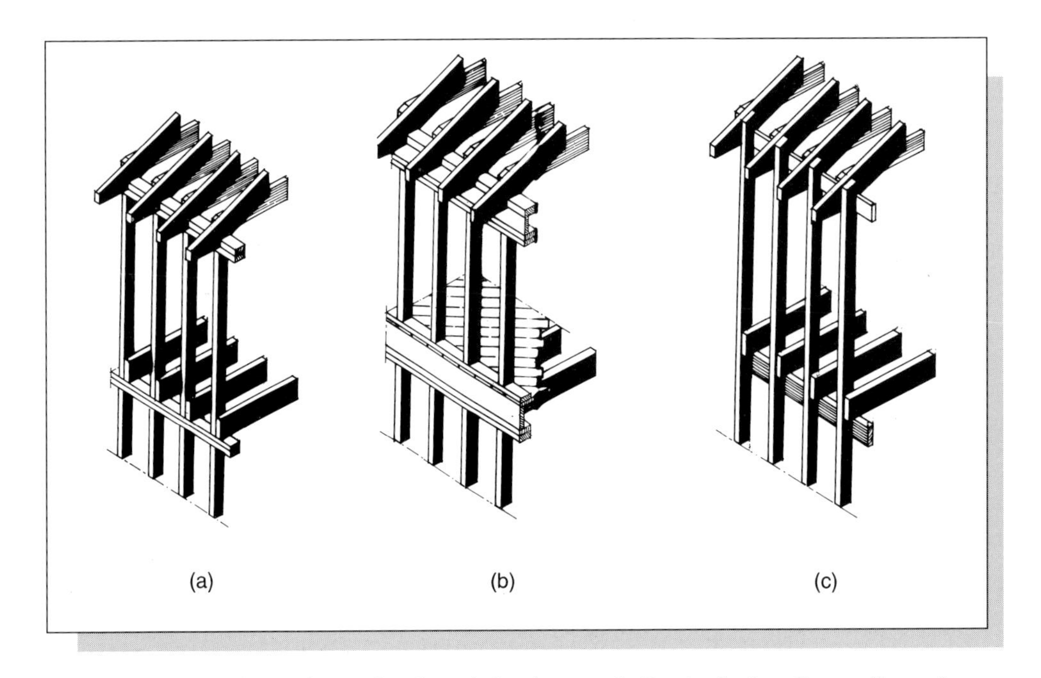

Figure 13.42: Different forms of modern timber framework. Bracing by boarding or diagonals.

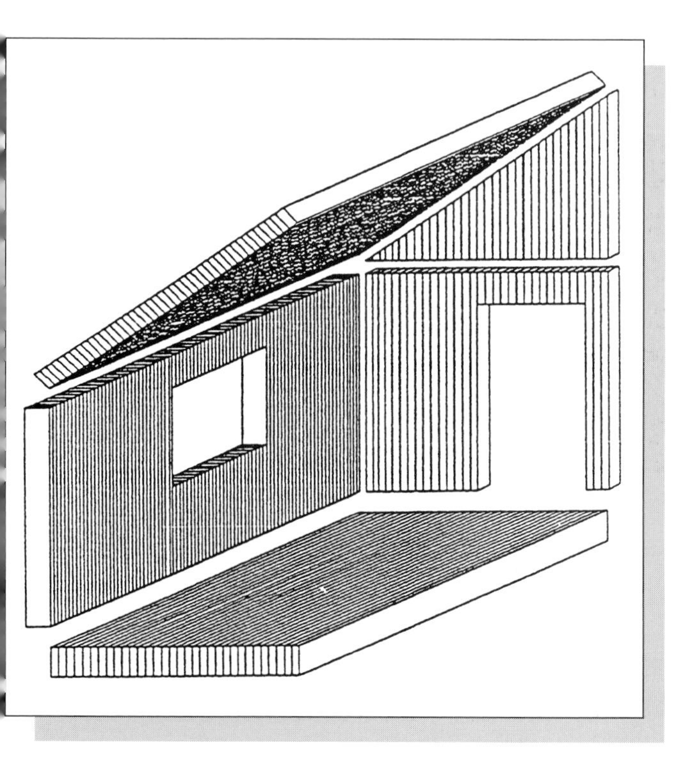

Figure 13.43: Principles for construction with massive timber elements.

is therefore very interesting in a resource perspective even if the volume of material used is high.

Roof structures

The use of materials for roof structures is almost the same as for floors. Many structural alternatives are available through combining components in different ways. Roofs fall into three main categories: single raftered, purlin and forms made of trusses, with a smaller group known as shell structures.

Shell structures

These structures are seldom used despite the fact that they use material very economically. The timber used in shell structures must have good strength properties. It is also an advantage if the timber is light.

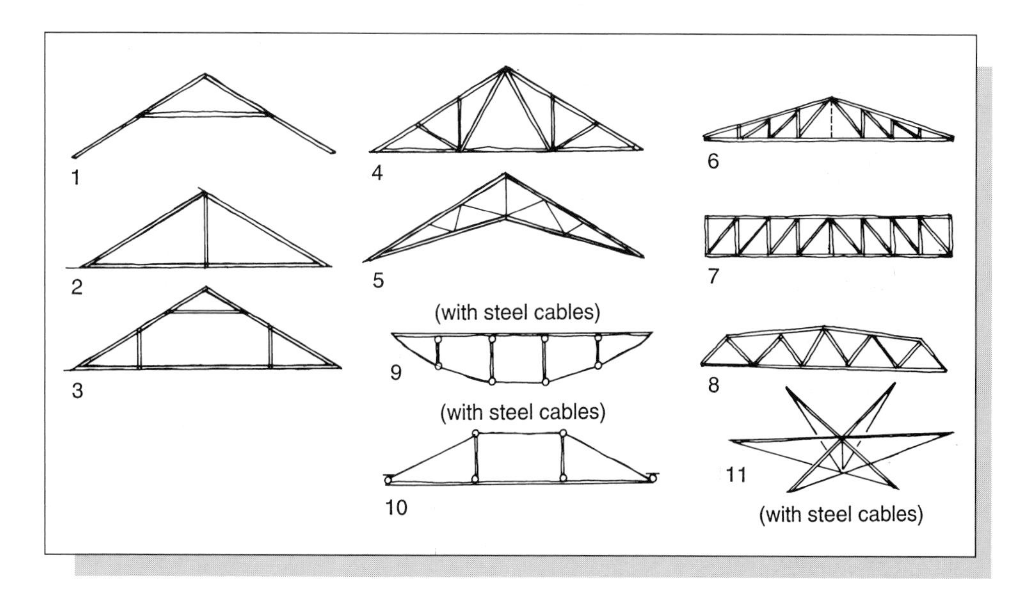

Figure 13.44: Roof trusses constructed in solid timber, some with steel cables.

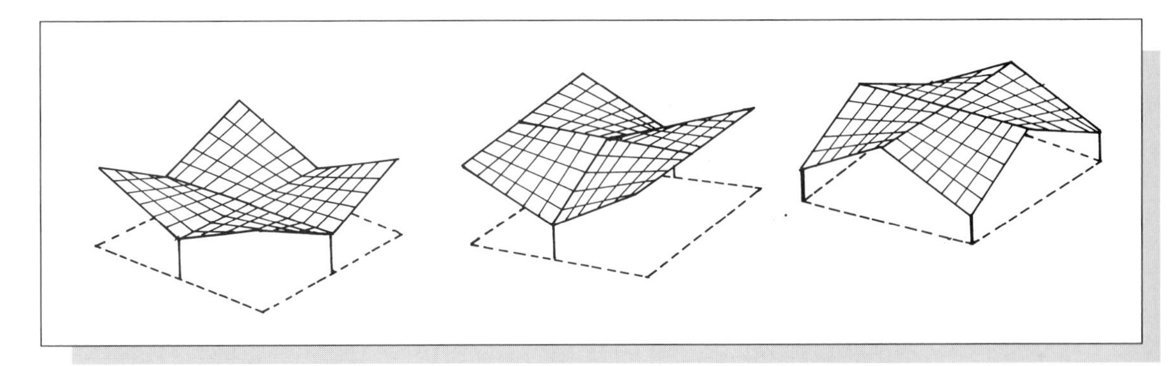

Figure 13.45: Possible combinations of double curved shells. Source: Schjödt 1959

Shell structures must cope with all weather conditions and penetrating damp, which really tests materials. Fir, spruce, larch, oak, ash, elm and hazel are best suited for this.

Shell roofs made of timber have existed for thousands of years, particularly in tent structures. They are very light and economical in material use, which has been a necessity for migrating nomads. There are two main types of shell roof: double curved shells and geodesic domes.

Double curved shells (hyperbolic paraboloid)

A compact version of the double curved shell started to appear in Europe at the beginning of the 1950s in buildings such as schools and industrial premises. Its span varies from 5–100 m. The shell is built *in situ* over a light scaffolding, and consists of two to three layers of crossed tongued and grooved boarding. The thickness of the boarding is approximately 15 mm. The shells are characterized by the fact that two straight lines can go through any point on the surface of the roof. The boarding is not straight, but the curving is so small that it can bend without difficulty. The shells are put together as shown in Figure 13.45, depending upon the position of the columns.

A lighter version, well suited for small permanent buildings, consists of a rectangular grid of battens. The battens are screwed together at all the intersections with small bolts. The shell can be put together in this way for transport. When erecting the structure permanently, the grid is fixed to a solid timber frame and the bolts are tightened. This structure can be used for small pavilions or bus shelters, for example.

Geodesic domes

The first geodesic dome was erected using steel in Jena, Germany, in 1922. Timber is a possible alternative. The method is a simple prefabricated system based on triangles, always constructed in the shape of a sphere. In this way a stable structure is produced which tolerates heavy loading. The spaces between the grid can be filled with thermal insulation. These domes are used as houses in the northern parts of Canada. The most common use of them in Europe is for radar stations, although there are reports that their waterproofing is questionable.

Figure 13.46: A traditional Icelandic dwelling made of peat.

Peat walls

Structural walls of peat were once more widespread in Ireland, Scotland and Wales. There are still a few peat houses in Iceland, and this building technique spread to Greenland during the eleventh century. Building in peat was also undertaken by immigrants in North America, especially amongst the Mormons, who worked a great deal with this material after 1850.

Peat is no easy material to build with, and most of the alternative building materials such as timber, stone, concrete and earth are more durable and stable, but the question of economy and access to resources is also important.

A well-built peat house can have a life span of approximately 50 years, when the decomposition of peat will be beyond its critical point. Peat has a higher strength in a colder climate and with special climatic conditions such as those on Iceland, and good maintenance, some examples have had a much longer life span. One advantage of peat is its high thermal insulation. Icelanders worked with two qualities of peat which they call *strengur* and *knaus*.

Strengur is the top 5 cm of the grass peat and is considered the best part. It is cut into large pieces that are laid in courses on the foundation walls. This method is particularly suitable for dwellings. *Knaus* is of a lower quality. These are smaller pieces of peat, 12.5 cm thick, which are laid according to the 'Klömbruknaus' method (see Figure 13.46).

A serious problem with peat walls is the danger of them 'slipping out'. This risk can be reduced by stiffening the corners with stone or short timber dowels which can be knocked through the layers as the building progresses.

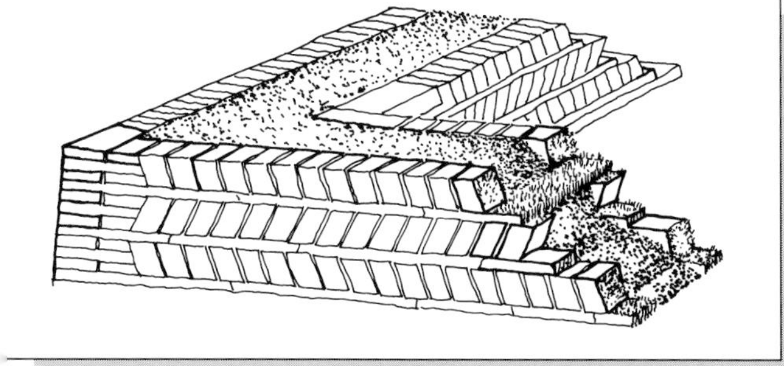

Figure 13.47: A peat wall contains layers of peat with earth between them. In the corners, strengur *peat is used; the rest of the wall is laid with* knaus. *Souce: Bruun 1907*

The energy and material used by different structural systems

Every structural system has its own specific use of material, depending upon its strength. Solid structures of brick and concrete are highly intensive in their use of material, whereas timber and steel are usually more economical, but each material can have different structural methods using different amounts of material.

Figure 13.47 shows structural alternatives to columns and beams. This example shows steel components, but the same principles apply for timber. The lattice beam is the most effective use of material, and the most economical is the lattice beam with radial lattice work.

One aspect of material–economical structures is that they are often more labour intensive than simple structures. The lattice beam with many joints costs more to produce than the equivalent laminated timber beam, even if the use of material is twenty times less. In some cases, the extra cost of transport and more

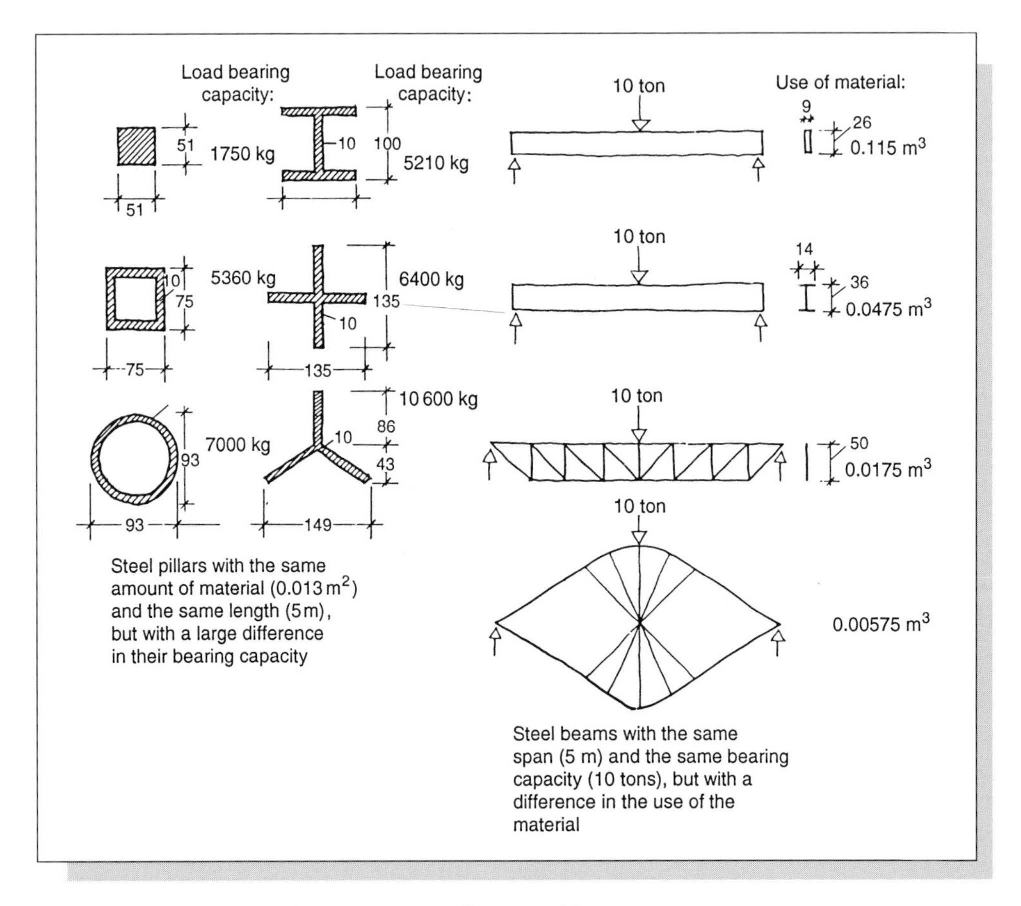

Figure 13.48: Structural alternatives to columns and beams.
Source: Reitzel and Mathiasen, 1975

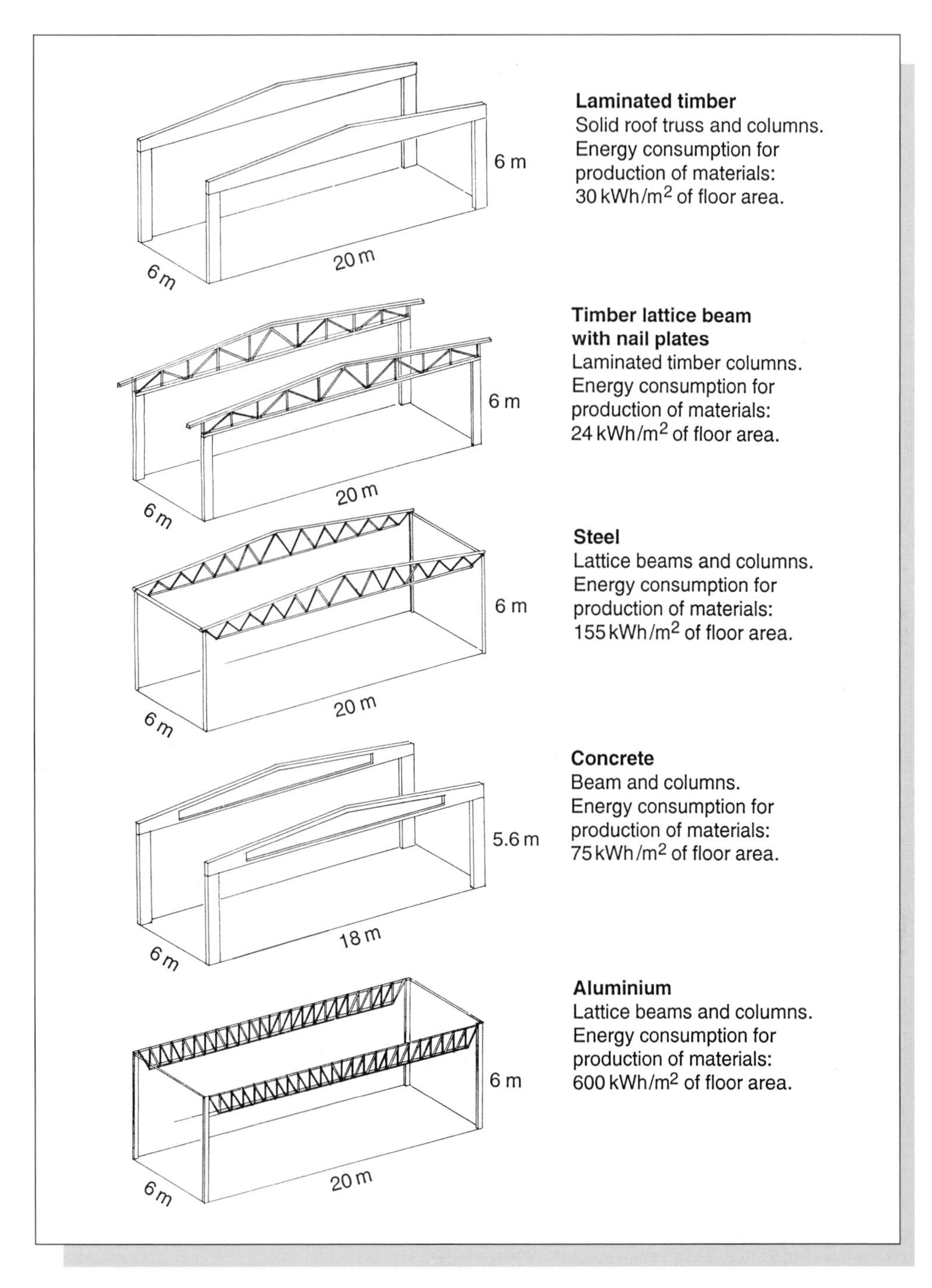

Figure 13.49: Comparative calculation of the use of primary energy when using different structural materials. Source: 'Report no. 302520', Norwegian Institute of Timber Technology, 1990.

Table 13.5: Environmental profiles of structural materials

Material	*Compressive strength (kp/cm²)*	*Tensile strength (kp/cm²)*	*Quantity of material used (kg/m²)*
Horizontal structures:			
Aluminium beams, 50% recycling	4300	4300	15
Steel beams, 100% recycling	5400	5400	40
In situ concrete[(2)]	150–700	7.5–35	400
Precast concrete[(2)] (normal concrete)	150–700	7.5–35	380
Precast aerated concrete[(1),(2)], good insulation	30	4–5	130
Precast light aggregate concrete[(1),(2)], good insulation	30	4–5	190
Softwood beams	450–550	900–1040	40
Pine beams, pressure impregnated	470	1040	40
Spruce, laminated timber	450	900	35
Hardwood beams	400–620	800–1650	35
Vertical structures:			
Aluminium studwork, 50% recycling	4300	4300	5
Steel studwork, 100% recycling	5400	5400	30
In situ concrete[(2)]	150–700	7.5–35	350
Concrete blockwork[(2)]	150–700	7.5–35	260
Aerated concrete blockwork, good insulation[(1),(2)]	30	4–5	150
Light aggregate concrete blockwork, good insulation[(1),(2)]	30	4–5	220
Lime sandstone[(2)]	150–350	7.5–17.5	220
Granite, sandstone, gneiss	200–2000	100–320	500
Gabbro, syenite, marble, limestone, soapstone	200–5000	160–315	500
Well-fired solid brick	325	33	220
Well-fired hollow brick	75–150	7.5–15	170
Low-fired solid brick	150	15	200
Earth, without fibres added	40	Up to 6	800
Softwood studwork[(3)]	450–550	900–1040	1
Pine, pressure impregnated	470	1040	1
Spruce, laminated timber columns	450	900	1
Hardwood studwork[(3)]	400–620	800–1650	1

Notes:

(1) Structural materials with high thermal insulation; need little or no extra insulation

(2) Inclusive of reinforcement

(3) A comparison has recently been done by the Norwegian Building Research Institute between timber framed and log buildings. This has shown that log buildings are slightly better than timber framed buildings in use of resources and pollution effects over a period of 50 years. The log building also has a better potential for re-use.

(4) Advancing to '2' if in brickwork specially prepared for re-use.

Effects on resources			*Effects of pollution*				*Ecological potential*		*Environmental profile*
Materials	*Energy*	*Water*	*Extraction and production*	*Building site*	*In the building*	*As waste*	*Re-use and recycling*	*Local production*	
3	3	3	3	1	2	2	✓		3
2	1	2	2	1	2	2	✓		2
2	2	2	3	3	2	1		✓	2
2	2	2	3	1	2	1	✓	✓	2
2	3	2	3	1	2	1	✓		2
2	3	2	3	1	2	1	✓		2
1	1	1	1	1	1	1	✓	✓	1
2	1		3	2	3	3	✓	✓	3
2	1	1	2	1	1	2	✓	✓	2
1	1	1	1	1	1	1	✓	✓	1
3	2	3	3	1	2	2	✓		3
2	1	2	2	1	2	2	✓		2
2	2	2	3	3	2	1		✓	2
2	2	2	3	1	2	1	✓	✓	2
2	3	2	3	1	2	1	✓		2
2	3	2	3	1	2	1	✓		2
2	1	1	2	1	2	1	✓		1
1	1	1	2	2	1	1	✓	✓	1
1	1	1	1	1	1	1	✓	✓	1
1	3	3	3	1	1	1	✓	✓	3[4]
1	3	3	3	1	1	1	✓	✓	3[4]
1	2	3	3	1	1	1	✓	✓	3[4]
1	1	1	1	2	1	1	✓	✓	1
1	1	1	1	1	1	1	✓	✓	1
2	1		3	2	3	3	✓	✓	3
2	1	1	2	1	1	2	✓	✓	2
1	1	1	1	1	1	1	✓	✓	1

intensive use of raw materials, especially when mounted on site, can change the economic equation quite drastically.

The primary use of energy during the production of structural materials is dependent upon the quantities of material produced and the material used. A comparison of the use of primary energy of different structural systems in different materials is given in Figure 13.48. In conclusion, a timber lattice beam is most economically efficient compared with a laminated timber beam or a steel or concrete beam (Norsk Treteknisk Institutt, 1990).

Environmental profiles

Table 13.5 and further tables at the end of the following two chapters give environmental profiles of materials. They are organized in such a way that each functional group has a best and a worst alternative, The evaluations rate the best as 1, the next best as 2 and the worst as 3. Different materials may be given the same evaluation and in some cases only first and second placings are given.

The evaluations relate to the present-day situation. The ecological potential column gives an idea of the product's or material's future possibilities within the aspects of re-use/recycling and local production and thereby adjusts the final environmental profile. These evaluations are based on information given in Table 1.3 (Effects on resources), and Table 2.5 (Effects of pollution) in Section 1, and on the more qualitative evaluations in Sections 2 and 3.

The amount of materials is given in kg/m^2 for a normal well-insulated building. The loss factor for material that disappears during transport, storage and building is not included. This is given in the third column of Table 1.3. The loss factor has to be used when calculating the quantifiable environmental damage for individual products. This is done by using Table 1.3.

References

BJERRUM L *et al*, *Jordhus*, NBI, Oslo 1952
BRUUN D, *Gammel bygningskik paa de islandske Gaarde*, FFA, Oslo 1907
BUGGE, *Husbygningslare*, Kristiania 1918
DRANGE T *et al*, *Gamle trehus, Iniversitetsforlaget*, Oslo 1980
GRUNAU E B, *Lebenswartung von Baustoffen*, Vieweg, Braunschweig/Wiesbaden 1980
KOLB J, *Systembau mit holz*, Lignum, Zurich 1992
LAURITZEN E *et al*, *De lander på genbrug*, Copenhagen 1991
LIDÈN H-E, *Middelalderen bygger i sten*, Universitetsforlaget, Oslo 1974
LINDBERG C-O *et al*, *Jordhusbygge*, Stockholm 1950
NORSK TRETEKNISK INSTITUTT, *Energiressurs-regnskap for trevirke som bygningsmateriale*, NTI rapp. 302520, Oslo 1990
ORTEGA A, *Sulphur as building material*, Minamar 31, London 1989
REITZEL E, *Energi, boliger, byggeri*, Fremad, Köbenhavn 1975
SCHJÖDT R, *Dobbeltkrumme skalltak av tre*, NBI, Oslo 1959
SHADMON A, *Mineral Structural Materials*, AGID Guide to Mineral Resources Development 1983

14 *Climatic materials*

Climate regulating materials control the indoor climate, and are mainly orientated towards comfort. They can be subdivided into four groups:

- air-regulating
- moisture-regulating
- temperature-regulating
- noise-regulating.

Air-regulating materials are usually composed of a thin barrier over the whole of the outside surface of the building and resist the incoming air flows. They are also used in internal walls between cold and warm rooms, where there is a chance of a draught being caused in the warm room.

Moisture-regulating materials are primarily used for waterproofing foundations, and as an inner vapour barrier to stop moisture from inside the building penetrating the wall and damaging it. They include materials that can regulate and stabilize air moisture in permanent absorption and emission cycles.

Temperature-regulating materials mainly include thermal insulation materials built into the outside surface, but also materials that stabilize temperature relationships through their warmth-regulating properties. A subgroup for internal use are surface materials that can reflect, absorb or carry heat radiation through their structure and colour.

Noise regulation is necessary to reduce and transfer sound of different qualities in and between rooms, and to guarantee a good acoustic climate. External sources of noise, such as road and air transport, necessitate good insulation in both walls and roof. Noise-regulating properties are dependent upon the material used, its design, placement and size. Treatment of sound in building technology is otherwise seldom discussed and will only be touched upon briefly here.

Certain climate-regulating materials have qualities that put them in two or three groups. A thermal insulation material can also be airtight, regulate moisture and even stop noise. Different functions can be combined, e.g. timber can be a moisture-regulator while acting as a structural and surface material.

Thermal insulation materials

The thermal insulation of a building can be done in two ways: as static or dynamic insulation. There are even materials that reflect thermal radiation, thereby affecting the heat loss of a building and which should be considered as representative of a particular method of insulation of their own.

Static and dynamic insulation

In static insulation the insulation value of static air is used. The principle requires the use of a porous material with the greatest possible number of air pockets. These have to be so small that no air can move within them.

In dynamic insulation air is drawn through a similar porous insulation material. When the fresh air is led from outside through the surface of the wall, rather than through small ventilation ducts, it picks up heat loss flowing out of the building. Besides achieving a pre-warmed fresh-air flow into the building, the heat loss through the surfaces is reduced to a minimum. The optimal materials for such a wall should have an open structure with pores across the whole width, plus good heat exchange properties. A high thermal capacity is also an advantage, so that sudden changes in the outside temperature are evened out. Dynamic insulation is still being introduced into construction and has been used in only a few buildings.

The main part of this chapter considers the properties of different materials in relation to static insulation.

The technical demands of an insulating material (excluding the reflective layer) are usually as follows:

1. High thermal insulation properties
2. Stability and long life span
3. Fire resistance

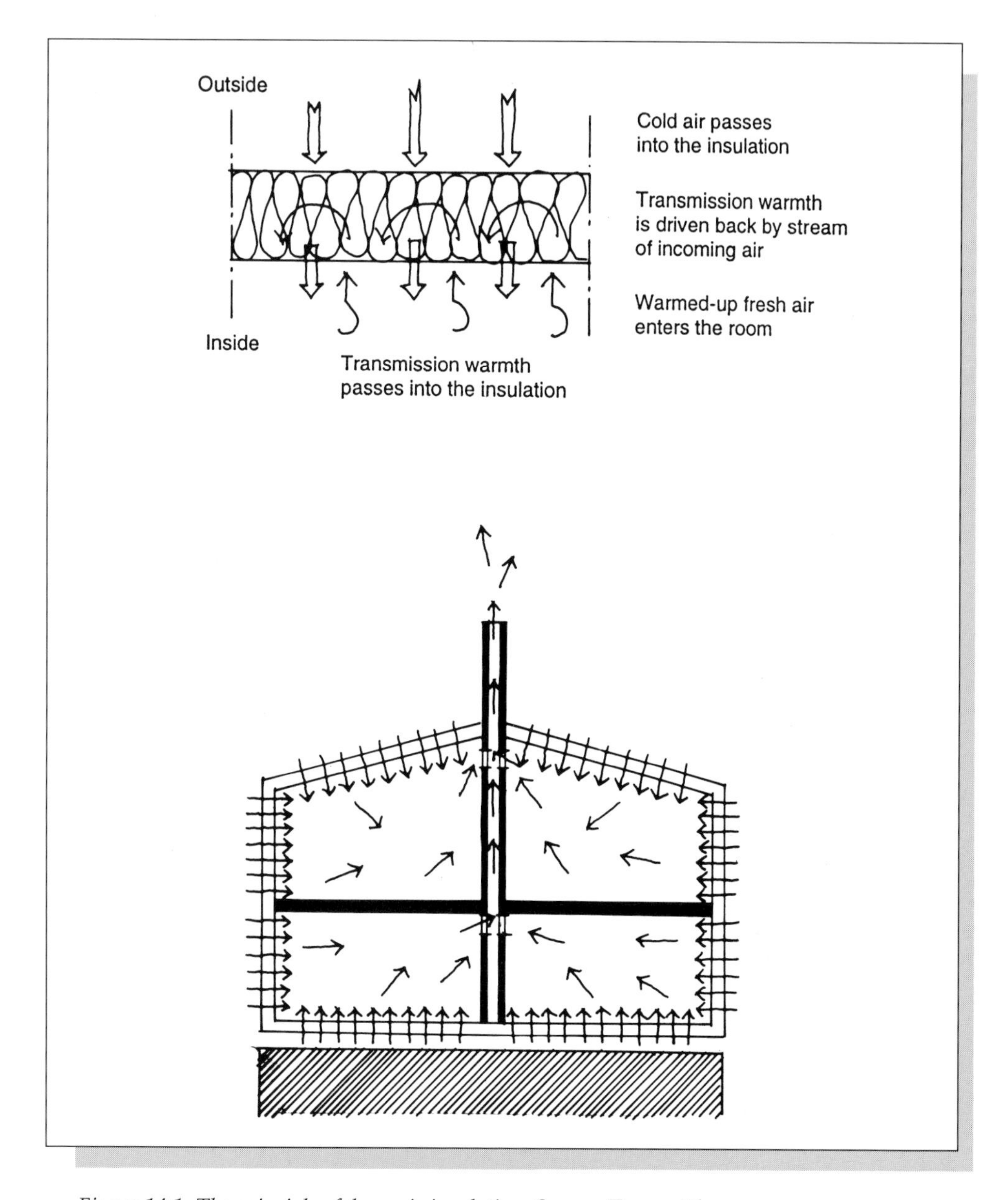

Figure 14.1: The principle of dynamic insulation. Source: Torgny Thoren.

4. Lack of odour
5. Low chemical activity
6. Ability to cope with moisture
7. Good thermal exchange properties (for dynamic insulation)

Figure 14.2: The McLaren Leisure Centre, Callander: a healthy building materials specification, with dynamic insulation in the ceiling. Source: Howard Liddell

The thermal insulation property for static insulation is usually called lambda (λ) and can be measured with special equipment:

$$\lambda = W/(m^{\circ}k)$$

Mineral wool has a lambda value of 0.04, while a woodwool slab has a value of about 0.08. This means that a double thickness piece of woodwool gives the same insulation value as a single thickness of mineral wool.

Calculating the value of insulation

The example quoted above comparing the thermal insulation of two materials is the traditional method of calculation, making the assumption that there is a linear relationship between the lambda value and insulation/heat loss. There are limitations to the lambda values. They give no indication of the material's structure, moisture properties or reaction to draughts (which every wall has to a certain extent). It takes no notice of the material's thermal capacity. In buildings that are permanently heated, as in hospitals for example, there is a great energy-saving potential and improved comfort if materials with high thermal capacity are used. The same is true for buildings where there can be wide and rapid changes in the inside temperature, for example, when opening the windows. The thermal insulation value of a material is reduced when damp. In frozen materials the ice conducts warmth three to four times better than water. This is important if using hygroscopic materials. Even if such materials seldom freeze, a lower insulation value is assumed during spring because of the higher moisture content.

Age can also affect insulation value. Certain products have shown a tendency to compress through the absorption of moisture and/or under their own weight, while others have shrunk (mainly foam plastics). The thickness of the layers of insulation needs to be appropriate for the local climate. Too much insulation can cause low temperatures and thereby hinder drying in the outer layers, which can lead to fungus developing in organisms in the insulation or inside adjoining materials.

Insulation materials are sold either as loose fill, solid boards or thick matting. The latter two can result in a damaged layer of insulation, because temperature or moisture content changes can cause changes within the structure. This is especially the case with solid boards, which need to be mounted as an unbroken surface on the structure and not within it. Loose fill insulation is good for filling all the spaces around the structure, but it can settle after a time. The critical factors are the weight and moisture content of the insulation. The disadvantages of hygroscopic materials become apparent here because they take up more moisture and become heavier. Settling can be compensated for by using elastic materials which have a certain 'suspension' combined with adequate compression. Structures with hygroscopic loose fill as insulation need topping up during the building's life span.

Thermal insulation materials usually occupy large volumes, but they are light and seldom take up more than about 2 per cent of the building's total weight. Many insulation materials do, however, have a high primary energy use and use of material resources, and produce serious environmental pollution during manufacture, and use, and even as waste. The waste material of forms of insulation must be specially treated. Only in exceptional circumstances is it possible to recycle or re-use insulation materials.

Warmth-reflecting materials

By mounting a material that has a low reflection rate for short-wave warm radiation on the building's south façade, solar energy can be used very efficiently, while a sheet of highly reflective material on the inside of the wall will reduce heat loss. This is especially utilized in modern window technology.

Table 14.1: The approximate reflection factor of solar radiation on different materials

Material	*Reflection*
Shiny aluminium	0.70
Aluminium bronze	0.45
Brick	0.14
Timber	0.14
White paint	0.70
Light paint	0.70
Black paint	0.01

Moisture-regulating materials

Moisture should not be able to force its way into a building's structure without being able to come out again. Apart from the danger of mould and rot within organic materials, the damp can freeze and cause the breakdown of mineral materials when frost occurs. Damp also reduces the insulation value of the material drastically.

Moisture can enter the structure in six ways:

- As moisture from the building materials
- As rain
- From the ground
- As air moisture from inside or outside
- As moisture from installations which leak, e.g. drainage, water supply or heating system
- As spilled water

The last two points do not need to be discussed, as the correct use of a material should prevent such circumstances, or at least minimize them.

Moisture within building materials

During construction a new house carries about 10 000 litres of water within its building materials. Drying time is strongly dependent upon the structure of the material. There is an unnamed relative material factor, s – the drying capacity of a material increases when the factor value falls. Lime mortar has an s-factor of 0.25, brick 0.28, timber 0.9, lightweight concrete 1.4 and cement mortar 2.5.

To avoid problems with moisture, building materials should be dried according to standard practice, and concrete, earth and timber structures should be allowed to dry before they are used with moisture-tight materials. Good ventilation design is important for an enclosed structure.

Rain

External cladding and roof coverings, discussed in greater detail in the following chapter, cope with rain. There is also a need for special components, partly to protect exposed parts of the building such as pipes going through the envelope, partly to carry the water away from the building. Such components are often made of metal sheeting, and are either built on site or prefabricated in a factory and transported to the building.

Ground moisture

The site for the building should be dry and well-drained. It is advisable to keep the natural level of the water table and keep all rainwater within the site without using the public drainage system. There is little need to overload the public system unnecessarily, and a stable water table is necessary to keep the local flora and fauna in a state of balance. Topography, soil and other site conditions can easily come in conflict with this strategy, but it is important to find a foundation system that suits the site.

Perimeter walls and slab foundations of concrete will always be exposed to moisture. This can be reduced by a layer that breaks the capillary action of water from the ground plus a watertight membrane, but it is always difficult to stop a certain amount of moisture entering the fabric of the building. Concrete slabs directly on the ground are problematic. There have also been a whole series of damp problems with organic floor coverings such as timber, vinyl sheeting, etc., laid directly onto the concrete, even where there is a plastic membrane in between. As insurance against such problems, concrete slabs on the ground should have mineral floor coverings such as slate or ceramic tiles.

All structures normally have a damp-proof membrane between the foundation and the rest of the structure, usually consisting of bitumen felt.

Air moisture

Air moisture is almost entirely produced inside the building by people, animals and plants, or from cooking and using bathrooms; this can damage the structure. Air moisture tries to penetrate the external walls and condense there.

Air's moisture content and condensation risk

The lower the temperature, the less water vapour air can hold. At 20°C air can hold 14.8 g/m^3 of water vapour, while at 0°C it can only hold 3.8 g/m^3. If the internal air at

20°C only holds 3.8 g/m^3, it can pass through the wall to outside air at 0°C without any condensation being formed, but if the air is saturated with 14.8 g/m^3 then there will be condensation within the wall of 11 g/m^3. In a normal situation, a room contains about 5–10 g/m^3 water vapour, while a bathroom, in short periods, can reach almost 14.8 g/m^3.

Big condensation problems can occur with open-air leakage or cracks in walls and roofs. At the same time, moisture diffusing through materials normally occurs without large amounts of condensation being formed inside the wall. A wall completely free of small cracks is unrealistic, so it is necessary to take certain precautions using one of the following principles:

- Vapour barriers
- The absorption principle
- The air cavity method

Vapour barriers

The use of vapour barriers has become the most widespread method in recent years. The main principle is that water vapour is totally prevented from entering the wall by placing a vapour-proof membrane behind the internal finish. The air and its vapour is then ventilated out of the building. This method has certain weaknesses. The only usable material for this purpose is plastic sheeting or metal foil. How long plastic sheeting will last is not really known. During the building process, rips, holes and such like will inevitably be caused. At these points small amounts of vapour will creep through, and after a time condensation will occur in the wall.

A more moderate and less vulnerable solution is a vapour check that limits vapour diffusion. This is not as absolute as vapour-proofing, but reduces penetration considerably. Materials used for this are high-density fibreboards and different types of sheeting. The choice of material is determined by the type of wind-proofing used on the outside of the wall. A rule of thumb is that the resistance to vapour diffusion on the inside must be five to ten times higher than the wind-proofing layer on the outside to give the vapour a direction (NBI, 1989). It is important to note that the windbreak's resistance to diffusion is often heavily reduced if it is damp – down to 10 per cent of its original value in the case of a porous wood fibreboard. It is therefore often possible to use the same material on both sides of the wall.

The absorption principle

Some materials used for walls are very hygroscopic and resistant to rot, and can absorb vapour. As the condition of the building changes in terms of its temperature and vapour content, the stored moisture is, after a while, released back into

Figure 14.3: Solid, untreated timber has very good moisture-regulating properties.

the room. Untreated wooden panelling in a bathroom is an example. When the bath is being used, the panelling absorbs a great deal of water vapour. Afterwards, if the window is opened or the ventilation is increased, the air dries out quickly. The panelling then releases the absorbed moisture back into the air of the room. In comparison with the traditional vapour barrier, this method will retain less acute damp in the room and strong ventilation will be less necessary. It also has energy-saving potential. A similar situation is created when the occupants of a house go to bed or leave for work – the moisture content in a living room with absorptive walls will be stabilized. Even if the temperature often falls during this period, the process still continues.

Untreated timber panelling, rammed earth and lightweight concrete are examples of materials that absorb and release moisture rapidly.

Hygroscopic materials and the regulation of climate

Hygroscopic materials form a cushion for damp in the same way as a heavy material is a cushion for temperature, and this exerts a positive influence on the internal climate. A moderate and stable moisture situation will reduce the chances of mites and

Figure 14.4: Bourne House, Aberfeldy (interior view). Surfaces with moisture-regulating properties. Source: Howard Liddell

micro-organisms growing. The deposition and emission cycles of dust on inside surfaces will be reduced. Water vapour carries various gas contaminants combined with water vapour molecules, which also penetrate the wall. Hygroscopic walls will therefore have a moderate air cleaning effect for nitrogen oxides and formaldehyde. This is only effective as long as the gases stay in the material or are broken down inside it. Hygroscopic materials lose their moisture-regulating properties if they are covered with diffusion resisting materials such as plastic wallpaper, varnish, etc.

It is also an advantage if sealing and insulation materials in the wall are hygroscopic. Condensation is no problem when the amount of condensed moisture is low compared with the material's potential capacity for holding moisture (below the threshold for rotting), as the water that is stored during a damp period can evaporate during the rest of the year. This applies under normal circumstances to brick, earth, timber and other natural fibres.

Constructions with insulation materials such as foamglass and mineral wool are not hygroscopic and should be insulated from internal moisture by a vapour barrier. Otherwise there is a risk of absorption in the structure, and this can be too much, even for timber. A wind-proof membrane with a large capacity for moisture absorption and permeability can compensate for this to a certain extent. Sheets of gypsum or porous fibreboard glued with asphalt or untreated are well suited for this, as long as their surfaces are not treated with less permeable materials.

The air cavity method

The final method is based on ventilating out moisture that has penetrated the wall. Moisture needs to be taken care of before it can condense. The dew-point, where the temperature is so low that saturation can occur, needs to be identified. This problem is most likely to occur in rooms that have a very high moisture content, or where there are materials of low moisture capacity combined with high watertightness in the wind-proofing membrane. There are ways of calculating this, but they have proved to be unreliable in practice as the climate is not very predictable. The air cavity can be either narrow or wide. One solution is cavity wall construction where the cavity is of a large volume, with a low temperature function, such as a conservatory or storage space.

Damage due to damp

Damage due to damp can be recognized through mould or the smell of mould. Other odours can also be caused by damp, because damp can cause gases to be emitted from glue, paint, mastics and other products.

Mould in organic materials can occur at a relative humidity of 90 per cent. Timber with a 20 per cent moisture content is easy prey for different micro-organisms. Materials that are not hygroscopic are often covered with a thin film of water in a damp atmosphere. The organic glue additives and oils in mineral wool can suffer strong attacks. Traces of mould can reach inside through cracks in the vapour barrier.

When the damage is done, the damaged area has to be removed and all the materials changed. The smell of mould can linger even after the damage has been repaired. This can be removed by ozone treatment. Ozone is, in fact, quite damaging to health, because it corrodes the inhalation routes in the body, and the gas will destroy plastic materials in the building, including the vapour barrier.

Air-regulating materials

Wind-proofing a building takes place in two areas, topographical and other wind breaking effects in the surroundings, and a wind-proofing membrane forming part of the building's outer skin.

Adjusting to the climate and external windbreaks

Nearby buildings, fencing, mounds, plains, mountains and vegetation regulate effect of wind on buildings. If the average wind speed around a building is reduced by 1 m/s, it is possible to reduce the energy requirement by 3 per cent. In the Norwegian coastal town of Kristiansund, where the average wind speed is 22 km/h (Beaufort scale 4), the loss of heat for an unscreened building through infiltration is 40 per cent greater than for a screened building. In a standard house, 30 per cent of the heat loss happens through infiltration. But the air around a building should not be completely still; 1–2 m/s is optimal. Heat radiation has a greater effect when there is no wind.

There are three main methods of reducing the infiltration of wind into the main body of a building:

- Windbreak
- Turbulence membrane
- Airtight membrane

Windbreak

A windbreak is perforated and should preferably be on battens at a good distance from the outer wall, so that a useful storage area is available in the space between. By using about 30 per cent perforation a minimal difference of pressures between the front and the back of the screen is achieved. The formation of eddies is thus reduced, and wind and rain are effectively slowed down. Suitable materials include climbing plants, trellis work, timber battens or metal ribbing.

Turbulence membrane

A turbulence layer is mounted directly on the main wall, and is usually made of different types of roughly-structured surfaces which cause innumerable small air movements in the material – a sort of air cushion. The wind is stopped dead instead of penetrating further into the wall. Materials suitable for this are roughly-structured render, cladding made of branches or a living surface of plants. None of the methods are 100 per cent efficient; there is always the possibility of weak points, and some wind will force its way through. The turbulence layer has no effect on infiltration as a result of suction, and usually needs to be complemented with an airtight layer.

Airtight membrane

Suitable materials for an airtight layer include sheeting, boards, paper sheeting and mastic, as well as external cladding. Holes and gaps in the structure, e.g. around windows and other building components, should be closed.

An airtight membrane for a whole wall often consists of paper sheeting, wood fibreboard or plasterboard sheeting, which can be improved by waterproofing. This is placed behind the external cladding and is well ventilated.

Wind breaking membranes should not let through more air than 0.1 m^3/m^2 with a pressure of 10 Pa. In extremely windy conditions such as heavy storms or hurricanes it is very difficult to prevent wind penetrating the building. In exposed locations it would be best to use heat insulation materials with good wind-proofing properties as well, e.g. well-compressed cellulose fibre.

Diffusion of gas and breathing walls

Internal climate usually needs a flow of fresh air equivalent to half to three changes of the whole air volume per hour, depending on the room's function.

In buildings which have airtight vapour barrier membranes in their walls, the flow of fresh air depends upon specific openings for ventilation such as windows. In a building with dynamic insulation, the flow of fresh air enters through the external surfaces. At the same time, contaminated gases in the internal air will be drawn out through the surfaces by gas diffusion. Gases have the particular property of always wanting to spread themselves evenly in the surroundings. The flow through the walls will therefore travel in both directions, and is permanent, though the pressure and the particular gas and molecular weight decide the speed. This also depends upon the material's capacity for letting through the different gases, i.e. the resistance to gas diffusion.

In principle there will also be substantial gas diffusion through materials that are initially far too dense to be used for dynamic insulation. For example, a 20 cm thick brick wall with an area of 10 m^2 lets about 90 litres of oxygen through each hour under normal pressure. This is the equivalent of one person's use in the same period. An equivalent calculation for concrete gives about 11.25 l/hour. The conditions for this calculation are that the oxygen content of outside air is 20 per cent and for inside air it is 15 per cent. It also assumes that conditions are ideal without complicated variations in pressure around the walls, ventilation intakes, etc.

Little is known about how walls breathe in practice. Researcher Lars Möllehave at the Hygienic Institute in Århus in Denmark has measured the diffusion of freon gas through material in walls in rooms with no cracks, which clearly shows that the process exists and is very active.

Snow as a climatic material

The thermal insulation of dry snow is equivalent to that of rockwool. This is reduced with increased water content.

Over large areas in Northern Europe, dry snow settles every winter and remains for six months, helping with insulation just when it is most needed. So it is quite clear that this snow should be conserved. There are six ways of retaining snow on a building:

Table 14.2: Properties of climatic materials and their use

	Temperature regulation			*Moisture regulation*		*Air regulation*
	Thermal insulation	*Thermal capacity*	*Thermal reflection*	*Sealing*	*Even moisture*	*Sealing*
Snow	Limited use					
Metal foil			Limited use	Limited use		
Lightweight concrete	In general use	Limited use			Limited use (aerated or with hygroscopic aggregate)	
Expanded minerals	Limited use	Limited use				
Expanded clay	In general use			In general use (expanded clay can be used as a capillary break)		
Foamglass	In general use			Limited use		
Foamed concrete	Limited use			Limited use		
Mineral wool	In general use					
Plasterboard				Limited use	Limited use	In general use
Porous brick		Limited use			Limited use	

Rammed earth	Limited use	Limited use		Limited use	Limited use
Asphalt/bitumen			In general use		
Plastic sheeting			In general use		
Foamed plastics	In general use		In general use		
Plastic-based mastic	In general use		In general use		
Plastic sealants					In general use
Building paper from plant fibres/ cellulose			Limited use		Limited use
Boarding from plant fibres/cellulose	In general use (wood fibreboard)		Limited use (wood fibreboard)	Limited use	Limited use
Matting from plant fibres	Limited use (flax and cellulose)			Limited use	Limited use
Loose fill from plant fibres/cellulose	In general use (cellulose fibre)			In general use (cellulose fibre)	In general use (cellulose fibre)
Building paper from woollen fibres			Limited use		Limited use
Matting and loose fill woollen fibres	Limited use			Limited use	

- A sloping roof of not more than 30°, preferably less
- A roof covering made of high friction material, e.g. grass
- A snow barrier along the foot of the roof
- An unheated space under the roof, or very good roof insulation
- Windbreaks in front of the roof
- Reduced sun radiation on the roof, e.g. a single-sided pitched roof facing north.

Many of these conditions have disadvantages. But the thermal insulation of snow should certainly be seriously considered when designing in areas where white winters are standard.

Snow is free, and is an efficient and environmentally-friendly insulating material. Zones with mild winters do not need 'snow-planning'; the same goes for sites exposed to wind, but in many cases well-planned placing of snow drifts can provide excellent protection from wind. This can be done using special snow fenders with an opening of approximately 50 per cent in the grid, and also with the help of planted hedges and avenues. Snow will settle on the lee side in areas of turbulence.

Metal-based materials

The type of material dominant in this kind of work is metal sheeting. The sheeting is used on exposed parts of the building's external skin, such as between the roof and building parts that go through the roof such as chimneys, ventilation units, vent stacks and roof lights, and on valley gutters and snow barriers at the foot of the roof. Not all metal products are usable, as some corrode. Combinations of different metals can create galvanic corrosion.

Stainless steel sheeting

This is usually an alloy of 17–19 per cent chrome and 8–11 per cent nickel. In aggressive environments one uses an alloy of 16–18.5 per cent chrome, 10.5–14 per cent nickel and 2.5–3 per cent molybdenum. Stainless steel can be used in combination with other metals. When corroding, chrome and nickel leak into the groundwater and soil.

Galvanized steel sheeting

This needs about 275–350 g/m^2 zinc. The material should not be used with copper. Gutters are often coated in plastic.

Aluminium

Aluminium normally has 0.9–1.4 per cent manganese in it. The products are often covered with a protective coating through anodizing. They can also be painted with special paint. They should not be used in combination with copper or concrete.

Copper

Copper is produced in a pure form without any surface treatment or other alloyed metals.

Zinc

Zinc is usually used in an alloy of zinc, copper and titanium. This should not be used in combination with copper. Its surface is painted with a special paint.

Lead

Lead is soft and malleable. It should not be used in combination with aluminium.

In terms of raw materials the use of metals should be reduced to a minimum. These details of the building are very much exposed to the climate and therefore to deterioration. Zinc corrodes quickly in an atmosphere containing sulphur dioxide, which is common in towns and industrial areas; the spray of sea salt also causes corrosion, so it is best used away from the coast. The zinc coating on galvanized steel is exposed to the same problems, but its durability is better in the long run. In particularly aggressive atmospheres even aluminium, lead and stainless steel will begin to corrode.

Metals have a high primary energy consumption and a polluting production process. For the people using a building, metals are neutral, even though a high percentage of metal is assumed to strengthen the building's internal electromagnetic fields. Metal ions may also be released into the soil around the building. This could cause an environmental problem, depending on the amount and type of metal in question – lead and copper are the most troublesome. Metal can be recycled when it becomes waste.

The use of metals should be reduced to a minimum and alternatives used where possible. Guttering, for example, can be made of PVC or wood (see Figure 14.5). The use of metal sheeting can be reduced or avoided in many cases by choosing other materials.

Materials based on non-metallic minerals

Many loose mineral materials contain natural pores which make them useful as thermal insulation. Examples are fossil meal, perlite and vermiculite.

Figure 14.5: A wooden gutter, well worn after decades of service.

Materials such as cement, magnesite and lime are bad insulators, but they have potential as binders for different mineral aggregates, to make them into blocks, slabs etc. In the same way expanded clay pellets, pumice, wood shavings and woodwool can be bound.

Aluminium powder added to a cement mixture acts like yeast and forms gas within the concrete. This becomes a lightweight concrete with good insulation value. It is also possible to foam up a relatively normal mixture of concrete to a foam using air pressure and nitrogen.

Quartz sand is the main constituent of glass and has a very low thermal insulation value, but glass can be foamed-up to produce a highly insulating and stable foamglass. The mineral wool glasswool also originates from quartz sand. The sand is melted and drawn out to thin fibres in the form of thick matting or loose wool, which also has good insulation value. A similar material, rockwool, is based on the rock species diabase and lime, treated in almost the same way.

All these mineral materials, except for those containing a lot of gypsum or lime, have poor moisture-regulating properties. Cement products take up and release moisture very slowly. Drying out a concrete building can take years, and during that period damage can occur to organic material touching the concrete.

Table 14.3: The use of non-metallic mineral climatic products in building

Material	*Composition*	*Areas of use*
Fossil meal, loose	Fossil meal	Thermal insulation
Perlite, expanded, loose	Perlite (possibly with bitumen or silicon)	Thermal insulation
Vermiculite, expanded, loose	Vermiculite	Thermal insulation
Aerated concrete	Cement, water, lime, gypsum, quartz, aluminium powder	Thermal storage insulation balancing of humidity, construction
Lightweight concrete with mineral aggregate	Cement water, with fossil meal, expanded perlite, expanded vermiculite, expanded clay, pumice or expanded blast furnace slag	Thermal insulation construction
Lightweight concrete with organic aggregate	Cement, water, with wood chips and saw dust, hacked straw or cellulose fibre	Thermal insulation, thermal storage, balancing of relative humidity, construction
Lime-mortar products(1)	Lime, water, sand	Balancing of relative humidity, thermal storage, moisture barrier
Gypsum products	Gypsum, water (possible addition of silicones, starch and covered with a layer of thin cellulose cardboard)	Balancing of relative humidity, moisture barrier, wind-proofing, sound-proofing
Sulphur concrete(1)	Sulphur	Damp-proofing, construction
Quartz foam (Aerogel)	Calcium silicate, hydrochloric acid	Transparent thermal insulation
Foamglass	Quartz, boron oxide, aluminium oxide, soda, lime	Thermal insulation, thermal storage, vapour barrier
Glasswool	Quartz sand, phenol glue, aliphatic mineral oils	Thermal insulation, sound absorption
Rockwool	Diabase, limestone, phenol glue, aliphatic mineral oils	Thermal insulation, sound absorption, sound insulation
Montmorillonite	Montmorillonite (can be placed between two layers of cellulose paper)	Waterproofing

Note:
(1) These materials are discussed in other chapters.

Most mineral insulation products have weak wind-proofing qualities, and require a separate membrane or skin such as render, timber panelling, or the equivalent.

Montmorillonite is a clay mineral well-suited to waterproofing because of its high moisture absorption coefficient. Render containing sulphur also has a high waterproofing quality.

These climatic products are based on materials from resources with rich reserves. What they nearly all have in common is that their extraction causes a large impact on nature, damaging the groundwater and biotopes. The more highly refined products are, the more energy they consume in production, with associated pollution during the process. Most mineral-based climatic materials are often chemically stable in the indoor climate. However, in many cases organic material additives can cause problems by emitting irritating gases and encouraging the growth of micro-organisms. Some of the materials produce dust problems during the building process and even after the building is finished. Some raw materials include radioactive elements which lead to a high concentration of radon in the indoor air.

As waste, mineral-based climatic materials can be considered chemically neutral – the main problem can be their volume. Attention must be given to coloured products, as the pigments may contain heavy metals.

Clean loose aggregates can be re-used, as can blocks and prefabricated units. They can also be crushed into insulating granules, which are particularly well-suited to use as underlay for roads.

Cement products

Cement can be used as an insulating material in three forms:

- As foamed concrete
- As aerated concrete
- As binder for light mineral and organic aggregates

Foamed concrete

Foamed concrete has considerably better thermal insulation properties than normal concrete – as high as 0.1 W/mK for densities of approximately 650 kg/m^3. It consists of Portland cement and fine sand in proportions of about half and half. The foaming agent is either tensides or protein substances. The latter can cause considerable problems in the indoor climate if it reacts with cement. The use of tensides, however, causes no such problems. Foamed concrete is seldom used in building construction because of its relatively low thermal insulation and low load-bearing capacity. It is used nowadays mostly for the levelling of floors, sprayed onto horizontal surfaces or into hollow cavities from mobile tanks transported by lorry. The environmental aspects of this concrete are the same as *in situ* concrete (see 'The composition of concrete', p. 193).

Aerated concrete

Aerated concrete is produced by reacting finely powdered quartz (about 50 per cent by weight) with lime, gypsum and cement. A yeast constituent such as aluminium powder is added to a proportion of about 0.1 per cent. Aluminium reacts to release hydrogen. When the substance is almost stiff, it is cut into blocks and prefabricated units which are hardened in an autoclave. Prefabricated units of lightweight concrete are usually reinforced with steel. Aerated concrete is the only commercial pure mineral block with good structural properties *and* a high thermal insulation value. The material is very porous, and needs a surface treatment which lets out/in vapour – hydraulic lime render, for example. If the water content becomes too high the material will easily be split by frost. The production of this aerated concrete is dependent upon aluminium. The total contribution of aluminium in the external walls of a relatively large private house is 10–20 kg.

Aerated concrete normally has good moisture-regulating properties and does not have any negative effects on the indoor climate, although the steel reinforcement can increase the electromagnetic field in a building. Aluminium will have completely reacted in the finished product, and in practice aerated concrete can be considered inert and problem free as waste. Both prefabricated units and the blocks can be re-used, depending upon how they were laid and the mortar used. Strong mortars are used nowadays which make it difficult to dismantle the components without damaging them. More appropriate mortars are weak lime cement mortar and hydraulic lime mortar. Crushed aerated concrete can be used as insulating granules for road building, and also as aggregate in lime sandstone, different light mortars and light concretes.

Concrete with light aggregate

This is usually produced as blocks, slabs or floor beam units which are relatively strong. There is a difference between products that have an organic and a mineral aggregate. Mineral insulating aggregate in concrete can be light expanded clay, pumice, fossil meal and exfoliated vermiculite, perlite or slag. The first two and expanded perlite have the lowest moisture absorption coefficient, and are therefore best-suited to products used for insulation. The others have a very high moisture absorption coefficient and are best used as insulation for high temperature equipment.

Sawdust and chopped straw can be used as organic constituents in concrete. Blocks are also produced using broken up, waste polystyrene, and it is possible to produce lightweight concrete mixed with waste paper. With the exception of woodwool slabs, discussed later in this chapter under 'Timber', concrete with organic constituents generally has a low thermal insulation value compared to rival products such as aerated concrete. In light expanded clay blocks it is becoming more usual to cast in a thermal insulating membrane of expanded plastic, usually polyurethane, discussed later in this chapter under 'Plastics'. 'Woodcrete', which contains the maximum proportion of sawdust possible, achieves much higher thermal insulation values than normal concrete, and can be compared with a light expanded clay block, for example. Woodcrete should be a viable alternative because it provides a considerably warmer, softer surface than pure concrete. It is also a good sound insulator, and will not rot because of the high pH of the cement. The sawdust has to be treated in the same way as the wood in woodwool slabs before production (see 'Woodwool cement – production and use', p. 282).

Raw material for concrete with light aggregate is widely available. The pollution caused by the processes involved is the same as for concrete (see 'Concrete structures' p. 192). To attain acceptable thermal insulation levels, considerable thicknesses are necessary, and the primary energy consumption is high. Erecting a fully-insulated wall of light expanded clay block insulated with expanded polyurethane uses 75 per cent more energy than for an equivalent construction in timber (Fossdal, 1995).

Except for possible pollution from granules and the use of plastic sheeting, the production and use of concrete products usually causes no problems. The use of steel reinforcement with these products may increase the electromagnetic fields within a building.

Light expanded clay blocks are initially inert and the waste from them can be used as fill for road building, as ground insulation or as insulating aggregate in smaller concrete structures, light mortars and render. Lightweight concrete blocks can easily be re-used if they are held together by weak mortar, as can larger concrete units that have been bolted or placed without fixing. Lightweight concrete products can be produced in local small and medium-sized factories.

Gypsum products

Gypsum is used mainly for sound-proofing and wind-proofing boards which are also very good moisture regulators. The products are cast from 90–95 per cent gypsum which has fibreglass added (0.1 per cent by weight) as reinforcement. The following constituents are also added, to a total weight of 1 per cent: calcium ligno-sulphate, ammonium sulphate and an organic retardant. In the wind-proofing boards the additives include silicon (0.3 per cent by weight). The boards are often covered in cardboard which is glued with a potato flour paste or PVAC glue. Acoustic boards have a covering of woven fibreglass on the surface.

Gypsum is sourced from power stations as a by-product, or from nature. In both cases the raw material situation is good, even if it is hoped that polluting coal power stations become less active in future. The materials needed for the additives are renewable or obtained from fossil resources. The cardboard covering is produced from a minimum of 90 per cent recycled cellulose. Extraction of gypsum has a large impact on the natural environment, and the use of gypsum from power stations improves the waste situation.

Apart from dust, the use of gypsum has no particular problems, except when additives, e.g. the retardant diethyl triamine, are used. When silicon is added, methyl chloride is used. Once in the building, however, the products cause no problems.

Gypsum products are less well suited for re-use, but can be recycled through the addition of 5–15 per cent waste gypsum to new products. The gypsum industry is very centralized, which often makes recycling an uneconomic proposition. There is chance of sulphur pollution from demolition and building waste because of microbial breakdown.

Fossil meal products

Fossil meal is a sedimentary earth that can be used as fill or aggregate in cast cement blocks or insulating mortars. Fossil meal products have good thermal properties and a high moisture absorption rate, making them suitable for insulation of high temperature equipment such as kilns, kettles, hot water tanks, baking ovens and high temperature equipment in industry. It can also be used in walls between rooms as a fill. It has a powder-like consistency, and must be placed between paper sheets so as not to leak out into the room.

Fossil meal mortars are made by mixing fossil meal with a cement, or even with plant fibres up to 30 per cent by weight. Water is added and the ingredients are well mixed together. The mortar is then ready for use on hot water pipes, for example, preferably in several layers, each 1–2 cm thick. A canvas is bound over the last layer, which can be painted or rendered with lime.

Blocks of fossil meal can be made using cement as a binder. It can also be used as an insulating aggregate in brick products. Fossil meal contains large amounts of silicium dioxide and can be superficially considered dangerous with respect to silicosis. However, in fossil meal this substance is not the crystalline silicium oxides as in quartz, but an amorphic version which is completely harmless. Fossil meal is relatively widespread and causes considerable blemishes on the countryside when extracted. The waste phase does not cause any problems. Unmixed parts can be re-used or can even be left in the natural environment, covered with earth.

Perlite and pumice products

Perlite is a natural glass of volcanic origin mined by open-cast methods in parts of the world such as Iceland, Greece, Hungary and the Czech Republic. It is pulverized and expanded in rotating kilns at about 900–1200°C, which increases its volume between five and twenty times. Expanded perlite was first produced in the USA in 1953. It has the consistency of small popcorn and is used as loose fill and aggregate in mortars, render and lightweight concrete blocks. It is also used for the thermal insulation of buildings, the insulation of refrigerating rooms and high temperature insulation.

Because the material absorbs a little moisture there is the risk of a reduced insulation value and an increased settling problem within a wall. To avoid this, a moisture preventative is added to the mix before it is poured into the wall. Perlite mixed with silicon (about 1 per cent by weight) at 400°C is called Hyperlite. Bitumen can also be added in a proportion of about 15 per cent.

Using perlite as an aggregate in render and mortars can achieve an increase in the thermal insulation of a wall. For example, 15 mm perlite render is the equivalent of a whole brick wall thickness or 240 mm concrete. In this case the perlite is not impregnated.

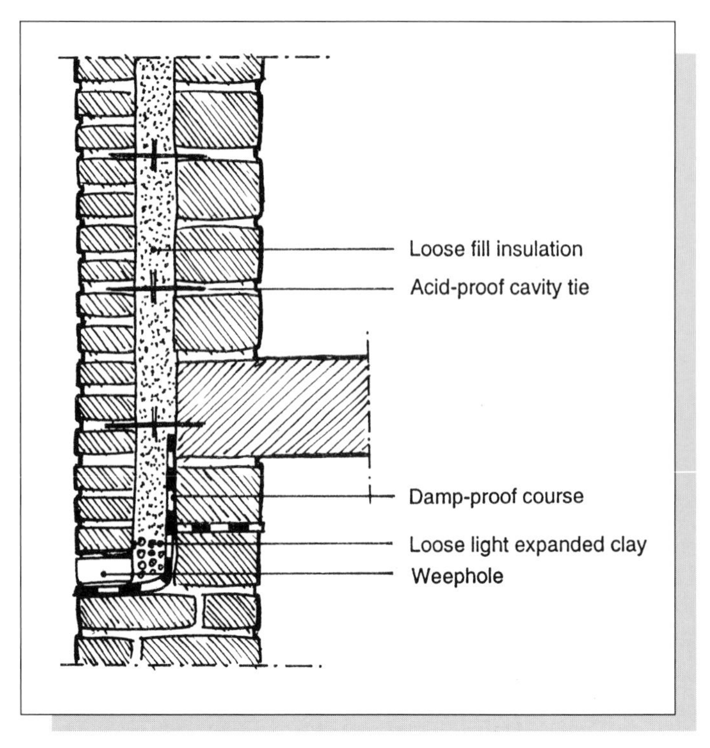

Figure 14.6: Principle for perlite insulation in cavity walls.

Lightweight concrete blocks with perlite can be produced in many different mix proportions. When perlite is exposed to even higher temperatures naturally, it expands and becomes a porous and monolithic rock called pumice. The pores in this stone are not connected, so the material does not absorb any water. Building blocks of pumice in combination with cement have almost the same properties as light expanded clay blocks.

Pumice occurs naturally and in large quantities in Iceland.

Perlite reserves are large. The only pollution risk related to perlite is possible irritation from exposure to its dust. The use of bitumen and silicon additives raises the question of oil extraction and refining in the environmental profile. Pure and silicon-treated perlite have no side effects once installed in a building. Depending upon how the bituminous products are incorporated, small emissions of aromatic hydrocarbons may occur.

As a waste product, bituminous perlite must be disposed of at special depots. Pure perlite is inert. The siliconized material is also considered inert. Recycling is possible by vacuuming the loose material out of the structure, compressing it and re-using it locally.

Vermiculite products

Vermiculite is formed through the disintegration of mica, which liberates lime and takes up water. When vermiculite is heated to 800–1100°C, it divides into thin strips. These release water, curl up like snakes and swell to become a light porous mass which can be used as an independent loose insulation or as an aggregate in a lightweight concrete in the proportions 6:1 vermiculite to cement. Other mineral binders can be used. Prefabricated slabs are made in varying thicknesses, from 15 mm to 100 mm.

As with the other mineral materials, vermiculite is particularly useful for high temperature equipment. It easily absorbs large amounts of moisture, even more than untreated perlite. As normal wall insulation it has a tendency to settle a great deal. This can be solved by applying compression up to 50 per cent, using a coarser form of the material. The environmental situation is approximately the same as for perlite.

Foamed quartz

By adding hydrochloric acid to a solution of waterglass (calcium silicate), silicic acid is formed in a jelly type mass. Its trade name is 'aerogel'. This is used as transparent thermal insulation, usually between two sheets of glass. It is best used in connection with solar heating. The sun's radiant energy penetrates the gel, while it prevents the loss of heat through convection and loss of long-wave radiation (see Figure 14.7).

A transparent layer of insulation on the south wall of a brick building, can provide much of the heat it requires because the warmth goes through the wall and into the building. Heavy brickwork will even out the temperature and prevent overheating or too much cooling.

This type of gel is at present not in general use, and has disadvantages: it does not tolerate water and has a tendency to crumble. But it has few negative consequences in relation to the environment and resource extraction.

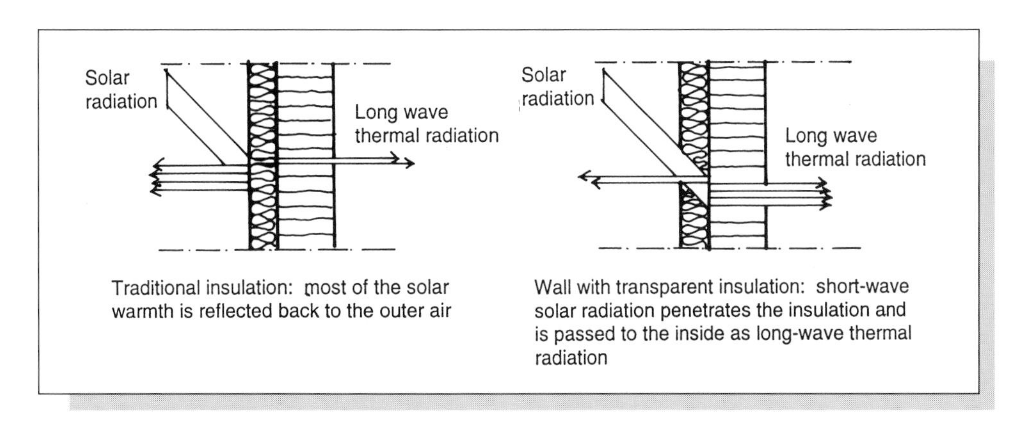

Figure 14.7: Transparent thermal insulation.

Foamglass

Foamglass is usually produced by adding carbon to a conventional mass of glass and heating it to 700–800°C until it starts bubbling. The product is usually made in the form of slabs. These are gas- and watertight with high thermal insulation properties, and they are mainly used as insulation underneath ground floors. The raw material is usually new glass, but a rougher product can also be made from recycled glass in the form of blocks or granules.

Blocks of foamglass not only have a high thermal insulation value, but also have structural properties similar to conventional lightweight expanded clay blocks. They are also easy to screw and nail into. They are usual cemented together with a bituminous mass. The granules are based on 95 per cent by weight recycled glass with added sugar, manganese dioxide and lime. They are used as light aggregates in concrete or as loose insulation.

Products based on new glass production use high levels of primary energy and polluting production methods (see 'Ecological aspects of glass production', p. 105). Products based on recycled glass are environmentally better, despite the high level of energy use when re-melting the glass.

Within the building these products present no problems. One exception is the use of bitumen as a jointing material and any metal reinforcement used can increase the electromagnetic field. These products have no moisture-evening properties. Extensive use of them in a building can lead to an indoor climate with rapid air moisture changes and, in certain cases, the possibility of damp in adjacent materials.

Components containing bitumen must be disposed of at a special tip. Blocks and granules can be re-used in building. Foamglass is inert and can be crushed and used as an insulating layer in road building. There is no other way of recycling this material.

Synthetic mineral wool fibres

Glasswool/fibreglass

Glasswood/fibreglass is made from quartz sand, soda, dolomite, lime and up to 30 per cent recycled glass. The mass is melted and drawn out into thin fibres in a powerful oil burner. Glue is then added to the loose wool and heated to form sheets or matting in a kiln. Phenol glue is commonly used in a proportion of about 5.5 per cent of the product's weight. To give a high thermal insulation value the diameter of the fibre should be as small as possible. The usual size is about 5 μm.

Rockwool

Rockwool is produced in approximately the same way as glasswool, starting with a mixture of coke, diabase and limestone. Basalt and olivine can also be used. The quantity of phenol glue is lower – about 2 per cent by weight. The diameter of the fibres varies from 1–10 μm.

Both types of mineral wool, especially rockwool, have aliphatic mineral oils added up to about 1 per cent by weight to reduce the dust. An emulsifier is often added in the form of a synthetic soap, for e.g. polyethoxylene, up to 0.2 per cent by weight, and a foam reducer, usually polymethylsiloxanol, up to 0.5 per cent by weight. Both glasswool and rockwool are usually made as matting, but both types are delivered as loose wool. Mineral wool products are light and have extremely good thermal insulation values.

When used as insulation both glass and rockwool need a vapour barrier of aluminium or plastic sheeting, partly to avoid dust and partly because the material cannot regulate moisture particularly well. Research has shown that in timber frame buildings, rockwool, and to a certain extent glasswool, increase rot and damage caused by damp on the timber framework, unlike the more hygroscopic insulating materials such as cellulose fibre (Paajanen, 1994).

Mineral wool products can also be criticized for other reasons. Many experiments indicate a connection between exposure to mineral wool fibres and skin problems, itching, eye damage and respiratory irritation. The latter has, in many cases, led to chronic bronchitis. It is also possible that these materials have carcinogenic effects. Acoustic panels functioning as sound insulation are normally the most common source of mineral wool fibres in the indoor climate (Bakke, 1992).

It has been shown that dampness in mineral wool can lead to the emission of vapours which can later enter the building. The problem is more acute when the wall becomes warm, e.g. through solar radiation. The type of gases released are aliphates, aromates and ketones. The aliphates in particular can affect air quality detrimentally. All of these gases irritate the ears, nose and throat (Gustafsson, 1990).

Damp mineral wool smells sour, which can imply the release of amines. Additives in mineral wool that contain nitrogen are very susceptible to mould. The amount of mould in an infected material can be 1000 to 50 000 times the amount in uninfected material (Bakke, 1992).

Raw materials are abundant for the main constituents of glasswool and rockwool. The production of glasswool occurs in relatively closed processes. The emissions from production are little and limited to formaldehyde and dust in addition to energy pollution. Large amounts of phenol, ammonia, formaldehyde and dust are released during the production of rockwool, and large amounts of waste are produced. Phenol can be washed out of rockwool waste. Unpolluted waste can be compressed and recycled for the manufacture of new mineral wool, although the industry is so centralized that this form of recycling is economically unrealistic.

Montmorillonite

Montmorillonite occurs mainly in bentonite clay, a very disintegrated type of clay made from volcanic ash. The minerals in montmorillonite not only absorb water on the surface, but also within the mineral structure. It therefore has the capacity of taking up large amounts of water and swelling to twenty times in volume. This absorption occurs quickly, and when the surroundings dry out again, the clay releases its moisture. It is therefore useful as an absorbent waterproof membrane on foundation walls made of brick and concrete. Bentonite clay can be purchased in panel form, packed between two sheets of corrugated cardboard: the clay is approximately 0.5 cm thick and the cardboard gradually rots away. The panels should be under a certain pressure, which can be achieved by a compressed layer of earth of at least 0.4 m.

There is an abundance of montmorillonite clay, but in very few places, so high levels of transport energy are needed. The environmental problems of this product are otherwise of no consequence.

Fired clay materials

Fired clay in the form of bricks is mainly a structural material and has a low thermal insulation value. However, it is possible to add substances to the clay which burn out during the firing and leave air pockets in the structure. The lighter product that results can be found in slab or block form.

Clay can also be expanded to light expanded clay pellets for use as loose fill, or it can be cast with cement to form blocks or slabs. By exposing light expanded clay to even higher temperatures, the light, airy granules cohere into a solid mass which can be used to form blocks known as Zytan blocks. This type of block is no longer in production because of the very high primary energy use required.

All fired clay products are chemically inactive. In the indoor climate there are no particular problems with these products.

Certain types of brick are good moisture regulators. The more developed the microporous structure, the better the moisture regulation. Low-fired brick and brick with a high proportion of lime give the best results. Because of their high primary energy use, all fired clay products should be recycled, preferably in their original undamaged state. Coloured and glazed clay products may contain heavy metal pigments, and as a result can cause problems when they are finally disposed of.

Fired clay

Blocks of porous clay are fired at temperatures of 1000°C or more. The organic ingredients in the block (sawdust, pieces of cork, etc.) are burnt away to leave an internal structure with isolated air holes. In one particular product, granules of polystyrene are used as the aggregate for burning out the clay. During the firing the polystyrene granules vapourize. The vapours from the polystyrene have a polluting effect, whereas the completed product is probably free from polystyrene.

Table 14.4: The use in building of fired clay climatic products

Product	*Areas of use*
Low/medium fired brick	Balancing of humidity
Brick with high lime content (15–20% lime)	Balancing of humidity
Bricks containing materials such as sawdust, peat, hacked straw and powdered coal, that are burnt out during the firing process	Thermal insulation
Bricks with fossil meal as an insulating aggregate	Thermal insulation
Expanded clay, loose	Thermal insulation, capillary break
Lightweight concrete	Thermal insulation
Zytan block	Thermal insulation

Insulating aggregate such as fossil meal can be added, and once fired the blocks have a relatively high thermal insulation value.

Fired clay blocks with fossil meal as thermal insulation

One part clay is mixed with 15 parts fossil meal into a homogeneous mass. It is also possible to add 25 per cent sawdust or pieces of cork before the mass is pressed into forms and fired. Hard blocks can be used structurally, while the blocks with sawdust or cork pieces are primarily used for insulation. In addition to these solid blocks, the material can be formed into blocks with holes.

Fossil meal which naturally contains the right amount of clay to enable formation directly into blocks is known as sandy clay. In Scandinavia this form of fossil meal occurs only at one site, Jylland in Denmark, and the sources are not very plentiful.

Figure 14.8: Highly porous bricks balance humidity in a bathroom. Hydraulic lime mortar is used to improve the possibilities for re-use. Source: Gaia Lista, 1990

Light expanded clay

Expanded clay can be used as loose fill or cast with cement into blocks or other structural units. It has a relatively high thermal insulation value. Loose expanded clay pellets can be used under the slab of a building as a capillary break. Light expanded clay and light clay thermal blocks have good structural properties, but they are poor moisture regulators because the pore structure is closed.

Blocks and prefabricated units of light expanded clay are well-suited for dismantling and re-use as long as they were originally fixed together with weak mortars or mechanical jointing, such as bolts. Loose expanded clay pellets around surface water piping and ground insulation can also be re-used if they have been protected from roots, sand and earth. All expanded clay products are inert and can be recycled for use as insulation under roads, etc.

Figure 14.9: A traditional buried dwelling in Tunisia.

Earth and sand as climatic materials

Earth has a relatively low thermal insulation value, but, as with most materials, a thick enough layer can provide adequate protection against the cold. In the animal world it is not uncommon for rodents or other wild animals to live in the earth and benefit from the warmth. Man has also used this to advantage, and there are examples of underground buildings in most cultures, including underground towns in China, Turkey, Tunisia (see Figure 14.9) and Mexico.

Underground buildings

A buried building can be defined as a house roof and at least two walls covered by layers of earth at least 50 cm deep. The insulation value of earth is about one-twenty-fifth of the value of mineral wool, so if the roof is thinner than 2–3 m, extra insulation is needed. By planting trees or bushes on the roof, heat loss is reduced. The building should preferably be on a south-facing slope to take advantage of solar radiation. The floor must be higher than the water table. The loading on the roof can be more than ten times that of a normal building and the pressure on the walls slightly greater than that on a normal basement wall. It is important to have good drainage from the roof, and that the earth is laid on a well-drained material with a high friction coefficient.

Today, houses are generally built above ground. There are probably cultural reasons for this move to the surface of the Earth, because, practically speaking, nothing is as sheltered as a buried house! People, it seems, no longer want to live like rats. But in the USA, 600 underground buildings were erected between 1978 and 1980, including many libraries, schools and office buildings. The cost of an underground building has been calculated at about 10–20 per cent more than that of a conventional building (Winquist, 1980). The main aim of these buildings is to save energy, and it is symptomatic that the sudden rise in popularity of these buildings came after the energy crisis of the early seventies, only to

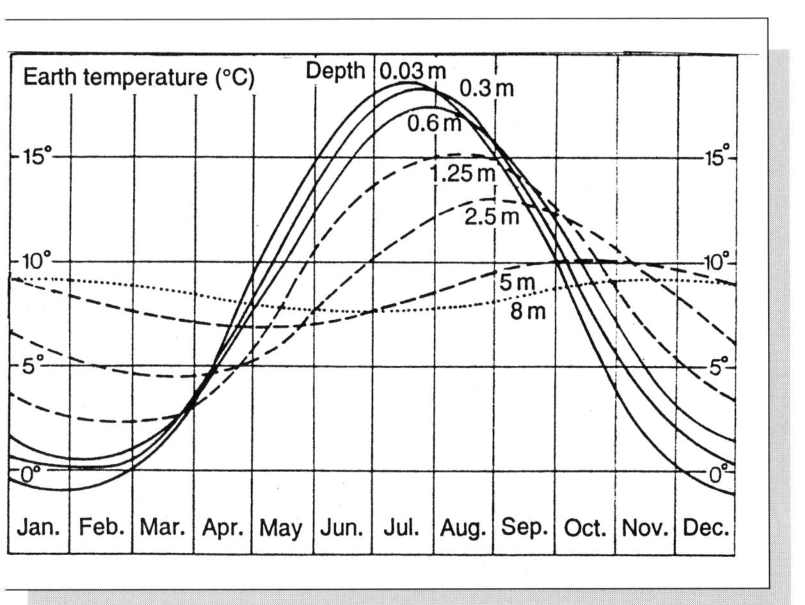

Figure 14.10: The temperature at different depths of the Earth throughout the year in southern Scandinavia. Source: Låg 1979

fall again once oil prices began to fall.

The American experience is that underground buildings have a reduced energy consumption, from 20–80 per cent of that of buildings above ground. Several factors influence this: the insulation of the earth mounds around the building, the warmth from the earth, the heat capacity of the earth mass and the protection from wind. Half buried buildings have better protection against noise, and the disturbance to the landscape is minimal. At 20 cm below the surface, the variations in temperature over 24 hours are hardly noticeable. This means smaller temperature changes in the fabric of the building and thereby fewer maintenance problems and a longer life span. These houses cannot, of course, be built where there is radon in the ground.

Figure 14.11: A cabin partly buried in a sensitive area along the south coast of Norway. The materials and structure have been chosen with respect to the climate, earth and water analyses. The aim has been to reduce the physical and chemical traces of the building to a minimum when it finally disappears. Source: Gaia Lista, 1997

Earth structures on the surface of the Earth also have interesting climatic aspects, particularly with respect to thermal insulation and moisture regulation. In northern Europe there are indoor swimming pools and moisture-sensitive libraries built with clay as the main material. A whole series of earth-based renders have been developed for concrete and hard fired brick in order to reach a more stable humidity within the building. To achieve reasonable thermal insulation, an insulating aggregate or another substance such as plant fibre is added to the earth. Earth has both a high heat capacity and good sound insulation properties. It is also wind-proof when compressed.

Water cannot usually penetrate a horizontal layer of earth more than 50 cm deep. The thick earth roofs found in Iceland are relatively safe from leakage. Earth containing a large quantity of clay is waterproof, even in thinner layers. The optimal clay is bentonite (see 'Montmorillonite' p. 269), which is waterproof at only 0.5 cm thickness. Normal clay needs thicker layers.

Two recipes for watertight layers of earth

In *The art of building*, Broch suggests the following recipe for waterproofing a brick and stone vault (Broch, 1848): first a 3 in thick layer of coarse sand on the vault, then a layer of finer sand, then 6 in of clay mixed with soil and finally a layer of turf. We have to assume that he was dealing with mausoleums and fortresses. The 'Podel' mixture, launched by James Brindley in 1764, was a method for damming water. The method is most interesting for external spaces: one part soil and two parts coarse sand are mixed, then stamped together or made wet until they do not let through any more water. The minimum thickness of the layer is 70–90 cm.

Clay as an infill between the joists in the floor space often has sound-insulating, moisture-regulating and, to a certain extent thermal-insulating properties. It can also affect the energy situation through its heat capacity and weight.

Filling with clay between joists

The clay is mixed with chopped straw, sawdust or similar material, and water is added, so that the mass becomes the consistency of porridge. This is used for the lowest layer, and should hinder leakage into the rooms below. When this has dried and stiffened, the cracks that have formed are filled by pouring a thin clay gruel over. The space up to the top of the joists is filled with dry clay.

Pure sand is often used as sound insulation in the floor structure. It is heavy and effective because it lies close up to the structure. Sand also has a considerable heat capacity.

All climatic earth materials are favourable from an ecological point of view. This includes all phases without exception, from its extraction as a raw material to its final disintegration. In the indoor climate earth is not a problem as long as it is not exposed to continuous and comprehensive damp conditions.

Bitumen-based materials

Bituminous products have good waterproofing qualities and are often used as damp-proofing on foundation walls and between foundation walls and the structure, etc. The first known building use of bitumen can be traced back about 5000 years to the Indus valley, where it was used to make a temple pool watertight. This fatty material often forms part of other building materials that are exposed to moisture, such as perlite, wood fibre wind-proof sheeting and different building papers, such as wind-proofing and roof covering. Coal tar was once used instead of bitumen. Such products are no longer in use.

It is usual practice to oxidize the bitumen mass by blowing air into it. The material is then warmed up and applied directly onto the surface, e.g. a foundation wall. Solvents can be added to give a more workable consistence. Mixing bitumen with crushed stone produces asphalt. Damp-proofing for foundation walls can be carried out with a strong building paper membrane impregnated with bitumen, or by applying 3–4 mm of asphalt reinforced with fibreglass. This can also be used underneath a bathroom floor or a timber structure. The joints are welded to make them watertight.

Bituminous mastic for making joints watertight consists of a solution or emulsion of bitumen with fine stone powder or synthetic rubber. The mixture contains high levels of solvents. Bituminous sheeting is often built up on a fibreglass or polyester base.

Bituminous products do not have a long life span if they are exposed to a combination of sunlight, wide variations in temperature and a lot of damp. They can also be attacked by acids found in soil. When protected from these conditions, they can be very durable.

Today, bitumen is based solely on oil, which is an extremely limited resource with a high pollution factor in its extraction and a potential for accidents. The production of bitumen-based materials is intensive in its use of energy and also has a high rate of pollution, but on a somewhat lesser scale than that of oil-based plastic products.

The heating of bitumen on a building site emits dangerous fumes – polycyclical aromatic hydrocarbons (PAH) amongst others, though the amount of PAH in bitumen is considerably lower than that in coal tar. Some of the products contain solvents. If bitumen products are exposed to heat or sunlight, fumes can be released into a building. Bitumen products cannot usually be re-used or recycled. Both bitumen and coal tar contain substances that are the initial stages of dioxin, which can seep out; waste products should therefore be carefully disposed of (Strunge, 1990).

Plastic materials

Many plastics have good water- and vapour-proofing properties and high thermal insulation properties when produced as a foam. As a sealant, plastic can take on many guises: paint, sheeting, paper, sealing strips and mastics.

Sheeting foils and papers

Three plastics are used for sheeting: polyisobutyl, polyethylene and polyvinyl chloride. Cellulose acetate is also usable, but is not produced for this particular purpose.

Polyisobutyl sheeting is produced in thicknesses of 1–2 mm and used primarily as damp-proofing for foundations. Polyethylene, the most-used plastic, is the only one used for vapour barriers, either alone or as a coating on paper sheeting. The sheeting is 0.025–0.2 mm thick. Polyvinyl chloride sheeting is not as vapour-proof as polyethylene, but it is used when higher strength is required.

Paper sheeting is made mainly of polyethylene and polypropylene and is used as a membrane in bathroom floors and as external moisture-proofing on foundation walls. The sheeting contains added stabilizers to increase its durability, and other additives such as a fire retardant and colouring.

Polyisobutyl and polyvinyl chloride contain large amounts of plasticizer. Paper plastics usually have fewer additives. Polyethylene foundation paper contains carbon as a ultraviolet stabilizer.

Building goods

The most common plastic in this case is PVC, mostly used as gutters and drainpipes. These are coloured and usually stabilized with cadmium.

Mastics

Apart from linseed-oil-based putty, the mastics available on the market today are plastic- or bitumen-based. A mastic has to fulfil the conditions of constant elasticity and durability. The plastics usually used are polysulphide, silicone, polyurethane, and various acrylic substances. The composition of these substances is complex and is usually based on at least five chemical substances with at least eight different additives. Mastics often have pigment and fibres added, usually fibreglass. Silicones are easy prey for mould in damp situations, and often have organic tin compounds added, about 0.05 per cent of the mastic. Polyurethane mastics contain 10–60 per cent phthalates. Plastics of polysulphide, polyurethane and polyacrylates contain chlorinated hydrocarbons as fire retardants and secondary plasticizers. Up to the end of the 1980s PCBs (polychloro biphenyls) were an important part of mastics for sealing between modules in prefabricated buildings.

Sealing strips

These are used mainly between the sheets of glass in windows and in window and door reveals. Important plastics used in sealing strips include polyurethane, polyamide, polyvinyl chloride, ethylene-propylene rubber, chloroprene rubber (neoprene) and silicone rubber. The products include different additives such as fire retardants, stabilizers and pigments.

Insulation materials

Different insulation materials are produced from polystyrene, polyurethane and urea formaldehyde. Foamed polyvinyl chloride and polyethylene were once used. The materials are foamed up using chlorofluorocarbons, pentane or carbon dioxide, and fire retardants and stabilizers are added.

Climatic products in plastic are based entirely on oil, which is an extremely limited resource with an extraction that is both polluting and carries a potential risk. Refining the products requires a great deal of energy compared to other materials. In all phases from production to use in the indoor climate and waste, the majority of plastic products can cause considerable pollution (see 'Pollution related to the most important building plastics', p. 149).

Sheeting and paper sheeting have very important roles in water- and vapour-proofing. Durability is therefore a decisive factor. According to existing documentation it is unlikely that plastic products have these qualities. In terms of pollution, products made of polyethylene and polypropylene produce lower levels. Goods made of PVC usually contain cadmium as a stabilizer against sunlight and other climatic influences, and as waste, cadmium has a high pollution potential (see 'Cadmium', p. 80).

Mastics must be applied when still soft. During the hardening process, the indoor climate can be badly affected by emissions of aromatic, aliphatic and chlorinated hydrocarbons. Chemical and physical breakdown of the material also occurs. At the Royal Theatre in Copenhagen, an unpleasant smell occurred after the use of a mastic. It could best be described as garlic or rotten eggs, and came from the sulphur compounds released on oxidation with the air (Gustafsson, 1990). There have also been many cases of serious mould growth on polymer mastics in bathrooms.

Mastics break down when exposed to weather and wind, becoming powdery. They then fall into or out of the joint. This process progresses much more quickly than was assumed during the 1960s when building methods with precast concrete elements began, and today a large number of buildings have considerable problems and high maintenance costs as a result. The decayed remains of the mastic also represent a toxic risk both inside the building and in the surrounding soil.

Sealing strips of plastic are already hardened by the manufacturer and are a lower pollution risk in the indoor environment. Their durability is much shorter than the products they are built in to, and they can be difficult to replace after a few years.

Insulation made of polystyrene and polyurethane is usually delivered as a readymade product from the factory; urea formaldehyde foam is sprayed in on site. The latter emits a lot of fumes during the hardening phase, particularly formaldehyde. Depending upon how the materials are built in, polystyrene can emit extra monomers of styrene while polyurethane can release small amounts of unreacted isocyanates and amines. Even if the level of emission per unit weight for these products is relatively small, large quantities of the materials are contained in buildings. There is also a great deal of uncertainty about how long plastic insulation materials will last.

The re-use of plastic-based climatic products is not particularly appropriate because of their short life span. Even the recycling of climatic plastic products is not very practicable, as most of them are fixed to other materials. An exception can occur in cases where pure insulating boards of expanded polystyrene (EPS) have been used. However, many of the plastics can be transformed to energy by burning them in special furnaces with smoke-cleaning systems. Ashes from the furnaces and plastic waste which is not recycled must be disposed of safely to prevent seepage into the ground water or soil.

Timber materials

Timber has many good climatic properties both in its natural form and when reduced to fine particles. Log walls have covered all the climatic functions in Scandinavian dwellings for hundreds of years. The narrow joints between the logs are usually filled with moss. Timber is wind-proof, it is a good regulator of moisture and it has a useful insulation value even if it does not quite achieve present standards, which can be reached by adding a little extra insulation on the outside.

When timber is reduced to smaller particles, it has insulating qualities. Sawdust, shavings and woodwool are available from different types of timber and in different sizes. These can be used directly as compressed loose fill. In Sweden, Finland and inland Norway this was the most widespread form of insulation in framed building up to the 1950s. Loose fill can also be made into sheets by adding cement, magnesite or glue. It is possible to make insulation boards bound by the glue from the wood itself, e.g. wood fibreboards.

Cellulose can be produced from wood pulp for use in corrugated insulation board and paper to protect against damp and wind. Thermal insulation made of loose fill cellulose was patented for the first time in England in 1893. This was made of shredded recycled paper, preferably containing a fire retardant and impregnated against moisture. This method is very widespread today; around 1980 this covered about 30 per cent of the insulation used in Canada. Even in

Scandinavia this method is becoming very popular, especially as a way of recycling printed paper. Predecessors of this method, piles of old newspapers and magazines in walls and floors, often fall out of old houses when they are demolished. Newspaper has relatively good moisture-regulating properties and thermal insulation properties. It works well as insulation as long as it is not exposed to water or condensation as a result of settling or leakage into the wall. Untreated newspaper, is however, a fire risk.

Some tree barks can also be of use as climatic material. Bark from cork-oaks is very suitable for thermal insulation, as is the bark from birch, which has been one of the most important waterproofing materials throughout history, especially as an underlay for roofs covered with turf.

Tar extracted from coniferous and deciduous trees can be used for waterproofing and impregnation.

All timber materials even moisture in the structure and indoor air. Wood fibreboards have good wind-proofing properties. Cellulose fibre, when well compressed into a wall, can have a wind-proofing effect, but a wind-proofing system cannot be based on cellulose fibre alone, as the fibre may well settle after a while.

Woodwool, wood shavings and shredded porous wood fibreboards can be used as sealing around windows and doors. They are pushed in between the building frame and door or window frame in the same way as linen strips, for example.

Timber resources are renewable. Many products are based on waste such as sawdust and cellulose, which in many parts of Europe is often burned or dumped. Additives in some products have a bad environmental profile, e.g. boron salts in cellulose fibre insulation and glues in some boards.

The primary energy used varies from product to product, but it is generally much lower than similar products in other materials. Exceptions include wood fibreboards which require high process temperatures and woodwool slabs which use a lot of cement.

The problems of pollution through the different levels of production, usage and waste are relatively small, except for a few additives in certain products. Boron salts in cellulose fibre can pollute the soil and ground water if they are not taken care of properly as waste.

Timber-based climatic materials can be generally considered extremely durable and stable. Hardboard products should be re-usable. This is principally the same for cellulose fibre and sawdust, which can be sucked out and then compressed again in another situation.

With the exception of woodwool slabs and cellulose fibre with boron salts (fire retardant), all products can become an energy source through burning. Pure timber products can be burned without specific smoke-cleaning systems, or they can be made into compost.

Sawdust and wood shavings

Figure 14.12: Thermal insulation made out of sawdust.

Loose fill of sawdust and wood shavings is timber in its most pure form, and can be used in walls, floors and ceilings. Sawdust is a fine-particled, hygroscopic material which takes up moisture and releases it into the air in the same way as timber, but at a slightly higher rate. It also has the same resistance to fungus and insects as timber.

Experimental buildings investigated by Professor Bugge at the Norwegian Institute of Technology demonstrated that after 30 years the sawdust was in perfect condition, with no sign of any deterioration. The buildings stood on the very damp west coast of Norway (Granum, 1951).

Thermal insulation of compressed wood shavings

Sawdust is well dried before use as a wall filler, preferably to less than 20 per cent moisture. Up to 5 per cent of slaked lime can be added to stabilize the lime and reduce the possibility of insects getting in, also making it less attractive to mice and rats. Using quicklime produces a continual drying process, as the lime absorbs plenty of moisture during slaking. This can be a useful solution if the moisture content of the sawdust is greater than 15 per cent, but quicklime is highly corrosive and reacts with moisture, emitting a lot of heat. Larger quantities of quicklime can therefore lead to fire.

To reduce the risk of fire, sand or pulverized clay can be added in proportions of 1:2 and 1:1 respectively. This is approved as non-flammable fill for floor construction with a thickness of 10 cm. Adding sand reduces the thermal insulation value. Alternative fire-preventing materials are soda, borax and waterglass. Borax, or a mixture of borax and waterglass in a ratio of 1:1 is used in a proportion of 5–8 per cent. In small buildings the need for fire retardants is not so great. Experience has shown that damage due to fire in sawdust-insulated buildings is no more likely than in other timber buildings, partly because the sawdust, due to its low weight, does not develop temperatures as high as timber (Granum, 1951).

Both sawdust and wood shavings can be rammed into walls. Loosely filled sawdust often forms gaps in the insulation, so it should be rammed in hard by hand, making 25 cm layers of loose fill at a time. Because of settling, refilling with sawdust is necessary every 20 years. Wood shavings, which are slightly more elastic, do not need refilling so often. Special design details are required, e.g. under windows, to make refilling simple. It is also an advantage if the vertical spaces within the framework are full height, e.g. in balloon framing, (see 'Structural framework', p. •••).

Table 14.5: The use of timber climatic products in building

Material	*Composition*	*Areas of use*
Timber panelling[1]	Untreated timber	Balancing of relative humidity
Woodchip	Woodchip, possibly with lime, sand, magnesium chloride, waterglass, borax, ammonia polyphosphate	Thermal insulation, balancing of relative humidity
Cork	Cork oak which can also be mixed with bitumen or gelatine	Thermal insulation, balancing of relative humidity
Woodwool slabs	Wood strands bound with cement or magnesite	Thermal insulation, thermal storage, balancing of relative humidity, sound absorption, sound insulation
Porous fibreboard	Mass of wood fibres with paper with or without bitumen	Thermal insulation, wind-proofing, balancing of relative humidity, sound absorption
Hard fibreboard	Mass of wood fibres, can have bitumen coating	Vapour barrier
Cellulose fibre loose or matting	Cellulose with borax or boric acid, and/or aluminium hydroxide	Thermal insulation, balancing of relative humidity, wind-proofing
Building paper/cardboard	Cellulose, glue, in certain cases bitumen, silicone or latex	Thermal insulation, balancing of relative humidity, wind-proofing
Bark from birch	Pieces of bark from birch	Waterproofing, balancing of relative humidity

Note:
(1) See 'Timber Cladding', p. 344.

Table 14.6 shows that sawdust has a lower thermal insulation value, the less dense it is, whereas the situation is the exact opposite with wood shavings. Differences between the degrees of compression are so large that it would be advisable to carry out test stamping and weighing before starting work.

It is also possible to insulate thermally with ground sawdust otherwise used for the production of wood fibreboard and building board. This fine-particled material can be blown into the structure and can produce thermal insulation values equivalent to those of mineral wool and cellulose fibre, i.e. approximately 0.04 W/m°C. These products often contain ammonium polyphosphate, in a proportion of about 8 per cent, as a fire retardant. This is a relatively harmless chemical which is also used as an artificial fertilizer. As a waste product it has no pollution potential and can be used to improve the quality of soil.

Table 14.6: The insulation factors of sawdust and wood shavings

Material	*Weight (kg/m³)*	*Insulation factor (W/m°C)*
Compacted sawdust	200	0.081
Compacted sawdust	120	0.071
Sawdust/sand (2:1)	750	0.100
Wood shavings (3–5 cm)	80	0.120
Compressed wood shavings	130	0.080
Well compressed wood shavings	150	0.070
Very well compressed wood shavings	180	0.060

(Source: Granum, 1951).

Cork oak

Cork oak cannot grow in northern or central Europe and therefore must be imported from southern Europe, mainly Portugal and Spain. The bark has probably developed to withstand the frequent forest fires that occur around the Mediterranean. The trees are ripe for peeling after 25 years and can then be peeled every 8 to 15 years. The material is used as thin boarding or crumbled for thermal insulation. Cork is built up of dead cell combinations of cork cambium and resins. It is usual to expand the cork to increase its thermal insulation value. It is then pressed at a temperature of 250–300°C. The cork's own glue components are released and bind the board together. Today it is usual to bind the boards with a bituminous material, gelatine and another glue in a cold process. In addition to its use as a loose material for filling walls, cork can be used in concrete for cork concrete blocks. Cork products are resistant to fungus and not easily penetrated by liquids. The material is easily flammable and burns with great intensity and heavy smoke. The waste of products glued with bitumen has to be specially treated.

Woodwool cement

Woodwool cement is usually produced as boards in thicknesses of 2.5–15 cm, but can also be produced as structural blocks. The board is used for sound insulation, and thermal insulation. Reinforcing the thickest boards with round wooden battens produces a material with good structural properties.

Woodwool cement is resistant to rot. It has a weak alkaline content of about pH 8.5; mould needs a pH of 2.5–6 to develop. Woodwool can therefore be used as

Figure 14.13: Woodwool slabs reinforced with round rods combine high thermal insulation values with structural integrity.

Source: Gaia Lista, 1990

foundation wall insulation. The woodwool should be laid on the inside, because running water in the earth will wash away the cement in the long term. The sound insulation qualities, when it is not rendered, are very good, and the boards are suitable for use as acoustic cladding.

A woodwool slab that has been cast into concrete or rendered has a lower insulation value, because the surface spaces will be filled with mortar. The effective insulation value of the rendered woodwool slab is the same as a 1 cm thinner board which has not been rendered. Boards with finer woodwool have a better insulation value than those with a coarser surface.

Woodwool cement consists of 65 per cent cement (by weight). To evaluate this material environmentally, the role of cement must be considered (see 'Additives in cement', p. • •). It is also used as part of some sandwich boards, glued or heated together with layers of polystyrene, polyurethane, rockwool or foamglass. These products have high insulation values, but have to be carefully handled as waste if they contain plastic.

Pure woodwool cement products cannot be recycled as material or burned for energy recycling. Boards which are mechanically fixed to a surface can, in principle, be re-used. Waste is almost inert and can be used as loose fill.

Woodwool cement boards – production and use

Wood with too much tannic acid, such as oak, cannot be used. Spruce is best, preferably waterlogged, but even this can be unsuitable in parts because of large quantities of resin and sugar. Particularly unsuitable wood can be sorted out in the following manner: a piece of the wood is put in cement mortar. If it can be pulled out after two days, then the timber in question is unusable (Chittenden, 1975). Woodwool from a lime tree can also be used.

Figure 14.14: Dwellings built with woodwool blocks under construction in Italy.

A woodwool slab is made in the following way:

1. Timber is cut up into 50 cm lengths and planed to woodwool.

2. The active ingredients in the woodwool are neutralized. There are several methods for this: The cell contents can be washed out by boiling the wooden particles or the woodwool can be oxidized in fresh air for a year. As a final treatment the particles can have substances added which accelerate the setting of the concrete. Sodium silicate (waterglass), calcium chloride and magnesium chloride can be used. The wood is left to lie in a 3–5 per cent solution for a while.

3. Cement with less than 1 per cent aluminium sulphate is mixed with water in a mechanical or manual mixer.

4. The woodwool is poured into this and well mixed in.

5. The mixture is poured into moulds and pressure is applied while they set. At this point, wood reinforcement can be inserted to increase strength. This is often used for the thicker slabs.

6. After 24 hours the slabs are taken out and cured for two to four weeks before being sold.

The slabs can be nailed, screwed or cast into place. The joints should be covered with a strip of netting if cement mortar is to be applied.

The slabs can also be cast with magnesite mortar with a little magnesium sulphate added, but these products are less resistant to moisture than cement products, and cannot be used as insulation for foundation walls. Magnesite boards were on the market much earlier than woodwool cement; they were first manufactured in Austria in 1914 and are still in production under the name of Heraklit.

Wood fibre boards

The manufacture of wood fibre boards is described in the following chapter. The porous products, softboards, are used for wind-proofing and have bitumen added in a proportion of approximately 12 per cent by weight. Hardboards are used for internal resistance to vapour and as a waterproof membrane under roofing (exterior), the latter usually impregnated with bitumen in a proportion of 5.5 per cent by weight. The normal thickness for softboards is 12 mm, but it is possible to manufacture thicker and lighter boards. With thicker boards, drying out is a problem in the wet production process. Hardboards used as climatic products come in thicknesses of 3–5 mm.

These products have a relatively high use of primary energy. As waste, the products containing bitumen have to be specially disposed of.

Cellulose fibre

Cellulose fibre consists of torn-up recycled paper or pulverized pulp. The fibre is treated with fire retardants and is used on site as loose fill. The proportion of chemical additives is as high as 18-25 per cent. These are partly fire retardant, partly to hinder mould and partly binder. The most commonly used compound is boric acid and borax. The fibre also contains traces of silica, sulphur and calcium from fillers used in the newspaper.

More recently, cellulose fibre matting has been manufactured using pure cellulose glue. Cellulose strips have also been manufactured for filling the space between window and door-frames and the building fabric. Building mats and filling strips are made from fresh cellulose fibre, and their production requires tree felling and a higher use of primary energy.

Loose fill cellulose fibre has been used as building insulation since the 1920s, and the material's durability is good as long as it has been placed in the walls or roof space in the correct way. This involves applying a high pressure when blowing in the fibre, to avoid settling later on.

Figure 14.15: Cellulose fibre.

In the production process workers can be exposed to dust made up of paper and fine particles of boric salts. There are no records of serious dangers from breathing in dust from paper, but it is generally advisable to be careful with very fine-particled dust because of its potential to irritate the lungs. Exposure to dust can occur at all stages from production to installation on site, but once installed correctly the fibre should cause no problems for those using the building. Cellulose fibre products have good moisture-regulating properties and are much less susceptible to mould than the mineral wool alternatives.

The products can be re-used and recycled, but cannot be burnt for energy recycling because of their fire-retardant nature. As waste boric salts and printing ink can seep into the earth or ground water. The effluent also contains eutrophicating substances which require special waste disposal.

Cellulose paper and boards

Cellulose building paper is usually manufactured from recycled paper and unbleached sulphite cellulose. It can also contain up to 20 per cent pulp. Boards are manufactured by laminating the sheets of paper together to 2–3 mm thickness, with PVAC glue (about 3 per cent by weight).

Cellulose building paper is used for covering joints, sound insulation in internal walls and for surrounding loose fill insulation. The boards are used for weather-proofing and are usually covered with black polyethylene on a moisture-resistant coating of natural latex. Thermal insulation panels are also made using sheets of corrugated cardboard which are laminated to different thicknesses. These were very popular between 1945 and 1950, and were often impregnated with bitumen to prevent damp.

The basic raw material of these products is environmentally positive, ignoring the consequences of the relatively small amounts of polyethylene, PVAC glue and bitumen. The same can be said of the manufacturing process except for the production of sulphite cellulose; depending upon the factory's cleaning technology, this can release huge amounts of eutrophicating substances. With the exception of bituminous products, they are relatively free of problems once in the building.

Durability is relatively good. Pure cellulose paper and laminated weather-proofing boards (with natural latex) can most probably be recycled into new cellulose products. The other products are best suited as a low quality cellulose fill in asphalt, etc. The materials can be burned for energy recycling, provided that the effluent gases are filtered from products containing plastic. Bituminous products that are not burned have to be safely disposed of, as do those containing small amounts of plastic. Dumping cellulose products will lead to an increased level of nutrition in the water coming from the area. Pure cellulose should be composted under controlled conditions.

Birch bark

The bark of birch trees has been widely used as a waterproof membrane under turf roofs. It has to be kept permanently damp to prevent it cracking. The pieces of bark were taken from large birches, conveniently known as roof birches. A roof had between three and twenty layers of bark, depending upon the required durability. The bark is very resistant to rot and can be used as waterproofing in other potentially damp areas, e.g. foundation walls. Because it prevents damp and spreads moisture evenly, it is better than asphalt paper for protecting built-in beams. During the rebuilding of the Church of St Katarina in Stockholm during the early 1990s, 300-year-old birch bark was found at the end of inbuilt beams. They were exceptionally well-preserved. The same method was therefore used in the rebuilding. In 1948 the Danish engineer Axel Jörgensen wrote: 'Building traders should set up an import of birch bark from Sweden or Finland, so that we could once again use this excellent protective medium' (Jörgensen, 1948).

Bark should be removed as carefully as possible, so as not to damage the tree's layers. The tree can then continue to grow, though it may not produce more building-quality bark. Bark is loosest during spring, and the best time to take the bark is after a thunderstorm (Høeg, 1974). Bark has also been used as insulation in walls, especially cavity walls, where its considerable resistance to rot and its high elasticity produces a stable wall.

Peat and grass materials

Many peat and grass species have considerable potential as climatic materials, for thermal insulation and air and moisture regulation. Loose fill, boards, blocks and matting of bog peat and straw represent good thermal insulation materials. Many types of plants have good moisture-regulating properties, and some even have a high resistance to rot, such as flax, jute and moss.

Plant products often make suitable thermal insulation because, in a dried state, they contain air and have a stable structure that deters settling. In the case of straw, fibres or stalks are used after the leaves have been removed. Eelgrass, lichen, moss and peat can be used in a dried state. Parts of cocoa and maize plants contain cellulose, which makes good building materials.

Table 14.7: The use of peat and grass climatic materials in building

Material	*Composition*	*Areas of use*
Living turf[1]	Grass in soil	Thermal insulation, sound insulation, roof covering
Grass fibre loose fill	Straw (can be stabilized with clay)	Thermal insulation, balancing of relative humidity, wind-proofing of joints
Grass fibre bales	Straw baled and tied together with hemp (can be impregnated with waterglass and rendered with hydraulic lime render)	Thermal insulation for houses or temporary structures, balancing of relative humidity
Grass fibre matting	Straw fixed to galvanized netting, sewn into paper or pinned together	Thermal insulation, balancing of relative humidity, wind-proofing of joints
Grass fibreboards	Straw, possibly with glue and impregnated (can have outer layer of cellulose paper)	Thermal insulation, balancing of relative humidity
Loose peat fibre	Peat (lime can be added or other impregnating materials)	Thermal insulation, waterproofing, balancing of relative humidity, sound insulation, wind-proofing of joints
Peat fibre matting	Peat sewn into paper (impregnating materials can be added)	Thermal insulation, balancing of relative humidity
Peat fibreboards	Peat	Thermal insulation, balancing of relative humidity, sound absorption

Note:
(1) For more details see 'Turf Roof', p. 328.

For moisture-regulating and wind-proofing purposes these materials are usually used as fill in the gaps between windows, doors and the building fabric. It is therefore important for them to be resistant to rot and packed tightly. This is a critical part of the structure and needs high durability. Materials used in this way are flax, hemp, peat and fibres from nettles.

Climatic materials based on plants are very interesting, ecologically speaking. The insulation sector is particularly interesting because it represents such a large volume of material, and it would be to great advantage if this could be covered by renewable resources. With few exceptions, plants grown in the majority of European countries would be suitable.

Plant materials have no problems in relation to the indoor climate, and they often have good moisture-regulating properties. Impregnated and glued products should be avoided. Pure products can be either burned for energy recycling or composted when they have served their time in the building. Ordinary disposal can lead to increased nutrients in waste water which seeps into the surroundings. Certain jute products used for sealing joints are impregnated before transport.

Grass plants

Many different types of grass can be used as an insulation material in the form of loose fill, bales, matting or boards: e.g. wheat, rye, barley, oats, hemp, maize, reed and flax. Further south in Europe, straw is a more common roof covering. A straw roof has good thermal insulation and moisture-regulating qualities. Straw roofs are discussed in the following chapter, p. 356.

Loose fill

This is pressed into the structure with lime added to repel vermin. Flax and hemp have a very high resistance to rot. Straw from corn rye is the most resistant to moisture. However, the durability of corn-based materials as insulation is relatively limited. Straw stabilized with clay is a better material. It prevents settling, increases alkalinity and improves resistance to rot.

'Leichtlehm'

During the 1920s in Germany a building technique called 'Leichtlehm' was developed. Leichtlehm is not structural and needs a separate structural system. A mix of straw and clay is rammed directly into the wall or produced as blocks, which can later be built up with a clay mortar. Straw mixed with clay needs a good protective surface treatment, and is given an extra skin for protection on very exposed sites.

Leichtlehm is produced as follows:

1. All clays can be used. The clay is dried and crushed and poured into a large tub (often a bath tub!) of about 200–300 litres, ten times as much water is added and mixed well in. A motorized mixer can be used or the work can be done by hand. About 2 per cent soda waterglass is added to reduce the surface tension, so that the water can more easily penetrate the clay particles. This reduces the amount of water required and makes the drying time shorter.

The clay should lie in the water for two hours. If using wet clay, it should be laid in water so that it is just covered and left for 24 hours.

2. The mixture is tested: 1 dl of the mixed clay gruel is poured evenly onto a piece of glass. The diameter is measured. If it is much less than 15 cm, it needs more water. If it is much more, than it needs more clay.

3. The clay gruel is poured onto the straw until it is totally drenched. Any type of straw can be used, but rye is best. The stalks are stiffer and thicker than most others, so the greatest amount of air is retained in the walls and therefore the best insulation.

4. The mixture is put into moulds to form blocks or rammed into simple moveable shuttering on either side of a timber frame wall, 30–60 cm thick. The mixture must not be rammed hard. The middle is pushed down with the foot, while the edges are given a stronger pressure: they can be beaten down with a piece of wood. The more compressed the mixture is, the stronger the wall, but with a corresponding reduction in thermal insulation.

The different layers need to overlap each other when rammed within the shuttering. The holes left after removing the shuttering are filled with clay. Before ramming, the timber framework – the structural part of the wall – is covered with clay as a sort of impregnation. The drying time during the summer is between six and eight weeks, depending on the weather.

Figure 14.16: Wall construction in 'Leichtlehm'.

The fibres used to fill the joints between windows, doors and the timber framework must have a strong resistance to rot. The most suitable fibres are flax and hemp.

Straw bales

It is also possible to use bales of straw stacked on top of each other as thermal insulation. The size of a bale of straw is usually 35 × 35 × 60 cm, and it weighs about 20 kg, but both the dimensions and the weight vary somewhat, depending upon the baling equipment and the pressure used to put it together. Hard-pressed bales can even have a structural capacity. Building with straw bales was very popular in the USA until after the Second World War. They were used for everything from schools to aircraft hangars. The structure is usually placed on a damp-proof course on the foundation. The bales of straw must be properly compressed and dry (10–16 per cent) with no sign of mould or rot. They are stacked up on each other and coursed like normal brickwork. Between the courses, 70 cm-long stakes are pushed into hold them together. Extra reinforcement is used at the corners, against the openings, etc. After two to four weeks the walls are clad with chicken wire and rendered

Figure 14.17: Straw bale building. Source: Howard Liddell

with hydraulic lime render. The walls can be rendered direct with three or four layers of a clay-based render, mixed with cow dung and even hacked straw. It is also possible to use the same treatment on the inside, giving a very smooth surface. Rendered straw structures are non-flammable.

Plain straw bales can be made fire- and rot-resistant by dipping in a solution of 5 per cent waterglass, thin runny clay or lime gruel. On exceptionally exposed sites, rendered surfaces must be protected by an extra outer skin such as timber panelling.

During the 1980s, when a 75-year-old school built of straw was demolished in Nebraska, the straw was undamaged and fresh enough to be used as cow fodder. Such relatively unexpected experiences have led to a renaissance of straw bale building in both Canada and the USA, and in recent years straw bale building has begun in Europe, mainly in France.

Matting

Matting can be produced by binding fibres or stalks together with galvanized wire or by gluing or by 'pinning' them together. The latter method is used in the production of linen mats. The flax fibres are beaten into soft strands, then mixed with waterglass and boron. They are then filtered together on a special brush of nails to make them into an airy, effective insulating mat of various thicknesses. Denser felt products and strips for sealing joints are produced in the same way. Similar products based on hemp-fibre are in the pipeline. Stinging nettles can probably be treated in the same way.

One traditionally much older building material, reeds, can also be bound together with galvanized wire. Reed mats are used as thermal insulation and as reinforcement in concrete walls and prefabricated units. When rendered, the mats can also be used as false ceilings or a base for infill of party floor construction.

Strawboards

Straw is laid in a mould with the stalks lying at right angles to the direction of the board, forming the width of the board. They are then exposed to pressure and heat. This causes the straw to release its own form of glue that binds the whole board together. Porous boards have a thermal insulation equivalent to woodwool slabs. Under damp conditions they will be exposed to attack by fungus. Straw boards can also be produced as hardboards (see 'Production of straw boards', p. 359).

The first insulation boards made of straw were produced as early as the 1930s. They were made in thicknesses of 5–7 cm, under low pressure, reinforced with crosswires and covered with paper.

Flax boards are made of flax fibres boiled under pressure for several hours. The material is highly durable and non-flammable, and is used in some fire doors.

Linseed oil putty

Linseed oil comes from the seeds of the flax plant. Putty is a product of the working of a mixture of linseed oil and stone flour, such as chalk, heavy spar, powdered fired clay, powdered glass, etc.

Linseed oil putty is the only alternative to plastic-based mastics and window putties. It is environmentally much sounder than the alternatives, with no negative effects during production or use. As a waste product, it can be used in fill as long as no additives (e.g. lead) have been mixed in, to improve its elasticity. The elastic qualities of the putty can be preserved for a long period by painting with oil-based paint. Despite this the putty will eventually harden and begin to crumble. Linseed oil based putty must not be used in contact with damp lime or cement surfaces.

Bogpeat

Peat has been used a great deal as an insulating material and moisture regulator in its natural state or as a loose material, granules, mats or boards. In the past it has been used in Germany, Ireland, Scandinavia and Scotland. Today, insulation products of peat are again being produced in Sweden.

Peat usually consists of decayed brushwood, plants from marshes, algae and moss. For building, the most important moss is found in the upper light layer of a bog and has not been composted. Older, more composted peat can be used in certain circumstances, but it has a much lower insulation value. Totally black peat is unusable.

Peat is a good sound insulator, because of its weight, and could in many cases replace heavier alternatives such as sand in floor construction.

The same pigment substances that are in our skin are also found in peat. It will therefore probably protect us from certain frequencies of electromagnetic radiation. Matting made of peat can filter and absorb emissions of radon from building materials and foundations.

A peat-bog can contain many different sorts of moss, but this does not matter as far as insulation quality is concerned. Moss can be used as a sealing material between logs, for example, and for sealing joints between doors, windows and the building fabric.

There is very little risk of insect and fungus attack in dry peat, as long as it is not built into a damp construction. Peat has small quantities of natural impregnating toxins such as alcohol. It also has a low pH value (3.5–4), so it retards the spread of bacteria and protects against fungus. In certain products, however, it is still advisable to use impregnation.

Peat natural blocks

Peat consisting of moss can be collected from the bog as blocks and used as thermal insulation. It is dried and easily trimmed by sawing for building in walls. In framework and brick cavity walls the peat is built up with lime mortar in the joints. The acids in peat will attack the usual cement mortar, but this method can be used with sulphur concrete (see 'Sulphur concrete', p. 196).

Figure 14.18: Suspended ceiling insulated with compressed peat fibre constructed in Sweden in 1993.

Peat loose fibres

Peat fibres can be used as loose thermal insulation in floor construction and walls and is made from dried, ground peat with a little lime added, about 5 per cent. This is blown into the structure the same way as cellulose insulation.

As late as the 1950s in Scandinavia it was usual to insulate simple factory buildings with peat in empty concrete sacks. The leaves of the wall are built up on either side of the 'sack-wall'. Internal lining is superfluous. The sacks make sure that the organic acids in the peat do not come into contact with the concrete before it has set. Peat placed in the wall in this way is liable to settle (see Figure 14.19).

Peat as external waterproofing

A special form of denser bogpeat, rose-peat, has been used a great deal as the sealing material in dams. It is probably suitable as a moisture barrier for foundation walls. Its sealing ability is due to the fact that it can absorb and hold large quantities of water.

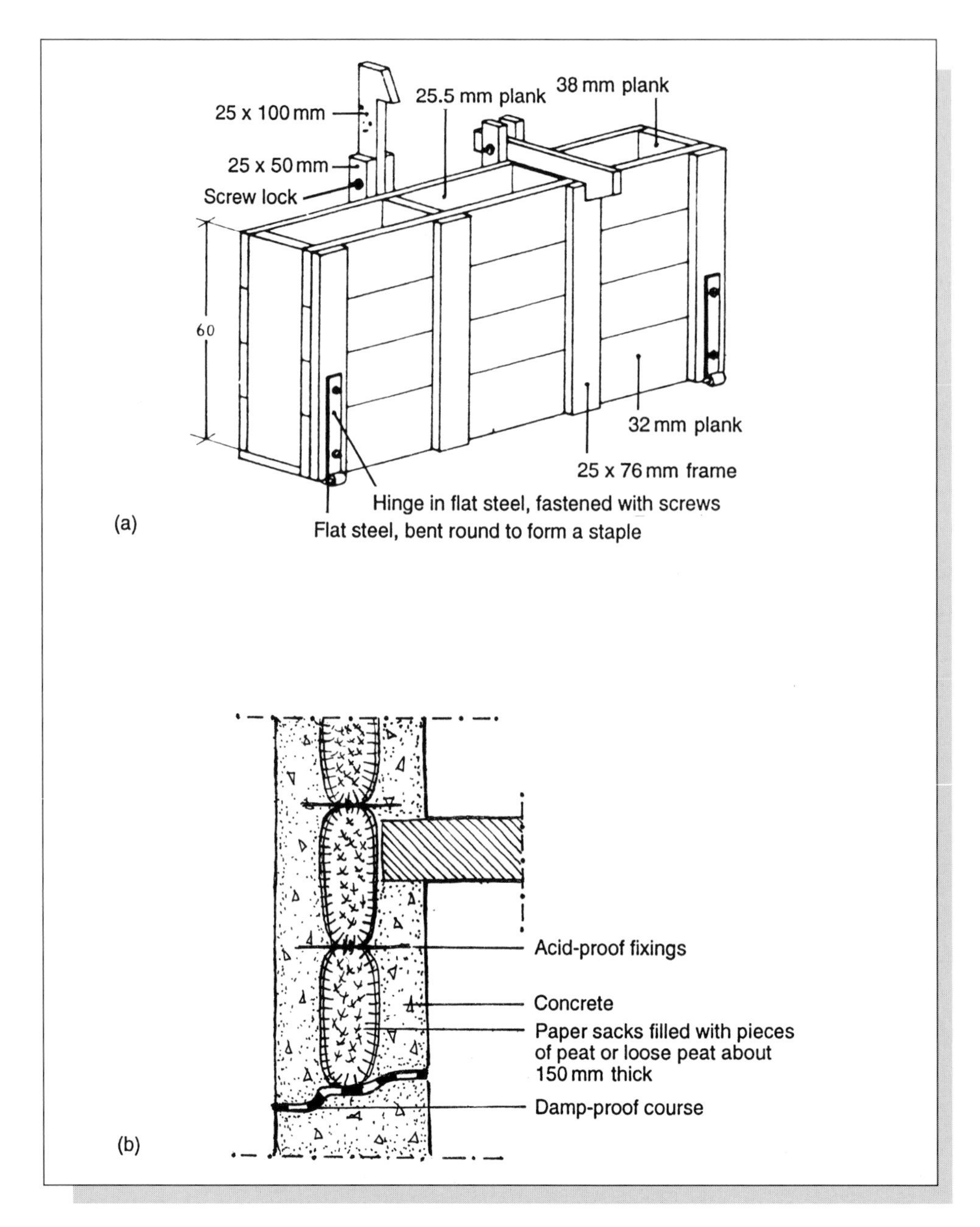

Figure 14.19: The 'sack-wall' made of cement sacks filled with peat: (a) constructing a box to hold the sacks; (b) building them in. Source: Haaland 1943

This peat, which is dark brown and available in most bogs, consists mainly of rotten leaves, and is found below the level of the roots in the bog. The plant fibres have to be visible, but the structure broken down. Rose-peat has to be free from roots and branches. When a piece

Figure 14.20: Log wall sealed with moss.

is rubbed between the fingers, it leaves a thick fatty, layer on the skin, a bit like butter. If the peat contains a lot of fibre, it feels rough. If it contains too little fibre, it feels smooth, like soap. The peat is cut out in cubes of 12 × 12 × 12 cm, often going down several layers before water fills the hole.

This special peat must not dry out and should be used as quickly as possible, but its properties can be preserved for up to a week in damp weather by covering it over with leaves and pine needles. It must be used in frost-free situations.

Extraction is simple, but somewhat heavy. The use of energy is minimal and its durability high. At the silver mines of Kongsberg in central Norway, dams of this peat are still watertight after being in use for 100 years.

Peat matting

This consists of peat fibres sewn into paper.

Peat boards

Peat boards are made in thicknesses of 20–170 mm. Their thermal insulation is very good and can compete with mineral wool or cellulose fibre. The most widespread method of production begins with the peat being taken to a drying plant where it is mixed and placed in warm water. It is then removed from the water, which is allowed to run off, leaving a moisture content of about 87–90 per cent. The mass is then put into a mould in a drying kiln to dry up to 4–5 per cent. To achieve different densities, different pressures can be applied. The whole process takes about 30 hours.

A dry production method can also be used. The peat is pressed into moulds so that the damp is driven out of it. By warming it to 120–150°C with no air, its own binders and impregnating substances are released. This is equivalent to its charring temperature, so the boards become fire resistant. There is also no need for added binders. The material has a strong resistance to fungus and insect attack. Its absorption of water is very low, and its stable moisture content is around 10 per cent. As the contents of the board are stable, there is no chance of it settling. This dry method of pressing came into use between 1935 and 1940 in the former Soviet Union. The method requires a relatively large amount of energy for the drying and setting processes, but this can be reduced to a certain extent by using solar energy for the warming process.

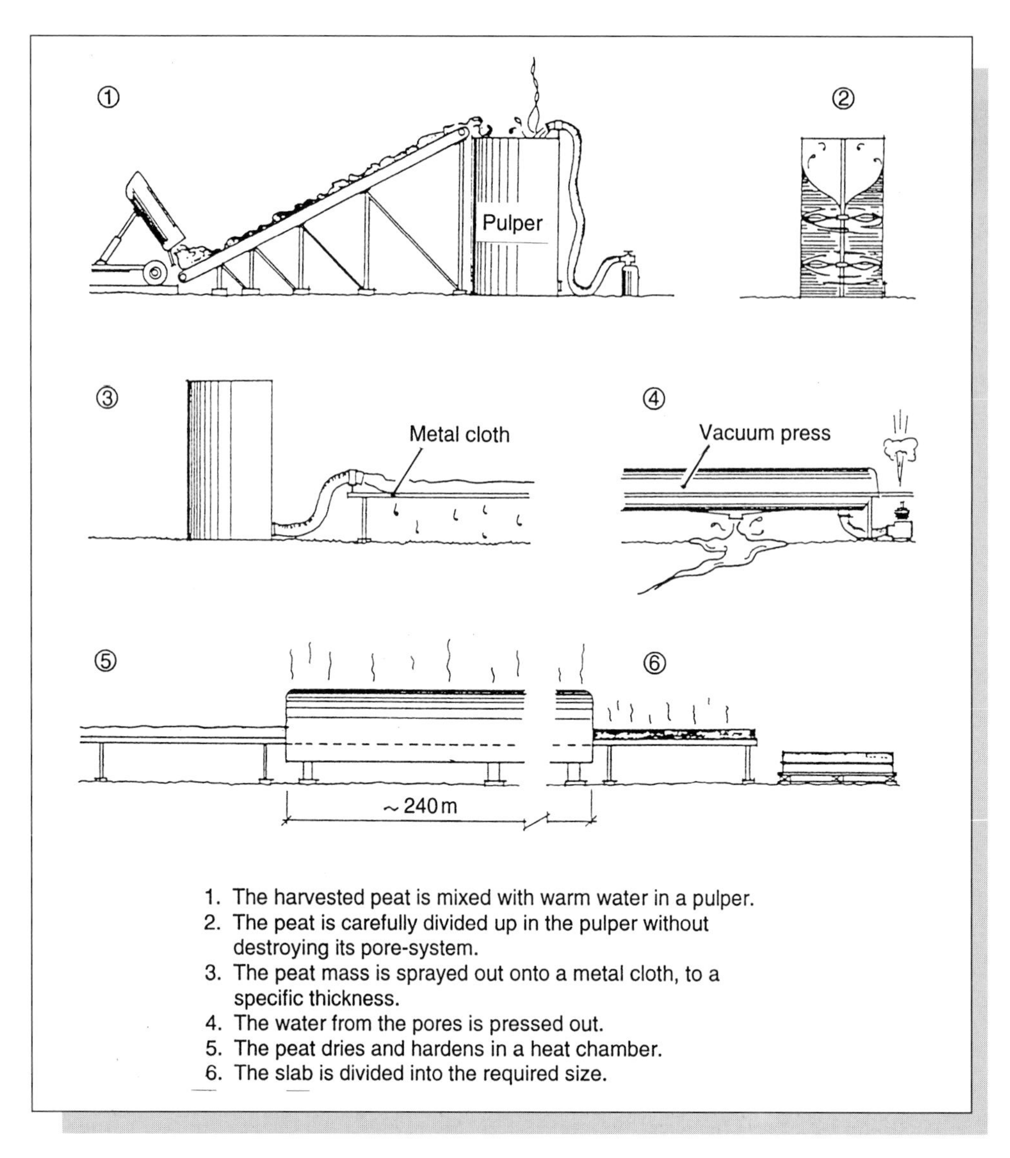

Figure 14.21: Pressing peat slabs using the widespread wet production method. Source: Brännström 1985

Moss

Moss has been used to seal the joints in log buildings for hundreds of years; between the logs, around doors and windows and in other gaps. The moss has to be put into all the gaps as soon as it has been picked, because it hardens and

Table 14.8: Climatic materials from animal products

Material	*Composition*	*Areas of use*
Woollen loose fill	Wool (can be impregnated against moths)	Sealant, thermal insulation
Woollen matting	Wool (can be impregnated against moths)	Thermal insulation
Woollen sheeting	Wool (can be mixed with hair from other animals, plant fibres and impregnated against moths)	Sound insulation against impact noise, thermal insulation, sealant
Woollen building paper	Wool (can be mixed with recycled paper)	Sound insulation against impact noise, balancing of relative humidity, covering of loose insulation
Asphalt paper	Woollen sheeting and bitumen	Roofing felt, sarking

loses its elasticity as it dries out. It can be boiled before use to reduce the amount of substances subject to attack from micro-organisms.

Moss stops air penetration when compressed, and prevents moisture penetration, as it is very hygroscopic and able to swell. It can absorb large amounts of damp without reaching the critical value for the materials next to the moss. These properties make moss useful as a substance that can absorb or regulate moisture on external walls.

There are two types of moss: *Hylocomium splendens* and *Rhytriadiadelphus squarrosum*. The latter is considered the best, as it can last up to 200 to 300 years as a sealing material in a log wall without losing its main functional properties. Sphagnum is a less durable moss.

Materials based on animal products

Climatic materials obtained from animals are hair, wool and hide. Reindeer skins have been widely used as insulation, especially amongst the Lapps. Animal fibres are high-quality thermal insulators and very good moisture-regulators.

The most widespread use of animals today is as the main ingredient for wool-based building papers and as thermal matting. It has also been used as an underlay for internal rendering and as sealing for joints.

Woollen matting competes with mineral wool as thermal insulation. The products often contain boric acid to prevent insect attack. In some cases, questionable

Table 14.9: Environmental profiles of thermal insulation

Material	*Specific thermal conductivity (W/mK)*	*Specific thermal capacity (kJ/kgK)*	*Quantity of materials used (kg/m² thermal resistance R = 3.75)*
Still air	0.024	1.0	
Water	0.50	1.9	
Dry snow	0.06–0.47		
Expanded perlite, untreated, 170 mm	0.045–0.055	3–4	13.5
Expanded perlite, with bitumen, 190 mm	0.055	3–4	15
Lightweight aggregate concrete blockwork (structural), 750 mm	0.210	1	560
Aerated concrete blockwork (structural), 400 mm	0.08	1	200
Foamglass boards, 170 mm	0.045	1.1	21
Foamglass granules, 350 mm	0.07	1	50
Mineral wool, 150 mm	0.04	0.8	3
Expanded clay pellets 430 mm	0.115		194
Expanded polyurethane 135 mm	0.035	1.5	3.8
Expanded and extruded polystyrene 150 mm	0.04	1.5	3.4
Expanded ureaformaldehyde, 180 mm	0.05	1.5	5
Compressed wood cuttings 200 mm	0.05–0.09	1.8	24
Porous fibreboard, unimpregnated, 200 mm	0.05	1.8	60
Wood wool slabs, 300 mm	0.08	1.9	69
Cellulose fibre, loose, 170 mm	0.045	approx 1.8	10.1
Cellulose fibre, matting, 150 mm	0.04	approx 1.8	11
Flaxen matting, 150 mm	0.04	approx 1.8	2.4
Slabs of peat, 150 mm	0.04	1.2	15
Straw bound together with clay, straw >100 kg/m³, 550 mm	0.12	1.2	330
Woollen matting, 150 mm	0.04	approx 1.8	3

Notes:
(1) This material also acts as a structural material, so no extra structure is needed.
(2) Thermal insulation varies a great deal with the different types of wood shavings/cuttings.
(3) If insecticide is added, much more care must be taken when this becomes waste.

chemicals are used for impregnation. To increase elasticity, polyester fibres are added to some products.

The more dense felt products usually consist of wool, but can also contain hair from cows and different plant fibres to keep the price reasonably low. Felt is used as sound insulation between floor joists and as thermal insulation around water pipes. It is also used, to a certain extent, as sealing around windows.

Wool-based building paper consists of a good deal of recycled paper, but the woollen content must not be lower than 15 per cent. Wool building paper is soft

Effects on resources			*Effects of pollution*				*Ecological potential*		*Environ-mental profile*
Materials	*Energy*	*Water*	*Extraction and production*	*Building site*	*In the building*	*As waste*	*Re-use and recycling*	*Local production*	
									1
1	2		2	2	1	1			1
2	2		2	1	2	3			2
3	3	2	2	1	1	1	✓		3(1)
2	3	2	2	1	1	1	✓		2(1)
2	3	2	3	1	1	1			2
1	2		1	1	1	1			1
2	2	2	2	2	2	2	✓		2
1	3		2	1	1	1	✓		2
3	3	3	3	1	3	3			3
3	3		3	1	2	3	✓		3
3	3		3	3	3	3			3
1	1	1	1	1	1	1		✓	1(2)
1	3	2	2	1	1	1			2
2	3	3	2	1	1	2	✓		2(1)
1	1	1	1	2	1	3		✓	2
1	2		2	1	1	3			2
1	1		1	1	1	1			1
1	2		1	1	1	1			1
1	1		1	1	1	1		✓	1
1	1		1	1	1	1(3)			1

and porous and is often used as floor insulation against impact noise. Wool is also used to make woven sealing strips known as 'textile strips'. They are relatively strong, but become hard when painted. Damp can cause them to shrink and loosen from their position in a building.

Wool is broken down at a temperature of over 100°C, by fat, rust, petroleum, alkalis and oil. The material does not burn, but smoulders when exposed to fire. The raw material for most woollen products is rejected wool from slaughterhouses, which would otherwise be thrown away. Woollen products can be considered problem-free as far as production and use is concerned. The use of

Table 14.10: Environmental profiles of joint filler

Material	*Quantity of material used* (kg/m^3)	*Effects on resources*			*Pollution effects*				*Ecological potential*		*Environmental profile*
		Materials	*Energy*	*Water*	*Extraction and production*	*Building site*	*In the building*	*As waste*	*Re-use and recycling*	*Local production*	
Mineral wool	20	2	2	2	2	2	2	2			2
Foamed polyurethane	35	3	3	3	3	3	3	3			3
Cellulose strips	150	1	2		2	1	1	2			1
Flax strips	150	1	1	1	1	1	1	1			1
Jute strips	100	1	2(1)		2(2)	1	2(2)	2(2)			2(2)
Coconut fibre strips	100	1	2(1)		1	1	1	1			1

Notes:
(1) Long transport routes from country of origin.
(2) The product is here assumed to be treated with fungicide. Without this, its position would be better.

Table 14.11: Environmental profiles of vapour barriers

Material	*Water vapour penetration (mg/m^2h Pa)*	*Quantity of material used (kg/m^2)*	*Effects on resources*			*Pollution effects*				*Ecological potential*		*Environ-mental profile*
			Materials	*Energy*	*Water*	*Extraction and production*	*Building site*	*In the building*	*As waste*	*Re-use and recycling*	*Local production*	
Plasterboard[1], 12 mm	10.5–14.2	12	3	3	2	3	1	1	2			3
Polyethylene sheeting, 0.15 mm	0.01	0.14	3	2		2	1	2	2			2
Polyisobutylene sheeting, 0.5 mm	0.004	0.5	3	3			1	2	2			3
PVC-sheeting, 1.0 mm	0.07	1.3	3	3		3	2	3	3			3
Hardboard[1], 3 mm	1.8	3.1	2	3	3	2	1	1	1	✓		2
Cellulose building paper[1], 0.5 mm	21.0	0.5	1	1	1	1	1	1	1			1
Cellulose building paper with aluminium lining, 0.5 mm	0.005	0.5	3	3	2	3	1	2	3			3

Note:
(1) Can only be used with insulation material that has good damp regulating properties.

Table 14.12: Environmental profiles of wind checks

Material	*Air penetration (m^3/m^2h Pa)*	*Quantity of material used (kg/m^2)*	*Effects on resources*			*Pollution effects*				*Ecological potential*		*Environ-mental profile*
			Materials	*Energy*	*Water*	*Extraction and production*	*Building site*	*In the building*	*As waste*	*Re-use and recycling*	*Local production*	
Plasterboard[1] with silicon, 9 mm	0.0006	9	2	3	3	3	1	1	2			2
Polyethylene sheeting, 0.15 mm	0.02	0.2	3	1		2	1	1	2			2
Polypropylene sheeting, 0.15 mm	0.017	0.2	3	1		2	1	1	2			2
Porous fibreboard[1] impregnated with bitumen, 12 mm	0.001–0.01	4.2	3	3	3	3	1	2	3			3
Cellulose building paper with bitumen, 0.5 mm	0.003–0.008	0.5	2	2	2	1	1	2	3			2
Laminated card[1] boarding, 2 mm	0.001	1.5	1	2	1	1	1	1	1	✓		1

Note:
(1) These products also have a structural function of windbracing.

Table 14.13: Environmental profiles of waterproofing membranes[2]

Material	Quantity of material used (kg/m^2)	Effects on resources			Pollution effects				Ecological potential		Environmental profile
		Materials	Energy	Water	Extraction and production	Building site	In the building	As waste	Re-use and recycling	Local production	
Glassfibre sheeting with bitumen	1.9	3	3	2	3	2	2	3			3
Bentonite clay[1]	4.8	1	1	1	1	1	1	1			1
Bitumen applied direct	3	3	2	2	3	3	3	3			3
PVC-sheeting	1.3	3	3		3	1	3	3			3
Polyethylene sheeting	0.6	2	2		2	1	1	2			2
Polypropylene sheeting	0.6	2	2		2	1	1	2			2
Polyester sheeting with bitumen	2	3	3	2	3	2	2	3			3
Wool-based sheeting with bitumen	1	3	3	2	3	2	2	3			3

Notes:
(1) Used for waterproofing tunnels, cellars and foundations.
(2) All materials cause high levels of environmental damage. The use should therefore generally be reduced, and water penetration should be hindered using other methods such as placing the bathroom on the ground floor and avoiding balconies on roofs of houses.

Table 14.14: Environmental profiles of external detailing materials[1]

Material	*Quantity of material used (kg/m²)*	*Effects on resources*			*Pollution effects*				*Ecological potential*		*Environ-mental profile*
		Materials	*Energy*	*Water*	*Extraction and production*	*Building site*	*In the building*	*As waste*	*Re-use and recycling*	*Local production*	
Stainless steel from ore	3.9	3	1	2	2	1	2	2[2]	✓		1
Galvanized steel from ore	5.2	3	1	2	2	1	2	2[2]	✓		1
Aluminium, 50% material recycling	2.4	1	3	3	2	1	2	1[2]	✓		2
Copper from ore	5.3	2	2	3	1	1	2	2	✓		1
Lead from ore	17	3	1	1	3	2	3	3	✓		3
Polyvinylchloride	2.7	2	2		2	1	1	2[3]	✓		2

Notes:

(1) All these materials have very negative enviromental effects. Saving of material has a greater positive effect than the choice of certain materials.

(2) These may have a surface treatment of varnish or paint, which causes a higher pollutiuon risk when dumped

(3) Colour pigments are added which have a strong influence on the pollution risk.

Table 14.15: Technical properties of secondary climatic materials

Material	*Technical properties* *Specific thermal conductivity (W/mK)*	*Specific thermal capacity (kJ/kg K)*	*Specific vapour penetration*[1] *(mg/m²h Pa)*
Metals:			
Steel	58	0.4	Vapour-proof
Non-metallic minerals:			
Lime sandstone	0.7	0.88	0.6
Lime render	0.9	0.96	0.4–1
Lime cement render	1.05		0.2–0.5
Cement render	1.15	0.92	0.03–0.4
Concrete	1.75	0.92	0.03–0.4
Stone:			
Granite	3.5	0.8	
Limestone	2.9	0.88	
Brick:			
Light/medium fired	0.65	0.92	1.4
Well-fired: solid	0.7	0.92	
perforated	0.6	0.92	
Earth (pisé and adobe):			
With fibre 10 kg/m³	0.96	1	Approx. 1.5
With fibre 40 kg/m³	0.615	1	Approx. 2
With fibre 70 kg/m³	0.420	1.05	Approx. 2
Timber:			
Pine/spruce perpendicular to the fibres	0.12	2.6	0.05–0.09[2]
Parallel with the fibres	0.35	2.6	
Oak/beech perpendicular to the fibres	0.165		
Parallel with the fibres	0.35		

Notes:
(1) This is the vapour penetration through a 1 mm-thick layer.
(2) A 15 mm thick piece of spruce panelling has a vapour penetration of 0.35 mg/m²h Pa.

poisonous additives, however, makes it necessary to dump the waste at special depots. The pure products can and should be composted, as normal dumping will lead to increased nutrients seeping into the ground water from the tip.

Materials based on recycled textiles

Products have been introduced in recent years based on unspecified recycled plastic-based and natural fibres. Melted down fibres of polyester are added (12–15 per cent by weight) to produce mats for thermal insulation.

The raw materials used are environmentally interesting, even if this is compromised somewhat by the fact that the added polyester is an oil-based product. In the building there is a possibility of emissions of the remaining monomer styrene. The material can probably be recycled into the same product again, but as waste it has to be specially disposed of.

Environmental profiles

Tables 14.9 to 14.15 are organized in the same way as the environmental profiles in Table 13.5 in the previous chapter.

References

BAKKE J V, *Mineralull og innemiljø*, Norsk Tidsskrift for Arbeidsmedisin nr. 13, Oslo 1972

BRÄNNSTRÖM H *et al*, *Torv och spon som isolermaterial*, Byggforskn. R140:1985

BROCH T, *Lærebog i bygningskunsten*, Christiania 1848

CHITTENDEN A E, *Wood cement systems*, FAO Doc no. 99, New Dehli 1975

FOSSDAL S, *Energi og miljøregnskap for bygg*, NBI, Oslo 1995

GRANUM H, *Sagflis og kutterflis som isolasjonsmateriale i hus*, NTI, Oslo 1951

GUSTAFSSON H, *Kemisk emission från byggnadsmaterial*, Statens Provningsanstalt, Borås 1990

HAALAND J, *Husbygging på gardsbruk*, Aschehoug, Oslo 1943

HØEG O A, *Planter og tradisjon*, Universitetsforlaget, Oslo 1974

LÅG J *Berggrunn, jord og jordsmonn*, NLH, AS 1979

PAAJANEN *et al*, *Lämmöneristeiden merkitys rakennusten biologissia vaurioissa*, VTT:r julkaisnja 791, Helsinki 1994

STRUNGE *et al*, *Nedsiving af byggeaffald*, Miljøstyrelsen, Copenhagen 1990

WINQUIST T, *Jordtäckta hus*, Byggforskingsrådet rapp. 10:1980, Stockholm 1980

15 *Surface materials*

The main purpose of surface materials is to form a protective layer around a building's structure. Through hardness and durability they must withstand wear and tear on the building, from the hard driving rain on the roof to the never-ending wandering of feet on the floor. Sheet materials can also have structural and climatic functions such as bracing, wind-proofing, moisture control, etc. Certain structures in brick, concrete and timber can have the same function as surface materials and therefore do not need them. Surface materials are otherwise used in roof covering, internal and external cladding, and on floors.

Because surface materials are used on large, exposed areas, it is important to choose materials that do not contain environmentally-contaminating substances which may wash into the soil or groundwater or emit irritating gases into the interior of the building. They should be both physically and chemically stable during the whole of their life span in the building or at least be easy to renew.

The roof of the building is its hat. The roof has to protect the building from everything coming from above, which sets requirements for how it is anchored, drained, and protected from frost, snow and ice. Most roof materials are used on the assumption that there is a material beneath them which helps to waterproof the building.

The external cladding has a similar task in many ways, but the demands are not as high, especially as far as waterproofing is concerned. In areas of hard rain and strong winds, durable materials are required.

Internal cladding has lower demands on it in terms of moisture and durability. The most critical factor is damage caused by the inhabitants of the building. Materials in ceilings do not need to have the same high standard as those in the walls. Internal surfaces should also have a higher level of finish to give a feeling of comfort and be pleasant to the touch. Cleaning should also be easier with these finishes. Thin layers such as wallpaper, stainless steel or hessian need a strong material to adhere to, but this material does not have to be of such a high quality. This is also the case with painted surfaces.

Table 15.1: The potential electrostatic charging of different materials

Material	*Electrostatic charging (V/m)*
Timber:	
treated with oil	0
varnished	–20 000
Fibre board	+50
Veneer	–110
Chipboard	–250
PVC	–34 000
Synthetic carpets	–20 000

Floor covering is the surface in the building which is most exposed to wear. It is also the part of the building with which the occupants have most physical contact, so comfort factors such as warmth and hardness must also be taken into account. Technical properties required in a floor material are:

- low thermal conductivity
- should not be too hard and stiff
- should not be slippery
- low risk of electrostatic charge

Table 15.2: Cleaning factors for floor materials

Material	*Cleaning factor*
Timber	5
Parquet flooring	4
Timber cube flooring	6
Concrete slabs	5
Terrazzo	3
Asphalt	5
Linoleum	4
PVC (vinyl)	2
Cork	7
Ceramic tiles	2
Stone slabs	3
Bricks	5

Note:
In the evaluation of the ease of cleaning different surfaces, the lower the cleaning factor, the more easily the surface is cleaned.

- should be easy to clean
- good sound insulation
- mechanical strength to resist wear and tear
- resistance to water and chemicals.

Many floor coverings need to be laid on a stable floor structure, e.g. linoleum and cork tiles, which cannot take any loading in themselves. The amount of moisture in the structural floor and its ability to dry out are critical: the quicker it dries out, the sooner the floor covering can be laid.

Flooring and damage to health

In the town of Steinkjer in central Norway, people complained of having aching feet after moving into new houses. Their wooden houses had burnt down in a fire and had

Table 15.3: The use of surface materials in building

Material	*Roofing*	*External cladding*	*Internal cladding*	*Flooring*
Metal	In general use	In general use	In general use in industrial buildings	In limited use in industrial buildings
Slate/stone	In general use	In general use in public buildings	In limited use in public buildings	In general use
Lime, in render		In limited use	In general use	No longer in use
Cement	In general use	In general use	In general use	In general use
Gypsum			In general use	
Fired clay products	In general use	In general use	In general use	In general use
Ceramic tiles		In general use in public buildings	In general use	In general use
Rammed earth				No longer in use
Bitumen	In general use in building paper			
Plastics	In limited use	In limited use	In general use	In general use
Climbing plants		In limited use		
Timber	In limited use	In general use	In general use	In general use
Grass plants	In limited use	No longer in use	In limited use in straw wallpaper	
Grass turf	In limited use			
Linseed oil				In general use in linoleum
Cellulose			In general use in wallpaper	
Wool			In limited use in wallpaper	In general use in carpet

been replaced with houses with concrete floors covered in plastic tiles. The complaints developed into minor damage to muscles and joints – the hard floors were the cause.

In the same way, over hundreds of years, horses used in the towns and cities suffered as a result of the hard surfaces under their hooves. They were put out to graze much earlier than country horses, used to working on a softer surface.

'Bakers' illness' was once a common problem in bakeries with hard concrete and tiled floors. These were in direct contact with the ovens, which warmed the floor by up to 30°C. The continual high floor temperature gave bakers headaches and feelings of tiredness. One way to avoid this was through wearing wooden clogs, as wood is a bad thermal conductor. A more common and serious problem today is high thermal conductivity in floors, which draw warmth out of the feet. A concrete floor will almost always feel cold.

Floors made of materials that are bad electrical conductors as PVC (see Table 15.1) create an electrostatic charge when rubbed which attracts dust particles out of the air. This is one of the most likely reasons for 'sick building syndrome'.

Metal surface materials

There are metal alternatives to all surface materials. Roof sheeting of galvanized steel and aluminium are increasingly being used as roofing in many building types, large and small. Different forms of metal cladding are also in use as external wall surfaces.

In industrial buildings the internal wall cladding is often made of stainless steel. This is easy to keep clean and particularly well-suited to premises that produce food. Flooring consisting of 6–8 mm-thick cast iron tiles with a textured surface is suitable for use in buildings used for heavy industry. Historic examples of the internal use of metal sheeting are limited. One example is the notorious lead chambers of Venice which were used for jailing particularly dangerous criminals such as the seducer Don Juan. The lead chambers were placed on roofs exposed to the sun, making them unbearably hot during the day and terribly cold at night.

Many metals can be used for roof covering and external cladding. Copper and bronze have been widely used on churches and other prestigious buildings. In the south west of England, lead from local mines is used as a roof material. In Iceland, walls and roofs covered with corrugated iron imported from England have been part of the established building tradition since the 1890s.

Modern metal sheeting is mainly made of galvanized steel, aluminium, copper, zinc and stainless steel. As far as internal use is concerned, stainless steel totally dominates the market. The products are often anodized with a thin surface layer or painted with special plastic paints. Linseed oil can be used to protect steel and zinc products. Certain metals cannot be used together because the combination causes galvanic corrosion. For example, when mounting sheeting, iron or zinc nails or screws must not be used to fix copper, and vice versa. Rainwater from a copper roof must not be drained over iron or zinc, as the copper oxide produced will soon destroy the iron or the zinc.

From an ecological point of view, the use of metals should be reduced to an absolute minimum. Metal products use a lot of primary energy and produce high levels of pollution during their manufacturing processes.

Once installed, metal products cause few problems. Their external surfaces can release metal ions when washed by rain which drain into the soil and ground water: lead and copper cause most problems in this case. The use of a great deal of metal in a building can also increase electromagnetic fields inside it.

Whole sheeting can normally be re-used. Many metals can be recycled, but as waste they must be disposed of at specific tips.

Non-metallic mineral surface materials

Mineral substances can be used to produce materials for all surfaces, either cast as a whole unit or as a component part, e.g. units for cladding, underlay for floors and other basic elements.

The first concrete roof tile was made in Bayern in 1844. Since the 1920s, concrete roof tiles have been in strong competition with clay tiles. Whether they can be as beautiful as clay tiles has always been a matter of great debate. As early as the beginning of the twentieth century the Norwegian engineer Bugge advised: 'Don't spend much time putting concrete tiles on dwellings because their form is usually unattractive, and their colours, in particular, are most ungraceful' (Bugge, 1918).

The colour of tiling has improved somewhat since then, to the extent that it can be difficult to tell the difference between concrete and clay tiles. The concrete tile has taken on both the colour and form of the clay tile, but the difference is more apparent when ageing – the clay tile is usually considered as having a more dignified ageing process.

In situ cast floors have a long history. They have been found in 7000–year-old ruins in the Middle East. Those mixes were of pure lime; today they are cement-based or made of concrete slabs.

Mineral surfaces consist of lime-, cement- or gypsum-based substances which have other constituents added, e.g. reinforcing fibres, which are then compressed to make sheets. They are most often used in situations where there is a need for highly fire-resistant materials, e.g. in walls between fire-cells and in external cladding.

Renders create a finished surface which often does not need further treatment. This is especially the case with lime renders, which can be given a matt or polished finish. The treatment of walls with render also dates back thousands of years. As well as its function as a surface treatment, render can also be considered a climatic material, as it can provide wind-proofing and moisture control.

The most common surface materials have rich reserves. Their common factor is that extraction of the raw materials entails heavy defacing of the environment, which can lead to changed water table levels or damage to biotopes.

These products usually present no problems in the indoor climate. The use of certain additives can incur a risk of unhealthy dust and fumes. If steel reinforcement is used, the electromagnetic fields in a building can increase. Many products can be re-used if they are easy to dismantle. They are usually inert and can be used as fill. Additives, such as metal colouring agents, can cause pollution when dumped.

Roofing materials

There are two types of concrete roofing: tiles and corrugated sheeting. Certain amounts of fibre must be added to give it the required tensile strength. The low weight of the sheeting makes it possible to produce it in a large format. More than any other concrete product, roofing needs particular care given to the proportions of the ingredients and the design of the sheeting or tile. One very important aspect is that the concrete used must have very low moisture absorption.

Concrete tiles and sheeting are usually made of Portland cement, but other hydraulic cements can also be used. The added fibres can be chosen from organic materials such as hemp, sisal, jute, reed, goat hair and cellulose, and from fibres of minerals such as silicate, steel, carbon, asbestos or mineral wool. Organic fibres are more easily decomposed. Research has proved that even when organic fibres have decomposed the sheeting has the same strength (Parry, 1981). The reason for this is partly that the fibres play their most important role during the setting process – it is during this period that the dangers of damage through shrinkage are greatest. Organic fibres used in concrete must be resistant to attack from lime. They also have to be free from any chemicals that can break down the cement. They can be treated the same as in woodwool slabs (see 'Woodwool cement boards – production and use', p. 284). It is also important that the fibres are easy to mix and bind easily with the mixture.

Roof sheeting was originally produced mainly with asbestos fibre, but this has now been replaced by cellulose fibre for health reasons, in a proportion of two per cent by weight. The sheets or tiles can usually also be applied to walls, either flat or corrugated.

Small scale production of corrugated sheeting
The mix for this sheeting is 5 kg cement, 15 kg sand and 0.2 kg fibre, mixed well with water. The mix is poured into a mould where it hardens over 24 hours. It is then placed in a damp, solar-warmed plastic case to cure for a month, or laid in water to cure for seven days. (The curing must not occur in dry air.) After curing the sheets are dried. (Parry, 1984.)

The Intermediate Technology Group (IPDG) in England have developed a production system for corrugated roof sheeting which is highly appropriate for small-scale production. The factory can produce 2000 tiles of 50 $\times$ 25 cm per week and needs four workers in a floor space of 25 m^2 with a courtyard of about 40 m^2. A factory that produces 20 000 tiles a week, employs 30 workers on a factory floor of 400 m^2 with a courtyard of 350 m^2. In this way one can produce roofing with low energy production costs and at three quarters of the price of corrugated metal sheeting. These have been produced for 20 years, and the life span of the sheeting is estimated at 50 years.

Environmentally speaking, cement roof sheeting can be considered better than the metal alternative. Roof sheeting is much more economical in terms of material use than roof tiles. All of the products can be re-used, but the sheeting can be more easily damaged under demounting and therefore has a lower re-use factor.

Floor coverings

Concrete

A normal concrete floor is highly durable and can cope with both water and chemicals, but on the other hand, it is unpleasant to walk on because it is hard and cold. This can be compensated for to a certain degree by adding sawdust, crumbled cork or light expanded clay. A concrete floor will produce a lot of dust through wear and tear unless it is treated with a waterglass solution, painted with a robust paint or covered by a strong floor covering. If the floor is to be covered with totally watertight material, the concrete must be completely dried out before the floor finish is laid, otherwise there may be alkaline reactions in the products with possible detrimental emissions into the internal air. Complete curing of the concrete is best guaranteed if it is well watered in the period after the concrete work.

Terrazzo concrete causes less dust problems than a pure concrete floor, and produces a much more hygienic surface. A terrazzo floor is a mixture of cement mortar and crushed stone of only a few millimetres in diameter, usually marble or limestone. For a harder floor, granite, feldspar or quartz can be used. The floor is cast in a 15–20 mm-thick layer on a concrete structural slab, and the surface is given a smooth finish by machine.

There are many types of floor tiles in terrazzo available on the market. They are usually 30 × 30 cm or 40 × 40 cm with a thickness of 4–6 cm. Pure concrete tiles are also produced as a floor finish; these are usually 30 × 30 cm square.

Concrete floors that have not cured properly are known to cause indoor climate problems. However, when the concrete product is properly cured and treated against dust, it is chemically stable and problem-free. Steel reinforcement can increase the electromagnetic field in the building.

Concrete and terrazzo tiles can be re-used if they are laid in a way that makes them easily removable. They can, for example, be laid in sand and given a weak lime cement mortar joint. *In situ* cast concrete floors can at the most be recycled as low quality aggregate or fill.

Figure 15.1: Floor of terrazzo slabs with marble tiles.

'Peatstone'

As a little curiosity, a floor tile made of 'peatstone' was in use at the turn of the century. Dry, hacked peat and sawdust were mixed with lime or dolomite. This was then mixed with wood vinegar and compressed to make slabs which were then dried. We know very little about the properties of 'peatstone' floors. It is perhaps the right time to experiment with this sort of flooring – it is very attractive in terms of energy and the environment.

Sheeting

There are three main types of mineral-based sheeting: cement-based, calcium silicate-based and gypsum-based. Apart from the binder, they often contain fibrous reinforcement. When they are mounted the joints must be filled. The filling material is, almost without exception, based on plastic binders, mainly PVAC glue or acrylate glue.

Cement-based sheets

Cement-based sheets are relatively new on the market. The first cement fibre sheets came in Japan in 1970. The sheets are non-flammable and are particularly strong. They can be used internally or externally without rendering as they will withstand frost. A binding of cellulose fibres or wood chippings from spruce or birch give the best results. The amount of wood chippings is usually about 25 per cent by weight. They are treated with a substance which reacts with lime (see 'Woodwool slabs – production and use', p. 284), and then mixed with Portland cement and water, after which the sheets are formed in a hydraulic press for seven to eight hours, then set in a special curing chamber.

Calcium silicate sheets

These are used as both internal and external cladding. They are non-flammable and strong. The sheeting is produced with up to 92 per cent by weight of quartz mixed with lime and a little cellulose fibre as reinforcement. Vermiculite can be used as aggregate.

Plasterboard

Plasterboard was first produced about 100 years ago. The usual sheeting products are used mainly for internal wall cladding, either covered by wallpaper or thin fibreglass woven sheeting for painting. Gypsum products also have an important role as climatic products (see chapter on 'Climatic materials'). The standard products are manufactured from 95 per cent gypsum with fibreglass reinforcement (about 0.1 per cent by weight). The following substances are also added to a total of about 1 per cent by weight: calcium lignosulphate, ammonium sulphate and an organic retardant. The sheets are covered with thin cardboard which is glued with potato-flour paste or PVAC glue. Pure gypsum sheeting is not particularly strong, but some sheets contain a large percentage of wood shavings, which increases strength.

The mineral sheets are based on raw materials with rich reserves. Gypsum as a by-product of power stations is used a great deal in the production of plasterboard.

The use of primary energy for calcium silicate products is low, but is much higher for gypsum and cement products.

Pollution from the production of sheeting is relatively low, calcium silicate sheeting causing the least. When built in, there are no problems with these materials, although asbestos may be found in older products. Calcium silicate and gypsum products are good moisture-regulators. The use of a filling between the sheets could result in emissions of monomers. The joints can also be covered by a timber strip or the products can be tongued and grooved to overcome the need for filling.

Products that do not have added filling can often be recycled. Pure plasterboard (gypsum sheeting) is too weak to be dismantled and re-used as is, but the material can be recycled as 5–15 per cent of new material. The gypsum industry is, however, very centralized, which makes it economically non-viable to recycle the products. Calcium silicate products can be crushed and recycled as aggregate in concrete. If it is finely ground, it can be used in mortars and render. The waste is inert and can be used as fill, as can pure mineral cement products. If there are high levels of organic substances in the products, when they become waste they may increase the amount of nutrients seeping into the groundwater. Sulphur pollution can develop from waste plaster through decomposition by microbes; this can be reduced by adding lime.

Render

There are several alternative renders, depending upon the surface to be rendered, climate, elasticity, etc. The usual binders are lime, cement, gypsum and sulphur or mixtures of these substances. Additives can make the render bind better or improve elasticity or thermal insulation; they include steel fibres, mineral fibres, perlite, hacked straw, or even hair from cows, pigs and horses. Pigments can be added; these should be fine grained and calciferous, usually metallic oxides. For external rendering or rendering in rooms such as bathrooms, waterproofing agents called hydrophobic substances are added, such as silicone products. Sand is also added, its grain size depending upon the surface quality required and how many layers of render are to be used. The final ingredient is water.

Rendering is labour-intensive work, but as a result it has a long life span. Well-applied lime rendering can last from 40 to 60 years, if it is not exposed to aggressive air pollution. Organic substances added to increase waterproofing and make application easier have a detrimental effect on the durability of the rendering.

The raw material availability of the different components of render is generally good and the environmental aspects of production are also favourable, especially for lime rendering. Pure rendering produces no problems within a building. Lime- and gypsum-based products have good moisture-regulating properties. Pure lime render can be recycled, in theory, by being re-fired, but this is impracticable in reality. Lime- and cement-based renders can be classified as inert, so their waste products can be used as fill. Pure lime render can be ground up and used to improve the soil. Dumping sulphur and gypsum waste can lead to sulphur pollution, but this can be reduced by adding lime.

Lime render

A normal lime render consists of slaked lime, sand and water. The proportion of lime to sand is 1:3 by volume. The render is put on in several layers until it is about 1.5 cm thick. It is most suitable for internal use, e.g. in bathrooms, but can also be used externally. For exterior use it should be protected against driving rain and continuous damp, otherwise it may be destroyed by frost because of its high porosity.

Nepalesian lime rendering

A render from Nepal should guarantee frost-resistance! The mixture consists of 15 kg lime, 6 kg of melted ox tallow and 36 litres of water. The tallow is for the waterproofing. The mixture has to be left for 24 hours at a low temperature. The water left on the surface is then poured away, and the creamy mixture at the bottom is mixed with 3 kg quartz sand. The render is applied in layers 3–5 mm thick. Curing takes weeks, and the surface must be protected during this period. The mixture is waterproof and weather-resistant, and is used externally on earth domes. (Minke, 1984.)

Lime rendering on earth walls

A condition for the use of lime render on earth walls is that the walls are well dried, and that the surface is even and without cracks. A thin clay gruel is applied to the wall and given a rough surface as a key for the lime rendering. The gruel consists of one part clay gruel and two parts sand with a grain size of around 4 mm. Pieces of hacked straw or hay 3 cm long are added and the mixture is then applied in two layers, straight after each other. The first layer is about 2 mm thick and the other is 5–8 mm thick. This is then left for two to three days.

The lime render is applied in two layers by trowel, without dampening the surface before application. The first layer consists of one part slaked lime, one part sand with a grain size of 4 mm and three parts hemp fibre or the equivalent, which is 5 mm thick. The next layer is 2 mm thick and consists of one part finely-sieved lime dough and three parts marble powder. In Japan, where this render originates, a small percentage of gelatine from seaweed is added. This makes the surface waterproof, although it is not vapour-proof.

For coloured render, pigment is added in the second layer (see Table 18.1). The surface is matt from the beginning, but a smooth shiny surface can be achieved by adding a third layer that is only 1 mm thick, consisting of one part fine-sieved slaked lime, one part white marble dust and one part pigment. The thin layer of render is put on with a trowel and smoothed out until it gels to a lustre. Then the surface is polished for one to two hours with the palm of the hand. This is obviously a very labour-intensive procedure.

Lime pozzolana render

A hydraulic lime or lime pozzolana cement gives a more weather-resistant render. It still needs to be applied in several layers to achieve a high durability. The first layer consists of one part hydraulic lime and two parts sand with a grain size of up to 7 mm. The second layer consists of one part hydraulic lime and three parts sand with a grain size of up to 5 mm. The third layer consists of one part hydraulic lime and three parts sand with a grain size of up to 5 mm, almost the same mix as the second layer.

Lime cement render

Lime cement render is used a great deal externally. It is somewhat stronger than lime render and more elastic than pure cement render. From 30–50 per cent of the binder is usually cement.

Cement render

This is mostly used as an external render in a retaining wall, tanks, pools, etc., and can be used on solid concrete walls, concrete blocks, lightweight concrete blocks, etc. First any cracks or damage to the surface should be smoothed out with a cement mortar of proportions 1:3, then the surface should be brushed with a cement gruel of the same mix proportions and finally rendered with a cement mortar of 1:1 on concrete walls or 1:3 on concrete block or lightweight block walls. The last treatment can be repeated, giving a surface which is as good as watertight.

Gypsum rendering

Gypsum rendering is mainly for internal use, especially as a moisture-regulating layer. This is a common plastering of buildings where brick or concrete block is the structural material. A mix of one part gypsum to two parts sand is usual. This sets in 10–30 minutes. Lime can be added to make the gypsum go further. For stucco work, a mix of three parts liquid lime and one part gypsum powder is used. More gypsum is needed for relief work in proportions of one part lime and two parts gypsum. A final coating can be one part lime and one part marble dust.

Sulphur render

This can be produced by melting sulphur at temperatures from 120–150°C. Sand, wood flour or the equivalent can be added. It is waterproof but cannot be used on materials with a high lime content.

Stone surface materials

Natural stone in the form of slate tiles is well-suited to many different uses, e.g. roof and wall covering and floors. Tiles cut from limestone, marble, syenite, sandstone and granite can be used as a floor finish, and as internal and external wall cladding.

Figure 15.2: A house with a recycled slate roof in Aberfeldy, Scotland. Source: Gaia Scotland 1993

Slate was used for roofing in France as early as the thirteenth century, on castles, palaces and churches. Since then, the material has spread over many parts of Europe, and to simpler buildings. Slate materials have generally been ignored during this century, partly due to the architect's attitude that slate is plain and uninteresting. Evaluated from an environmental and functional point of view, few materials can compete with slate. In highly exposed areas, it can successfully be used as wall cladding.

Cut and polished stone tiles have had a much greater use during the twentieth century, especially in public buildings. The products are not strongly layered and therefore need a developed technology to cut them to shape and divide them into layers.

The different types of tile are:

- *Roof tiles*
  Raw/rough tile, the oldest form, cut by simple splitting and dividing.
  Patchwork tile, which has the form of a drop and is usually made in small sizes, from 30 × 15 cm to 45 × 30 cm.
  Square tile, square with broken off corners, produced from slate in many sizes and thicknesses.
- *Floor tiles* of limestone and marble, usually produced in a thickness of 2–3 cm, while sandstone is around 8–10 cm thick because of its lower strength. Granite has a much greater variety of form and size. Round stones or square cobble stones from 5–12 cm can be used. All stones can be polished, which simplifies maintenance. Slate floors are often laid as tiles which are cut into squares or rectangles.
- *Wall tiles* of slate or other stones, produced in many different sizes. As they are not exposed to heavy loading, large dimensions can be used even with weaker types of stone.

The occurrence of slates and other stones for tiles is generally plentiful and well spread. The material is usually extracted from open quarries. This can change the local groundwater situation and damage local biotopes. The use of primary energy in extraction is initially quite low, but because stone is so heavy it is difficult to justify using it at long distances from its source.

Certain types of stone contain quartz and dust can be a risk during the working of the stone (see Table 7.3). Slate, limestone and marble have low radioactivity and therefore are no problem for the indoor climate. Certain types of granite can present a problem as a source of radon gas.

Stone floors are easily looked after, durable and resistant to spillage of water and other liquids, depending upon which stone is used. Marble cannot be used in men's toilets as it reacts with urine. Stone floors are hard and cold to walk on, unless floor heating is used.

Slate and stone tiles laid in a weak mortar can usually be taken up and re-used. Stone products that are fixed mechanically are easily re-usable. Over 90 per cent of the slates from an old roof can usually be re-used. It is necessary to ensure that they are high quality and not very porous with a high content of calcium carbonate. There is also a difference between stone that comes from coastal or inland regions. A coastal slate has usually been exposed to a more severe climate, with frequent changes between frost and mild weather. The same applies to stone tiles which contain lime or sandstone and have been exposed to a severe climate as wall cladding; these are not so easily re-used. All stone should therefore be carefully checked before re-use for strength and porosity. Dumping stone waste is seldom a problem.

Practical use of stone surface materials

Roof covering

Before laying, slate tiles are sorted into two, three or four groups of different thicknesses, unless this has already been done at the quarry. There is usually a timber board roof with felt on, if the roof is to be windproof, otherwise the felt can be left off. The slates are fixed onto battens, unless the site is very exposed to wind, when they are fixed directly onto the boarding.

The usual size of the battens is 25 × 50 mm. The distance between the battens depends upon the method of laying, the type of slate and its form, but mainly on the distance between the lower edge of the slate to the nail holes, minus the overlap.

The thickest slate is laid furthest down on the roof, to avoid large variations in thickness on the other courses. A slate hammer is used to split and shape the tiles. When breaking the corners, a special tool fixed to a wooden stump is used. The tiles are fixed with special slate nails which are 25/35 mm, 28/45 mm and 28/55 mm. The ridge is covered with rectangular slate tiles, timber boarding, zinc, copper or even turf.

Rough tiles

When laying rough tiles, holes are first bored or hacked in the tile with a drill or a special hammer. The tile is fixed to the batten with a strong galvanized nail or a slate tack. For large tiles wooden pegs made of ash or juniper can be used. As rough tiles do not always lie tightly on each other, they can be broken by heavy snow loads. One way of resolving this problem is to put lumps of clay under the end of each tile. If time is spent sorting the tiles so that they fit well together, then a roof of rough tiles can be as waterproof as any other.

Square tiles

The square tile is used for single-layer roofing. The overlap should be at least 45 mm for small tiles and 75 mm for large tiles.

Patchwork tiles

Patchwork tiles can be laid as a single or double covering. The following slopes are recommended:

Covering	Roof slope/climate
Double layer	Minimum of 18° everywhere
Single layer	Minimum of 22° in moderate climates or 27° in severe climates

For single laying the tiles must be at least 12 mm thick. For double laying they need only be 6 mm thick. The distance between the battens for double laying is somewhat less than half the length of the tile. An overlap of at least 50 mm is recommended.

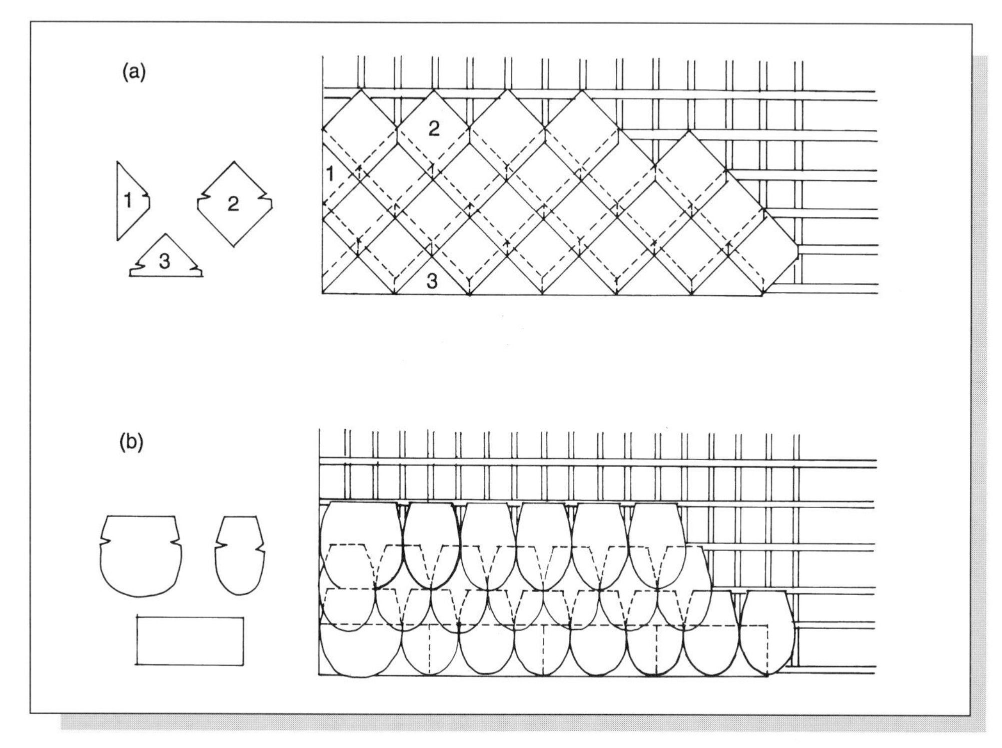

Figure 15.3: Laying of (a) square tiles and (b) patchwork tiles.

Patchwork tiles can also be used on rounded corners, and with some modification on cone, spherical and cylindrical forms. The main rule is that the size of the tile is reduced proportionally with the radius it is to cover. The tiles are nailed directly onto the rough boarding of the roof. To avoid the stone splitting because of the movement of the roof, each tile must be fixed to only one piece of boarding.

Wall cladding

Disastrous results can come from fixing a thin stone to a wall with mortar. The tiles can easily loosen or be broken, either through expansion when exposed to the sun or by the formation of condensation behind the tile, which then freezes and pushes it off. If the grain of the tile is vertical, there is a stronger chance of it being knocked off by frost than if the grain is horizontal. Hanging tiles can be prefabricated as a unit with the tiles pre-cast onto a concrete slab, which is then used as a wall cladding.

Cut stone cladding is mounted on special metal anchoring systems, with good ventilation behind the stone. The metal should be bronze, stainless steel or a copper alloy, which is bored into the structure. This cladding is very expensive and is usually used on offices or public buildings.

Walls can also be clad in the same way as a slate roof. All types of tile can be used, though patchwork and square tiles are the most appropriate because of their lightness. Only one layer is needed, and is mounted with slate nails, with good ventilation underneath.

Floor covering

A natural stone floor can be laid in several ways. It is usual to lay the stones in mortar directly on concrete. The concrete is primed with a mix of cement and sand, 1:1, while the mortar for laying the tiles is a mixture of cement and sand from 1:3 to 1:4. The mortar is laid to the necessary thickness and before laying the stone tiles, are given a coating on the underside with a cement and sand grout (1:1). The tiles are then knocked carefully into place with a rubber hammer. The joints are filled between three and seven days later with a grout of cement and sand (1:3).

For larger floor tiles hard deciduous wood can be used in the joints instead of mortar, or the mortar can be replaced with sand. The possibilities of re-use are then very good. In sheds,

Figure 15.4: Floor covering of slate, mixed with old roof slates. Source: Gaia Lista, 1990

winter gardens etc., it is often natural to lay the stones on earth or sand, without anything in the joints.

Marble is the only stone that needs proper maintenance. This is carried out with wax or polish.

Fired clay sheet materials

Fired clay can be used for a whole selection of surface materials for roof, walls and floors. These can be divided into two main groups: fired clay tiles and ceramic tiles.

Roof tiles of fired clay were used very early in the history of the Mediterranean countries. The principle used was that of 'nun' and 'monk' tiles (see Figure 15.5). The interlocking tile was first made in France in the mid-nineteenth century; it provides better waterproofing and increased fire safety. From around the end of the nineteenth century, all houses in small towns were ordered to have interlocking roof tiles or slates. Many clay tiles have been replaced with concrete tiles and metal sheeting, often given a profile to look like clay tiling.

Brick veneering of inner and outer walls uses bricks of standard sizes which are placed in mortar on solid concrete or timber frame structures in thin layers. Brick products can also be used as flooring, laid on sand or in mortar. Ceramic tiles are used on floors and walls. These are usually square or rectangular in form, but specially designed tiles of other shapes, e.g. triangular, octagonal or oval, are also available. Tiles can be glazed or unglazed; unglazed tiles are often coloured.

A better quality of clay is required for the production of roofing tiles and ceramic tiles than for bricks. There is, however, an abundance of raw material.

Ceramic tiles and fired clay products used as outside cladding, roof covering or untreated floor covering should have a very low porosity. This entails firing at high temperatures, something that results in high primary energy use and pollution levels. Lime cannot be added to reduce the pollution, as this would increase the porosity of the products. For brick veneers on inner walls, the waterproofing demands are less.

Fired clay products are an excellent material for the indoor climate. They are hygienic, do not release gases or dust, and are usually good moisture-regulators, if they are not highly fired and sintered. The jointing material for ceramic tiles usually have polymers such as epoxy and polyurethane as ingredients. These can cause health-damaging emissions into the indoor climate. In Sweden, mastics with organic constituents have lead to mould problems, especially in bath and shower rooms. Pure, biologically neutral combined cement and sand alternatives are far better for both floor and wall.

Fired clay products are very durable. They are not susceptible to aggressive gases and pollution in the same way as concrete and stone. Floor tiles, for example, are more durable than the grout between the tiles, and this may cause a problem. To take advantage of the material's durability it should be easy to dismantle and re-use. Roof tiles are no problem to recycle, but it must be remembered that stone from coastal climates has often been exposed to more frequent changes of temperature between freezing and thawing, making it more brittle.

A stone floor laid in sand is no problem to lift and re-use. The same can be said for internal brick cladding that is laid in a lime mortar or clay. However, if tiles or a brick veneer are laid on a cement-based mortar, it is almost impossible to remove them for re-use.

Crushed fired clay and ceramic products can be recycled as aggregate for smaller concrete structures, render and mortars.

Waste products from plastic-based mortars for jointing and colouring containing heavy metals are problematic. In cases where antimony, nickel, chrome and cadmium compounds are included, disposal at special depots or tips is required. No coding exists for coloured ceramic tiles, making it necessary to give all tiles the same treatment as a dangerous waste product.

Roof tiling

Production of roof tiles requires clay that has a high clay content, no large particles and a low lime content. Fired lime particles can absorb moisture in damp weather and destroy the tile.

Properties of different roof tiles

Type	*Properties*
Monk and nun	Moss grows on it in damp climate
Plain interlocking	Very good, fire resistant
Pantile, non-interlocking	Good, not so watertight at the joints
Pantile, interlocking	Very good, fire resistant

In addition some special tiles such as ridge tiles and hip-tiles. The weight of roof tiles varies from 30–40 kg/m^2. Tiles must be fired at a temperature approaching sintering, about 1000°C, to reduce their porousness.

There is a widespread belief that glazing increases a tile's resistance to frost. This is not necessarily true. A glazed tile can still absorb moisture. Apart from its visual appearance, the main purpose of glazing is to prevent the growth of fungus.

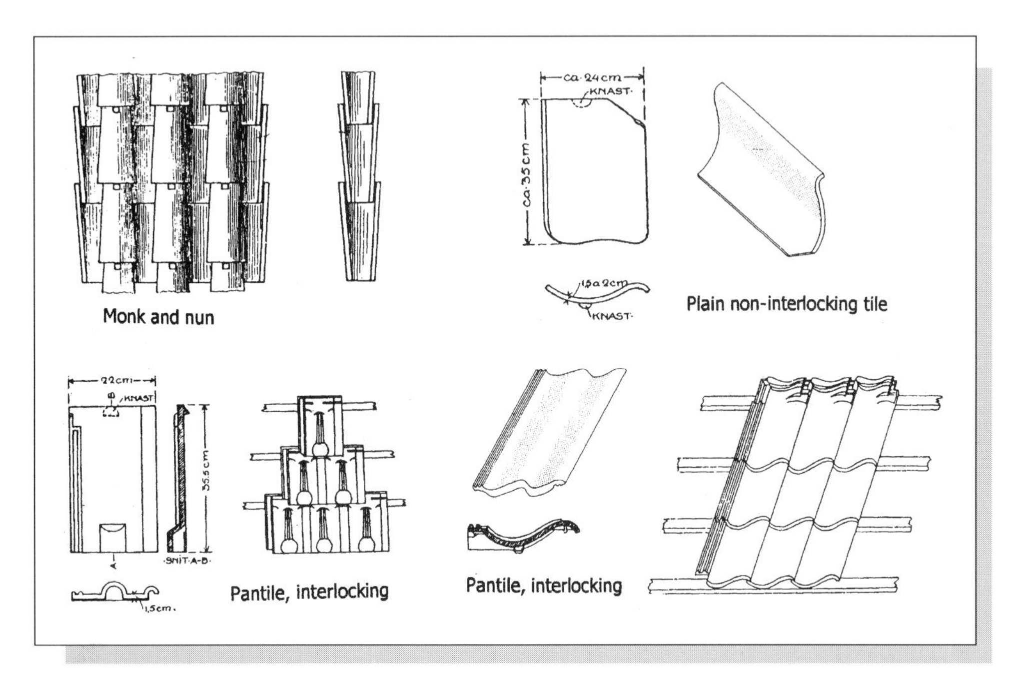

Figure 15.5: Types of roof tiling.

Ceramic tiles

There are many types of ceramic tile for many different uses. Tiles that are coloured all the way through are usually dry pressed and fired to sintering temperature. All ceramic tiles can be glazed.

Figure 15.6: Facial cladding with clay tiles.

Floor and wall tiles laid in mortar

The mortar is made of a mixture of cement and sand, in proportions of 1:4 or 1:5, and water, giving it the consistency of damp earth. It is laid to a thickness of 2.5–3 cm and evened out. A thin cement and sand grout (1:1) is then poured on and spread with a trowel. The maximum size of the grains of sand is 2 mm. The tiles are knocked in with a rubber hammer. The joints are then filled with a cement and sand grout (1:1–3), the maximum size of grain being 1 mm.

After jointing, a dry jointing material is spread over the whole surface in a thin layer. This lies in place until the laying pattern of the tiles becomes visible. The surface is then cleaned, and the floor is ready after four days curing.

Wall tiles are mounted in almost the same way with the same mix of cement and sand. It is an advantage if the back of the tile is textured and has a semi-porous surface. Laying floor tiles is relatively straightforward, but putting tiles up on a wall needs a well-trained professional.

Figure 15.7: In exposed coastal areas of Denmark, the roof tiles along the ridge and the gables are fixed with lime cement mortar to prevent them blowing off.

Floor finish of bricks laid in sand

A brick floor can be laid without cement using both well- and low-fired bricks. It is important to choose a brick with a smooth surface. A 3–5 cm-thick layer of sand is spread on a layer of stabilized insulating loose fill, and the sand is then dampened and compressed. The size of the grains must not be more than 5 mm and well-graded. The bricks are laid and knocked into place by a rubber hammer and sand is poured into the joints. The whole floor is then sprinkled with linseed oil, and this treatment is repeated twice at intervals of one week. This binds the sand in the joints and makes the brick surface easy to clean. It is also possible to treat just the joints with linseed oil, and treat the bricks with a soft soap. This floor surface can be used in both houses and public buildings.

Figure 15.8: Floor covering of bricks in sand, which are easy to remove and re-use. Source: Gaia Lista, 1988

Earth surface materials

An earth rich in clay can be rammed into a reasonably good

quality floor as long as it is given a smooth and dust-binding finish. Earth from the immediate vicinity should be used. It is rammed to the right consistency and the surface can be treated or covered with another finish. The use of energy is very low, and the floor returns to its original state when the building dilapidates, as it has not been chemically treated. For the users, earth makes a relatively warm floor, and it is soft and comfortable to walk on. Earth is the most widespread floor surface, world-wide, and the most ecological floor conceivable!

Laying an earth floor

The underlay must be well-drained, dry and firm, e.g. a 20–25 cm thick layer of light expanded clay fill. Light clay fill must be well bound with a lime–cement gruel. An alternative is a bound layer of crushed stone. Fine chicken net is placed on top of this. The floor should be rammed to a depth of 15–20 cm, in *lengths 1 m wide*, bordered by a plank, using the same technique and equipment used in wall ramming. Ten centimetres can be laid at a time. The earth should be the same quality as for Pisé building. The top layer must be well-sieved earth, and when it has been rammed, the surface should be evened out with a long-handled scraper.

If the floor is to be an underlay for a timber floor on battens, cork, linoleum, coconut or sisal mats, it has to dry out for a year before being covered.

If the floor is to be exposed it will be easier to maintain if it is rendered with an elastic mortar. In this case, fibres should be added to increase elasticity.

Plastic-based sheet materials

Sheet products in plastic are limited to building sheets, floor coverings, carpets and textile and wall coverings. Except for building sheets, the rest are discussed towards the end of this chapter.

The sheets are usually composite products consisting of sheets of paper sprinkled with a plastic, usually a phenol or melamine (about 25 per cent by weight), pressed together under high pressure and heated. These products are mainly used for wall and ceiling cladding, without any further treatment. A stronger sheet can be made with polyester and a mixture of stone particles reinforced with fibreglass.

Plastic products are based on oil, a very limited resource, the extraction of which creates high levels of pollution and is high risk. Their manufacture is energy-intensive and polluting. There is a strong chance that emissions from these products enter the indoor climate, depending upon how well the plastic has been cured.

These surface products and composite materials can seldom be recycled. Sheets with a high proportion of paper can be burned for energy recycling, but the smoke must be filtered. Sheets that contain minerals cannot be burned. Waste material left after demolition has to be dumped at special depots.

Living plant surfaces

Surfaces can be protected with living plants. These can be divided into two groups: roof coverings of turf and wall coverings of climbing plants.

Very positive environmental qualities result from the use of plants as living surface treatments. The exception is the waterproofing needed under a turf roof, which is usually either a plastic or bituminous product. Trelliswork for climbing plants should be made of high-quality non-impregnated timber.

Environmental advances with plant surfaces

Plant surfaces are an important factor in the environment of towns. Green plants bind and break down gases such as nitrogen oxide, carbon dioxide and carbon monoxide and produce oxygen. A combined leaf surface of 150 m^2 produces the oxygen needed for one person. A 150 m^2 roof that has 100 m^2 leaf surface per square metre supports the equivalent of 100 people. A wild, overgrown grass roof produces about 20 times as much oxygen as a well-looked-after lawn.

Planted surfaces bind dust, which is carried by rain to the ground. Well-planted areas also reduce vertical air movement. Over a conventional roof, vertical air currents of up to 0.5 m/s can be caused by solar heating of the roof material. On metal roofs the temperature can be as high as 100°C. This air movement can pick up dirt and form clouds of dirt over towns. A turf roof will reach no more than 30°C, almost totally eliminating the rising air movement.

Planted surfaces can provide good thermal insulation. Pockets of still insulating air are formed between the plants giving the same effect as a fluffy fur coat. Plants also reduce the effects of wind and the infiltration of air into the underlay. A turf roof gives an insulation of 46 dBA with 20 cm thickness and about 40 dBA with 12 cm thickness. This sort of roof is therefore particularly suitable near airports.

A large part of the year, the planted surface acts as a solar panel – turf roofs have a particularly high absorption coefficient. The plants develop their own warmth during the cold part of the year and prevent freezing. During the summer, dew will form on the roof in the morning. For every litre that condenses, an amount of warmth the equivalent of 0.65 kWh is emitted. The damp earth in the turf roof has a large capacity to store warmth. This can give the building a stable, warm, indoor climate during the winter, and a cool indoor climate in the summer. Walls covered in plants are cooled by their shade during the summer.

Turf roofs

Turf can be used as a cladding material on mounds along walls, but is more often used as a roof covering. Turf roofs are built up in several layers, the most

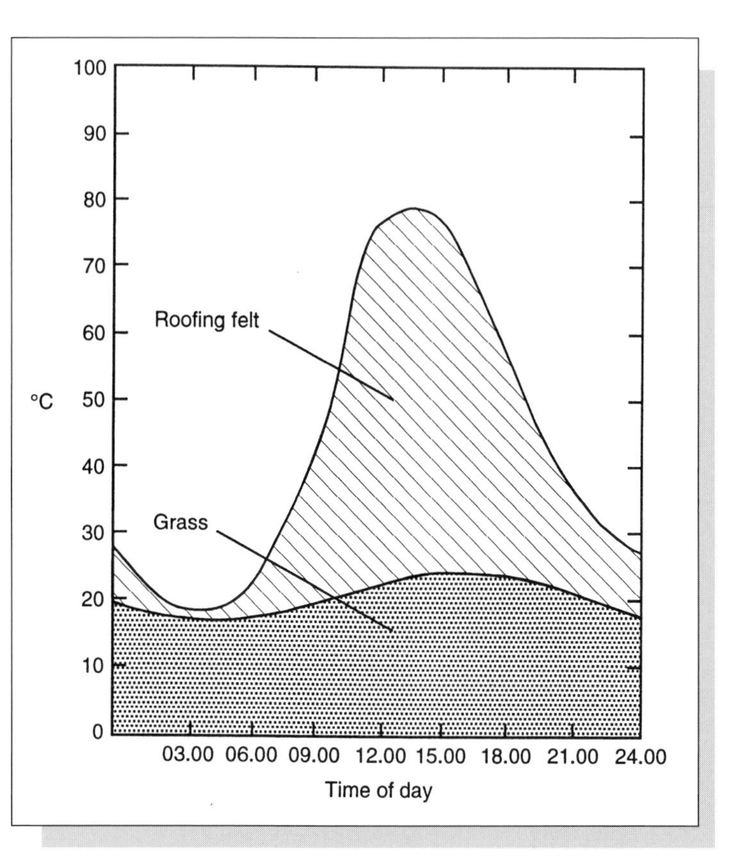

Figure 15.9: Comparison of the temperatures on roofs covered with bituminous roofing felt and grass during a period of 24 hours, on a clear summer day. *Source: H. Luz*

critical being the lowest waterproofing layer which prevents water from entering the actual roof structure. This was once done using birch bark, but is now achieved using bituminous products and plastic membranes. A normal waterproofing layer is built up in two layers with a polyethylene membrane of about 0.5 kg/m^2 on top of a polyester-reinforced bituminous felt of about 2 kg/m^2. Polyvinyl chloride products are also used. Bitumen-based glue and mastic is used for laying and jointing.

Turf roofs have dominated building history in northern Europe as long as can be remembered. Resources have been boundless and laying methods relatively simple, though labour-intensive. The high thermal insulation offered by turf roofing made it a strong competitor against slate, tiles and other materials that subsequently appeared on the market. The thermal insulation makes it common even in the tropics. There are houses in Tanzania which have a 40 cm-thick layer of earth with grass on the roof.

Climate has little effect on a turf roof, wherever it is. In very exposed, windy sites along the coast there are, however, stories of roofs of this type being blown off. With the demand for even better insulation and less labour-intensive methods the turf roof became less competitive. Today it is mainly relegated to Scandinavian summer cottages in the mountains. But during the last 10 years there has been a renewed interest in this roofing material, because of the ability of green plants to reduce air pollution noticeably by binding dust, breaking down gases and producing oxygen. It has been discovered that if 5 per cent of town roofs were covered with grass and plants, there would be a noticeable reduction in smog problems. These discoveries have led to heavily-polluted towns in

Europe, e.g. Berlin, experiencing a renaissance in the use of grass on roofs.

The insulating properties of turf roofs are difficult to assess – much needs to be taken into consideration: not only the earth structure but also the wind-proofing effect of grass, the collection of dew, the activity of the roots which develop warmth, its high capacity to store heat and its varying moisture content.

Figure 15.10: The roof garden of a large department store in Kensington, London. This type of roof garden has a very positive influence on the city climate.

Turf roofs are usually associated with folk architecture with just grass growing on the roofs. But other plants can be chosen, and the roof does not necessarily have to be sloping, it can be flat. The following plants are possible:

Plants	*Minimum depth of earth*	*Type of roof*
Grass	10 cm	Flat/sloping
Larger plants	10 cm	Flat/sloping
Bushes	25 cm	Flat/sloping
Small trees	45–80 cm	Flat/sloping
Vegetables	45–60 cm	Flat

Turf roofs have always been produced locally by people building for themselves. The methods are simple, and the grass and earth resources are infinite and can be used direct from their source.

Bituminous and plastic-based waterproofing layers reduce the otherwise favourable environmental qualities of this type of roof, both in terms of the extraction of the resource and the pollution related to them.

Earth in itself has unlimited durability – it is the waterproofing layer that decides the life span of a turf roof. Leakage problems and damage usually arise around flashings, where pipes, chimneys etc. penetrate the roof. Earth scraped off a damaged roof goes back to the soil and can later be used for a new turf roof. The waterproofing layer of polyethylene can, in theory, be cleaned and recycled, but it is doubtful that this would happen in practice. Bitumen and plastic can be energy recycled (burned) as long as there is a special filter on the smoke outlet. Waste material has to be deposited at special dumps.

Layering a turf roof

Flat turf roofs are made of several layers. The top layer has the planting with a soil layer underneath. Under is a filter layer which prevents heavier earth getting through, and beneath this is a further layer for draining away excess water. The waterproofing layer is furthest down and should prevent roots from growing through and water getting into the structure, which is preferably made of concrete. On a sloping roof of over 15° the filter or draining layers are unnecessary, but otherwise the roof is built up in the same way.

The plant layer

A wide spectrum of plants can be grown on roofs, some of which strengthen the network of roots and thereby the roof itself. They can stabilize it, retain moisture

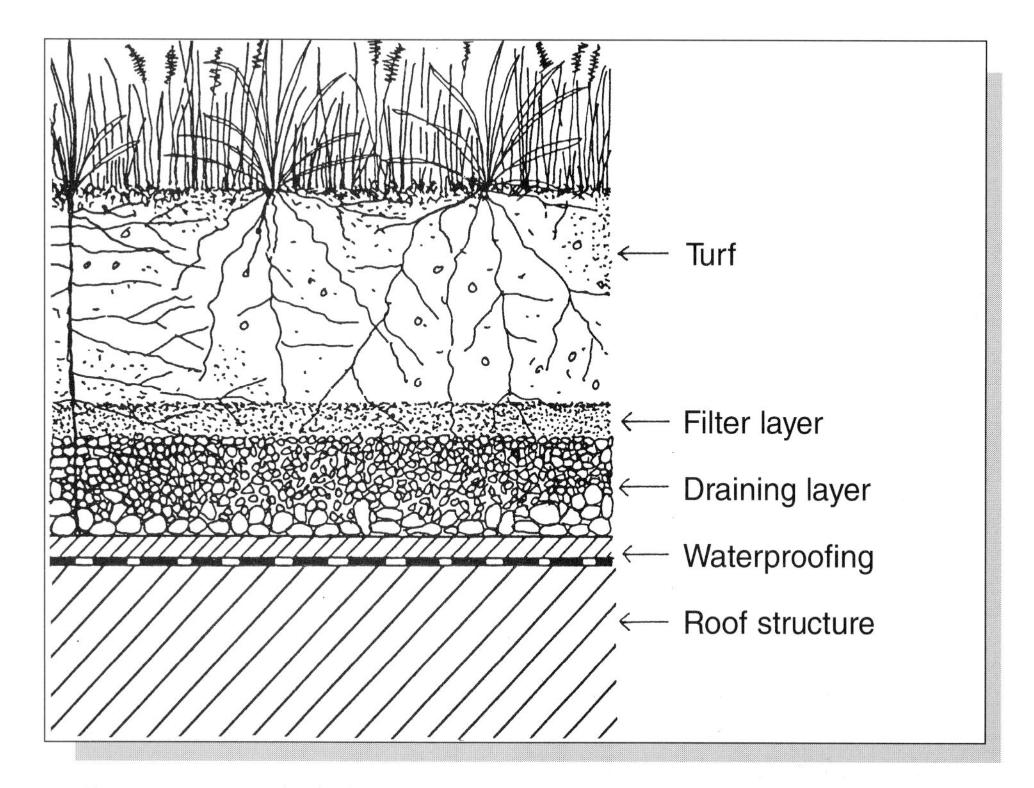

Figure 15.11: Principles for building up an almost flat roof using turf covering.

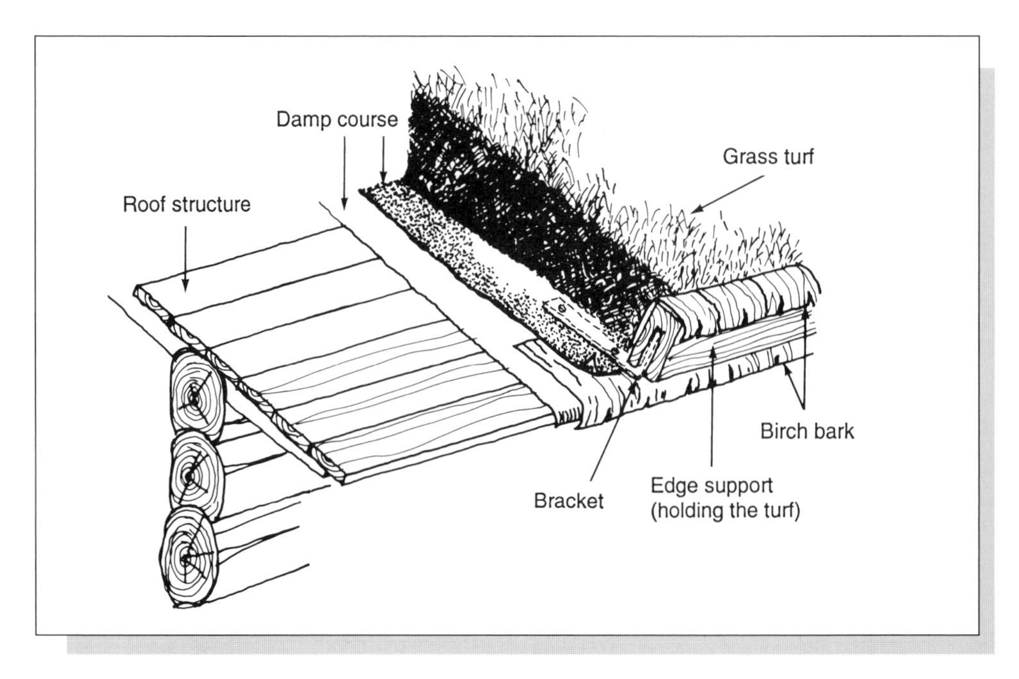

Figure 15.12: Principles for using turf covering on a sloping roof. Source: Norwegian Building Research Institute

over a dry period and even reduce fire risk. There are evidently many advantages to a varied flora on the roof (see page 161).

The earth layers

The usual turf for a roof comprises grass that is well bound by its roots, cut up into pieces 30 × 30 cm and about 10 × 15 cm thick. In Norway it is normal practice to use two layers of turf, the lower with grass downwards and upper with the grass on top. On the ridge, longer pieces of turf are used. Even loose earth can form a top layer, compressed to the same thickness as the turf. On a sloping roof, it is advantageous to lay a chicken net with 2–3 cm of earth on it before compressing the earth and sowing. For a roof with a slope of more than 27° it is necessary to lay battens to hold the turf in place (see Figure 15.13). These are not fixed through the roof covering but at the ridge, to each other, or resting on a batten at the eaves of the roof. The battens do not have to be of a very durable material, as they lose their function when the system of roots binds together.

The earth should have plenty of humus, which can be increased by mixing in compost or peat. A depth of at least 15 cm of earth is recommended. A thinner layer will dry out or erode easily. For sedum species, which are particularly

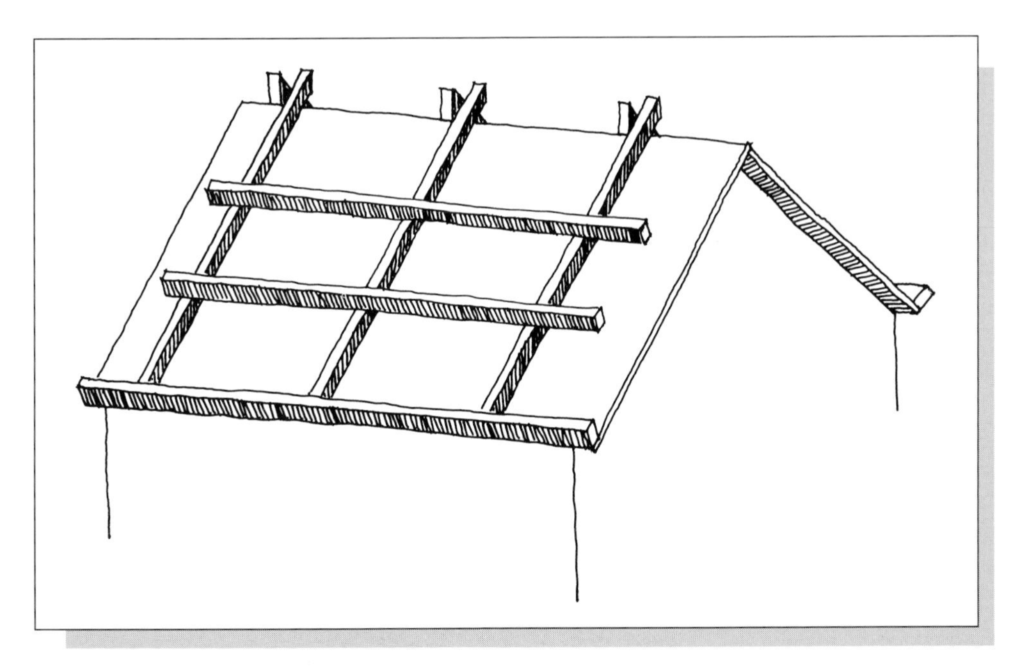

Figure 15.13: Battens for holding turf in place on steeper roofs.

resistant to dry periods, the depth of earth need only be 6 cm. On a roof with not much of a slope or a flat roof it is possible to use a layer of earth without turf for growing vegetables.

In Berlin around the turn of the century there was a method of covering courtyards with 20 cm building waste mixed with earth. A whole series of such courtyards exist in an area called Neu-Köln.

The filter layer

The filter layer, which is necessary on a roof with a slope of less than 15°C, can be rough sand or sawdust.

The draining layer

The draining layer, needed on a flat roof, can be rough or fine shingle or loose expanded clay pellets.

The waterproofing layer

This layer is necessary to ensure that excess water runs off the roof. There are different ways of achieving this, but the most common is bituminous or plastic-based solutions.

Birch bark

Bark from birch trees was the most usual waterproofing method until the mid-twentieth century. It is laid in six to 16 layers with the outside upwards, and the fibres following the fall of the roof to carry the water to the eaves. The more layers there are, the better the waterproofing.

The layer of turf over the bark layers must be at least 15 cm deep to prevent the bark from drying out and splitting. A roof angle of 22° is the lowest possible for this sort of waterproofing. This is a very labour-intensive technique and is dependent upon a limited resource.

Marsh-prairie grass

Marsh-prairie grass laid on thin branches was the usual waterproofing layer used by immigrants in the drier areas of the USA.

Tar and bituminous products

These have also been used, to a certain extent. In Germany during the 1930s a building with a flat concrete roof was coated with coal tar and then a 10–20 cm-deep layer of earth was laid on top. The roof has kept well through the years (Minke, 1980). Coal tar is not particularly good environmentally because of its high content of polycyclical aromatic hydrocarbons (PAH). Using a pure bituminous solution is a better solution, but there is little evidence as to how durable this would be. If using bituminous felt there should be at least three layers, but the durability is probably relatively low because of the acidic activity of the humus in the earth. A high proportion of quack grass in such a roof, *Agropyron repens*, would be inadvisable. Polyester reinforced bituminous felt is often used as an underlay for other plastic membranes. The material does not then come into direct contact with the earth.

Corrugated asbestos sheeting

This was used a great deal during the 1950s, but is no longer produced. This is due to the associated health risks and its limited life span.

Steel and aluminium sheeting

These cannot be used, because they are quickly eaten away by the acidic humus.

Slate and tiled roofing

It is actually possible to lay a turf roof on top of a sloping roof covered in slates or tiles, but it is unlikely to be an economical or resourceful use of materials.

Bentonite

Bentonite is a type of clay which expands when it comes into contact with water and becomes a tough and clay-like mass which prevents water penetration (see p. 269). This material is used in tunnel building and can also be used under a turf roof. The depth of earth must be at least 40 cm to give the clay enough pressure to work against. This restricts the use of this method to larger buildings with flat roofs. It would still need a layer of bituminous felt underneath.

Plastic

There are many different plastic materials on the market for this particular function, such as PVC or polyester sheeting with fibreglass reinforcement. The best product from an ecological perspective is polyethylene sheeting of about 0.5–0.7 mm thickness. This is an oil-based product but is relatively free from pollution when in use. When burnt it does not emit any poisonous gases. The polythene sheeting available today is mainly for sloping roofs. It has studs or small protrusions on its surface which stop the turf from sliding down, and is claimed to be resistant to humus acids. As the plastic is underneath earth, it is not affected by ultraviolet radiation or large changes of temperature, which have a tendency to break down plastics. The durability is unknown as there are no examples that have been in use for a long period. On flat roofs, reinforced PVC sheeting is the most common material. The plastic barrier is normally laid on top of a layer of bituminous felt.

Flashing

Flashings around chimneys and pipes that go through the roof are usually of lead or copper. The use of these materials should be kept to a minimum for environmental reasons. Slates can be used around chimneys on turf roofs (see Figure 15.14).

Climactic conditions affecting turf roofs

Sun

Strong solar radiation can cause the planted surface to dry out, especially if it is on a relatively steep roof facing south. If this angle is less than 20° there is no problem. For steeper roofs in drier climates the roof needs to be shaded or needs a thicker layer of earth giving a high water- and warmth-storing capacity.

Table 15.4: The uses of different waterproofing layers

Material	*Amount of work*	*Life span*	*Areas of use*
Bark from birch	Very high	Long (30–100 years)	Sloping more than 22°
Bituminous felt	Low	Medium/low depending on type of soil	All roofs
Corrugated asbestos sheeting	Low	Medium	Sloping more than 15°
Steel/aluminium sheeting	Low	Short	Sloping more than 15°
Slate/tile roof	Medium	Long	Sloping more than 20°
Bentonite clay with bituminous felt	Low	Unknown	Flat roofs
Polethylene sheeting with bituminous felt	Low	Unknown	Sloping more than 15°
Polyvinyl sheeting with bituminous felt	Low	Unknown	All roofs

Wind

The strength of the wind depends upon the height of the house and the local wind conditions. The stronger the wind, the slower the plant growth. Wind also has a cooling effect and can increase the drying rate, even causing physical damage in certain situations. For very exposed areas, planting should surround the building to protect it, with a thicker layer of earth on the roof, mixed with stones to give the roots a better hold.

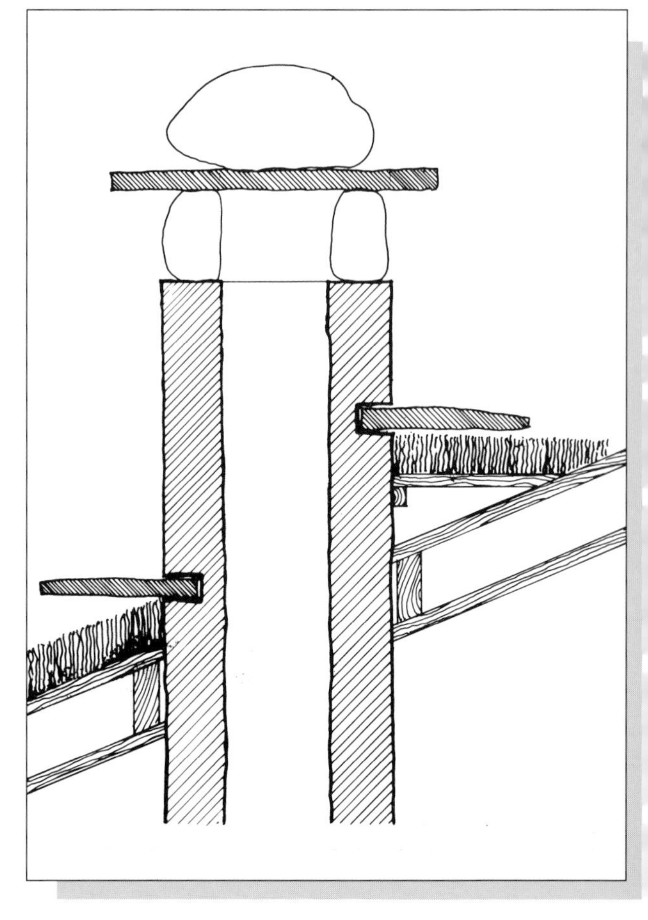

Figure 15.14: Slates used as protection from rain around the chimney.

Rainfall

Even if the earth in certain cases can be waterlogged, water is something that the planted roof needs in very large quantities. There is no groundwater reserve for them to draw on during a dry period. They are totally dependent upon the storage capacity of the layer of earth on the roof. A short dry period is no problem; after a little rain the plants can quickly recover. Shading can reduce solar penetration and a thicker layer of earth can store more water, especially if it contains more clay than sand. Automatic watering systems are necessary if vegetables are to be grown. Grey drainage water from the household can be used for extra fertilization.

Pollution

Green planting has a very positive effect on air pollution, but it can also be damaged by it. This can only occur in situations of extreme pollution, where there are strong concentrations of ozone, or dust that settles on the leaves and prevents photosynthesis. If the earth becomes too acid, lime can be added.

Erosion

Planted roofs do not receive any nutrition from the natural nutritional cycle, but are all the time losing humus, minerals, salts etc., as they are washed out. It is therefore natural to start with very rich earth. A little compost can be added occasionally, and autumn leaves should be left lying. The correct mix of plants can also add to the richness of the earth.

Wall cladding with plants

The qualities achieved by cladding walls with plants are somewhat similar to those of a turf roof, with increased wind and rain protection, extra thermal insulation and sound insulation, and better air quality.

There is a certain amount of scepticism as far as plant-clad walls are concerned, based on two main points: that the plants, especially ivy, eat into the wall, and that leaves can house all sorts of insects which can get into the building. However, as long as the materials used in the building are mineral, such as brick, and the render is of a high quality, then no damage will be caused by plants. In fact, they have the complete opposite effect, protecting the render from driving rain, drying out and large fluctuations in temperature. In Germany, rendered walls like these have lasted up to 100 years, while normal buildings have been re-rendered three to four times during the same period (Doernach, 1981).

Walls clad in timber panelling and other organic materials are less suitable for plants, but if they are planted, there must be plenty of ventilation between the plant and the wall. Ivy and other climbers that extend their roots into the wall should not be used.

Problems with insects have proved to be almost non-existent.

Climbing plants need no particular source of energy except a little fertilizer; the sun does the rest. The life span of these planted surfaces can be as much as 100 years, and ivy has been known to grow on a building for 300 years.

Orientation and planting

The different façades of a building offer different growing conditions for plants, just as plants can have different uses on different façades depending upon their orientation. On the south side plants that lose their leaves during winter should be grown to take advantage of solar radiation during the winter. In milder climates, fruit or vegetables such as grapes or tomatoes can be grown. On the east or west side it is better to have evergreens that form a thick green layer. Deciduous plants can be used if they have a dense growth of branches or have a hedge formation. On the north side it is necessary to find a thick layer of evergreen vegetation that is not dependent upon sunshine.

The planting has to be done during the spring or the autumn. The plants can be bought at a garden centre or found in the forest (e.g. honeysuckle, ivy, hops and blackberries). The plants are placed in the earth at 30–50 cm spacing and about 15 cm out from the wall. The depth of the holes should be between

30–50 cm depending upon the particular plant. The roots must have space to grow out from the building. Certain climbing plants are sensitive to high earth temperatures and prefer a shady root zone, which can be achieved by planting grass or small plants over them.

Apart from hedge and hanging plants, trelliswork is needed to help the plant on its way. Self-supporting climbers are quick to attach to walls, but others need more permanent trelliswork. This can be a galvanized steel thin framework or high quality timber battens. Timber battens are best placed diagonally. For fast-growing plants and heavy masses of leaves extra watering and fertilizing will be needed, especially at the beginning. Many of these plants must be pruned regularly.

Indoor plants

Russian and American space scientists have been working for years with so-called 'biological air cleaners' for use in space ships. These are plants with a high absorption capacity for organic gaseous pollution which is normal in modern interiors, such as vapour from solvents and formaldehyde.

Larger plants that do this are ivy (*Hedera helix*), the fig plant (*Ficus pumila*), devil's ivy (*Scindapsus aureus*) and the tri-leaf philodendron (*Philodendron* spp.), but potted plants such as the peace lily (*Spatiphyllum*) and the spider plant (*Chlorophytum comosum*) also do the same. The air-cleaning properties vary from species to species, and are also dependent upon the leaf area (see Fig. 15.15).

Timber sheet materials

Timber can be used in all the different situations where sheeting is needed: as whole timber, as one ingredient in sheeting and as cellulose for wallpapering. Wallpapering is discussed later in this chapter.

Timber can be used to cover roofs as shakes, shingles or planks. As cladding it can be used as panelling or wattle, and as flooring it can be used as boards, parquet tiles or timber sets. The sheeting is produced as fibreboard, cork, chipboard or veneer. The first two products have their own glue in the raw material which allows them to form sheeting; the latter two need added glue. This is usually urea formaldehyde glue added in a proportion of 2–12 per cent by weight. Laminate products are also made with chipboard in the middle and glued-on veneer or different types of plastic sheeting, often finished to look like timber.

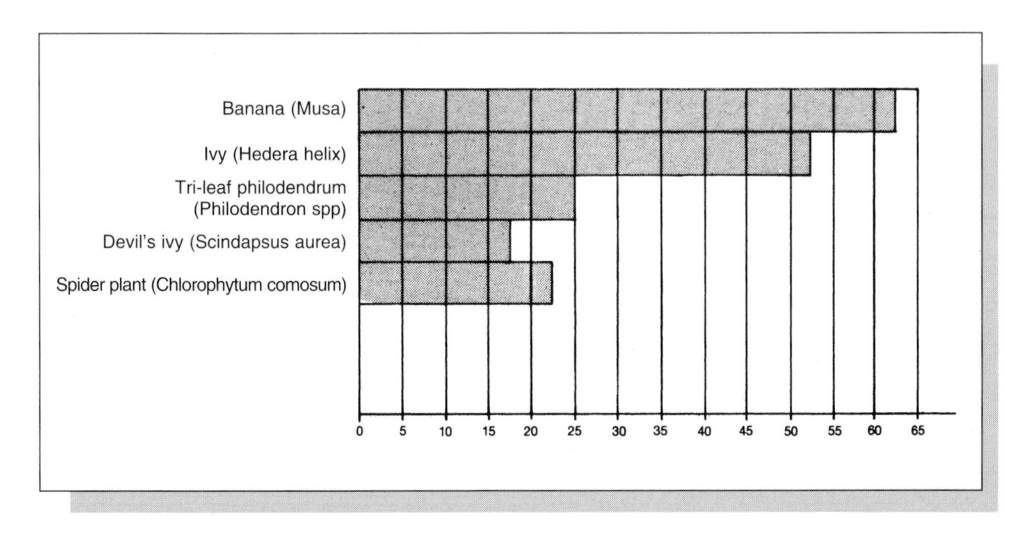

Figure 15.15: The absorption of formaldehyde by different plants, given in thousandths of a microgram per 24 hours, with a total leaf area of 0.54 m² per plant. *Source: Trädgard, 1989*

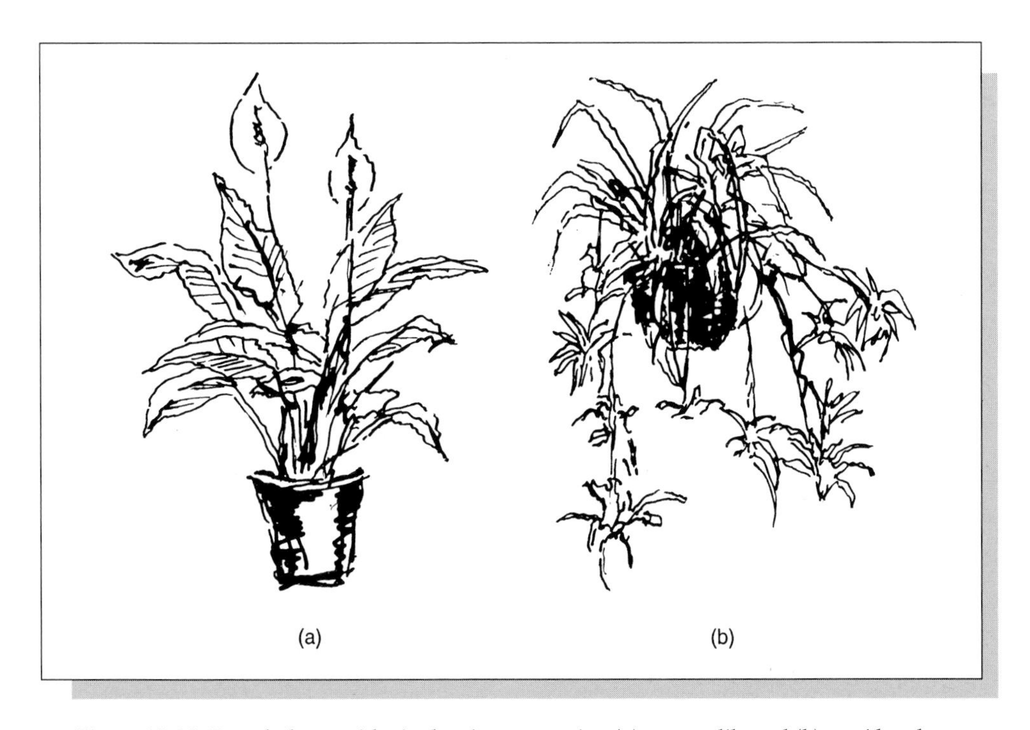

Figure 15.16: Potted plants with air cleaning properties: (a) a peace lily and (b) a spider plant.

All types of timber, both softwood and hardwood, are used for this sort of work, with very few exceptions. Products made of chipboard have no particular demands as far as quality is concerned and can even be made from wood shavings from demolition timber. The materials used for glue in parts of the production process and the impregnation materials used in external timber cladding come from questionable sources.

Timber is often a local resource, and all surface materials made of whole timber can be made locally. Timber is treated best at small mills. It is clear that it needs human attention, and there are limits to how mechanized a sawmill should be.

Durability is dependent upon the climate, the quality of the material and the workmanship, but is generally good as long as the timber is not over-exposed to damp. Artificially fertilized and quickly grown timber are undermining this opinion, and could lead to the down-grading of timber as a building material. Timber roofing is not suitable for damp coastal climates with a great deal of variation in temperature.

The primary energy consumption varies from product to product, but is generally low to moderate, with the exception of fibreboard.

There are generally no environmental problems relating to the production processes at sawmills or joinery shops. Wood dust can, however, be carcinogenic; this is particularly the case for oak and beech. The use of synthetic glue and impregnation liquids can pollute the working environment as well as the immediate natural environment, as effluent in either water or air.

Timber is generally favourable in the indoor climate, having good moisture-regulating properties, but these are often eliminated by treatment with varnish or vapour-proof paints. Untreated timber has good hygienic qualities. It proves to have far less bacterial growth on its surface than the equivalent plastic surface. Chipboard and veneer can emit gases from glues that have not set, mainly as formaldehyde. Pine can release smaller amounts of formaldehyde which can cause reactions in people who have very bad allergies.

Pressure-impregnated timber or timber treated with creosote should not be used in greenhouses or on roofs, where the rainwater passing over the timber runs into soil for cultivating food. Handling of creosote-impregnated materials can cause eczema on the hands and feet even without direct contact. Bare skin has to be protected. Creosote can also damage the eyes, and cause more serious damage to health.

Technically, all sheeting and boarding can be re-used when fixed so that removal is simple. Making all materials easy to dismantle would be a great advantage, especially in interior use. Re-use of exterior timber boarding panels or timber roofs would not be practical. These are surfaces that are exposed to all the elements and get worn out over the years, so there would usually be no purpose in re-using them.

Table 15.5: The use of solid timber as a surface material

	Roofing	*External cladding*	*Internal cladding*	*Floor*
Pine	x	x	x	x
Spruce	(x)	x	x	(x)[1]
Larch	(x)	x	(x)	x
Juniper		x		
Oak	(x)	(x)	(x)	x
Aspen	(x)	x	x	x[2]
Birch			x	x
Maple				x
Ash			x	x
Beech			x	x
Elm				x
Lime			x	
Common alder			x	
Grey alder			x	

Notes:
x Primary use.
(x) Secondary use.
[1]Better wearing when painted or varnished.
[2]Primary use; not so hard-wearing, but soft and warm.

Solid timber and fibreboard that is untreated, or treated only with natural products such as linseed oil, can be burnt for energy use in normal boilers or made into compost. Glued products have to be burned in incinerators or boilers with special filters in the chimney. Impregnated products cannot be burnt to produce energy, but have to be dumped on a special refuse tip. All-wool waste can lead to an increase in the nutrient level of the water seeping from the tip.

Roof covering

Spruce, pine, oak, aspen and larch can be used as roofing. Roofs can either be covered with cleft logs or planks, or with smaller units such as shingles. All the methods of timber roofing have one common requirement: they must prevent water gathering anywhere which would lead to fungus attack. This requires reasonably steep roofs and timber which has a mature quality, rather than fast grown timber. It may even be necessary to impregnate the timber.

The weight of a roof covering varies from 25–40 kg/m^2 according to how the roof is laid and the type of timber. The insulation value varies for the different types of timber, but is generally of no consequence. The use of timber roofs is

Table 15.6: The life span of different timber roof coverings under favourable conditions in a dry, cold climate

Type	*Life span (years)*
Shakes:	
no impregnation with steep roof	More than 100
maintained with tar, steep roof	More than 200
maintained with tar, shallow roof	More than 100
Cleft log roof	Probably very high
Plank roof, maintained with tar or linseed oil	30–50
Plank roof, pressure impregnation	60–80

often limited to small buildings in the countryside. This is because of the high risk of fire, especially when the roof is treated with tar. Thick materials usually give a better fire resistance than thin materials.

Any form of roofing has to be ventilated underneath. On non-insulated inland outhouses, the roof covering can be laid directly onto battens fixed to the roof trusses. On housing and in areas exposed to hard weather it is necessary to have a good roofing felt under the battens and a double batten system to allow water to run down under the battens carrying the timber roofing.

The materials for a roof need to be carefully chosen and the angle of the roof is critical. The steeper, the better. The stave churches have falls of up to 60°. Still much older shakes can be found on the wall than on the roof.

Timber is the roof covering with the least negative effect on the environment in terms of the use of resources and pollution during the production process, as long as it is not impregnated.

It is to the timber's advantage if the roof surface is treated with wood tar, preferably from beech, or linseed oil. Smaller timber components such as shingles and shakes can be put into a linseed oil bath and warmed to a maximum temperature of 70°C. In certain coastal areas, cod liver oil has been used instead of linseed oil. The oldest preserved shingles are to be found on the walls of Borgund stave church, Norway. They have been regularly painted with wood tar every fourth year since the late Middle Ages.

Liquids for impregnation based on poisonous mineral salts or oil- and coal-based poisons (see Table 19.3), will be washed out into local groundwater or soil.

The cleft log roof

This consists of half-cleft trunks laid over each other. This type of roof is very widespread in Finland and Sweden. Cleaving the timber gives a much more damp-resistant surface than sawing and chopping (see 'Splitting', p. 168). This roof has a longer life span than others, as long as drainage is adequate. The lower layer is often made of planks instead of

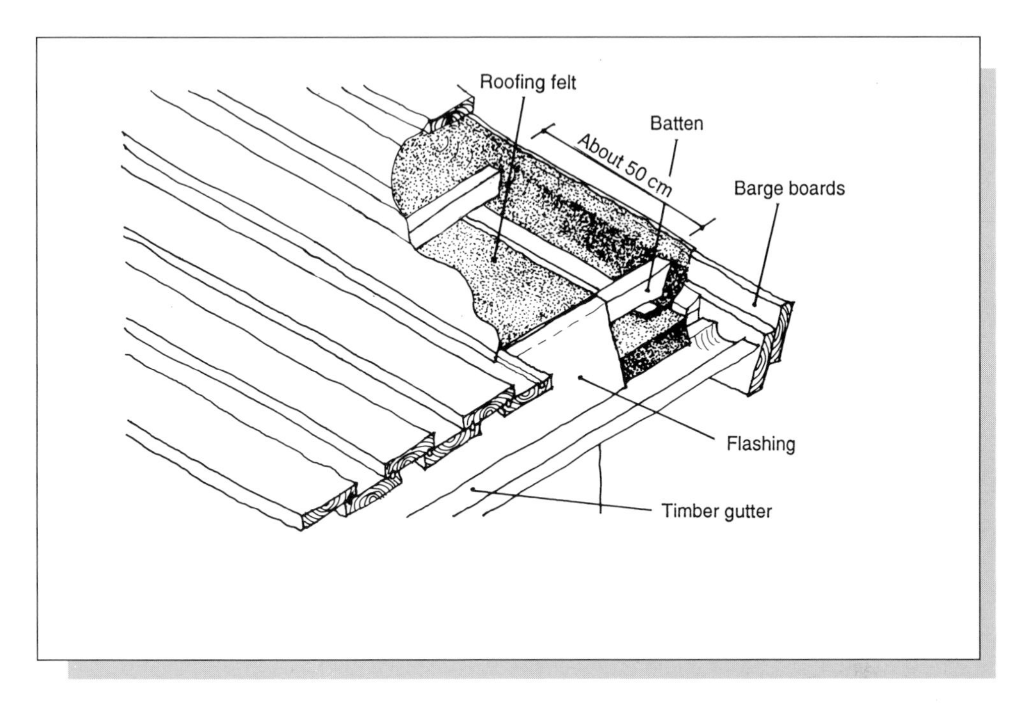

Figure 15.17: A plank roof. *Source: Norwegian Building Research Institute*

half-rounded timber and is therefore easier to lay, but the durability is probably not as good. If low quality half-round timber is used, the wood will swell and soon make the roof leak.

The plank roof

This is based on the same principle as the log roof, with planks lying on top of each other and running parallel to the slope of the roof. High quality pine should be used in less than 15 cm widths to reduce the chance of cracks forming. There should be grooves on the edges of the upper and lower planks for draining water. The planks are laid so that they press against each other when they swell in damp weather. The side with the inner grain of the tree must face upwards, especially in the case of the top planks. The root part of the log has the best quality and should lie on the lower part of the roof. The plank roof is often used as a base for other roof coverings.

The 'Sutak' roof

This is a method of roof covering that can only be used for steep roofs. Sutak roofs are usually found on small roof towers or ecclesiastical buildings and seldom in any other situation. The boarding is nailed onto the roof structure parallel to the ridge with about 5 cm overlap, with the inner grain facing upwards. This method was often used on the oldest stave churches.

Shakes

Logs that are to be used as shakes have to come from a mature tree and be well grown without any penetrating knots. The trunk is sawn up into 30–65 cm-long stumps and then split into quarters. The pieces are often boiled to reduce the chance of cracking when being cleft, but heating to over 70°C also makes the resin melts out, and impregnating effect is lost.

Cleaving is performed using a special knife which is 35 cm long and has a handle on each end. The sharp blade is usually placed radially on the end of the log and knocked in. As long as the blade is kept at right angles to the rings, it is possible to cut in at the side of the log. Rainwater is later taken off the roof in the perfectly formed annual rings. The shakes should be about 2–3 cm thick. It is also possible to cleave the shake with machine.

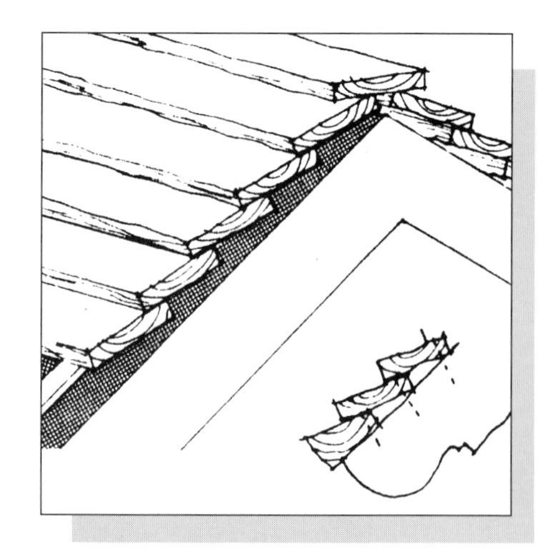

Figure 15.18: A Sutak roof. Source: Eriksen

The shakes are put on battens using the feather boarding principle with 2–3 mm between them to allow for shrinkage and expansion. A normal covering consists of two or three layers. They are nailed with wire staples so that the holes are covered by the next layer. Usually one staple per shake is enough. The staple should not be so long that it penetrates both the battening and the roofing felt. The laying details are shown in Figure 15.20. The shakes can be shaped in many different ways, the most complex being reserved for ecclesiastical buildings.

Archaeological discoveries show that shake roofs have existed since the early Bronze Age. Around 230 BC the majority of roofs in Rome were covered in shakes.

Shingles

Shingles are sawn by a circular saw. They are 40 cm long and 10–12 cm wide with a thickness of 1 cm at the lower end and 0.5 cm at the upper end. They are laid next to each other with a spacing of about 2 mm, usually in three layers, which means that the distance between the battens is about 13 cm. In the nineteenth century the majority of buildings in New York were roofed with shingles.

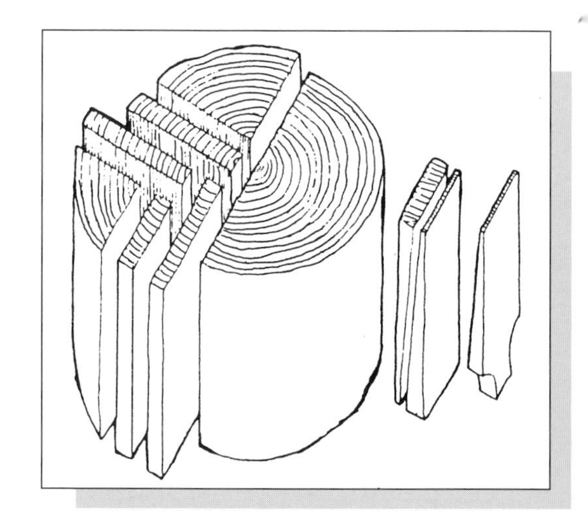

Figure 15.19: System for cleaving shakes by hand for softwood. Oak shakes are always cleft radially in the wood. Source: Vreim 1941

Timber cladding

Timber panelling has a long tradition as a cladding material, first as external wall panelling and later as internal wall and ceiling cladding. The different types of cladding have changed slightly in recent years, particularly to suit mechanical production. Special forms of panelling include cladding of shingles and shakes. Cladding with twigs and branches also has a long tradition in certain countries. Juniper

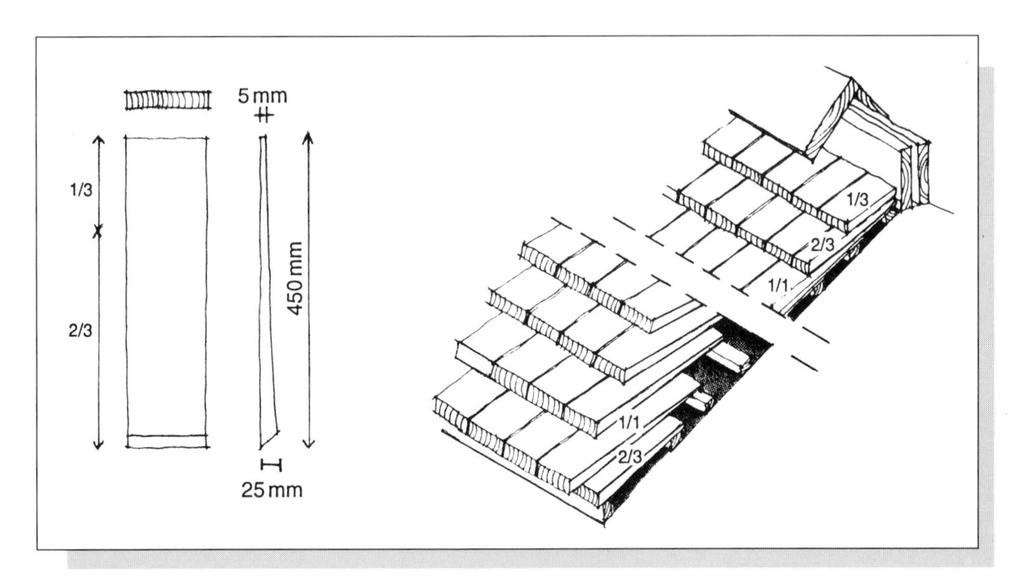

Figure 15.20: A traditional Norwegian technique for laying shakes in three layers. Source: Eriksen

is widely used and gives functional, long lasting protection against the elements.

Panelling for external walls should preferably be of high quality timber with no signs of rot. The planks should be sorted on site and the best ones placed on the most exposed façades of the building. Nailing through two planks should be avoided: they may split through natural movement. External cladding should be nailed at an upward angle to avoid water seeping in and staying there.

Timber panelling on an external wall is usually far more durable than the equivalent panelling on a roof. It is still important to choose the right system of panelling and use the correct form of chemical or 'constructive' timber treatment (see chapter on 'Paint, varnish, stain and wax' and 'Impregnating agents and how to avoid them', p. 429).

Interior wooden cladding has a very resilient finish compared with alternatives, and the surface has very good moisture-regulating properties if untreated, or treated with oil or lye.

Interior cladding materials can often be re-used, depending upon how they are fixed. There are building systems with standard components which make it possible to re-use materials several times over. External cladding is seldom re-used. It is therefore important to choose a surface treatment allowing burning or composting of the material. Impregnation of materials usually leads to having to dump them at special tips.

Different types of cladding

Figure 15.21: The construction of demountable internal panelling.

Exterior horizontal panelling

This is best used in exposed coastal areas. Driving rain runs off more easily and has more difficulty getting behind the panelling. The boards should be cut so that the stronger heartwood is facing outwards. When mounting the panelling the best quality boarding should be furthest down, where the panels are exposed to water and mud splashing from the ground.

Exterior vertical panelling

Driving rain can penetrate vertical cladding more easily so this type of cladding is more suitable for inland building. It is an advantage to have the heart side on the outside in all the panelling. It is also a good principle to lay the boarding the same way as it has grown, because the root end has the most heartwood.

Exterior diagonal panelling

This is very popular on the continent, especially in Central and Eastern Europe, because cut-off ends of boarding and shorter pieces of board can be used. In very harsh climates, diagonal panelling should not be used, as water does not run off as well as from other types of panelling.

Interior panelling

The strength of timber is not so critical for internal use, and of the softwoods, spruce is most economical. Quickly-grown timber serves the purpose, as do certain hardwoods. Birch is resilient. Aspen has a comfortable surface and a relatively good insulation value, and is often used in saunas. Because of its lasting light colour, it is also attractive as a ceiling. Other timbers appropriate for interior panelling are oak, ash, elm, lime and alder. Alder is particularly good for bathrooms, because it tolerates changes between very damp and very dry conditions. To reduce dust accumulating on the walls, it is better to have vertical boarding.

Interior panelling can best be re-used if it can be removed without damage, and should be fixed so that it is easily removable.

Shakes and shingles

Shake-clad walls have been and still are popular on the continent. Shingle cladding has been so popular in the USA that it dominated the building market, even in towns, around the turn of the century. The method for mounting on walls is the same as for roofing. The problem of water gathering is eliminated, and the life span is therefore much longer than the equivalent roof covering.

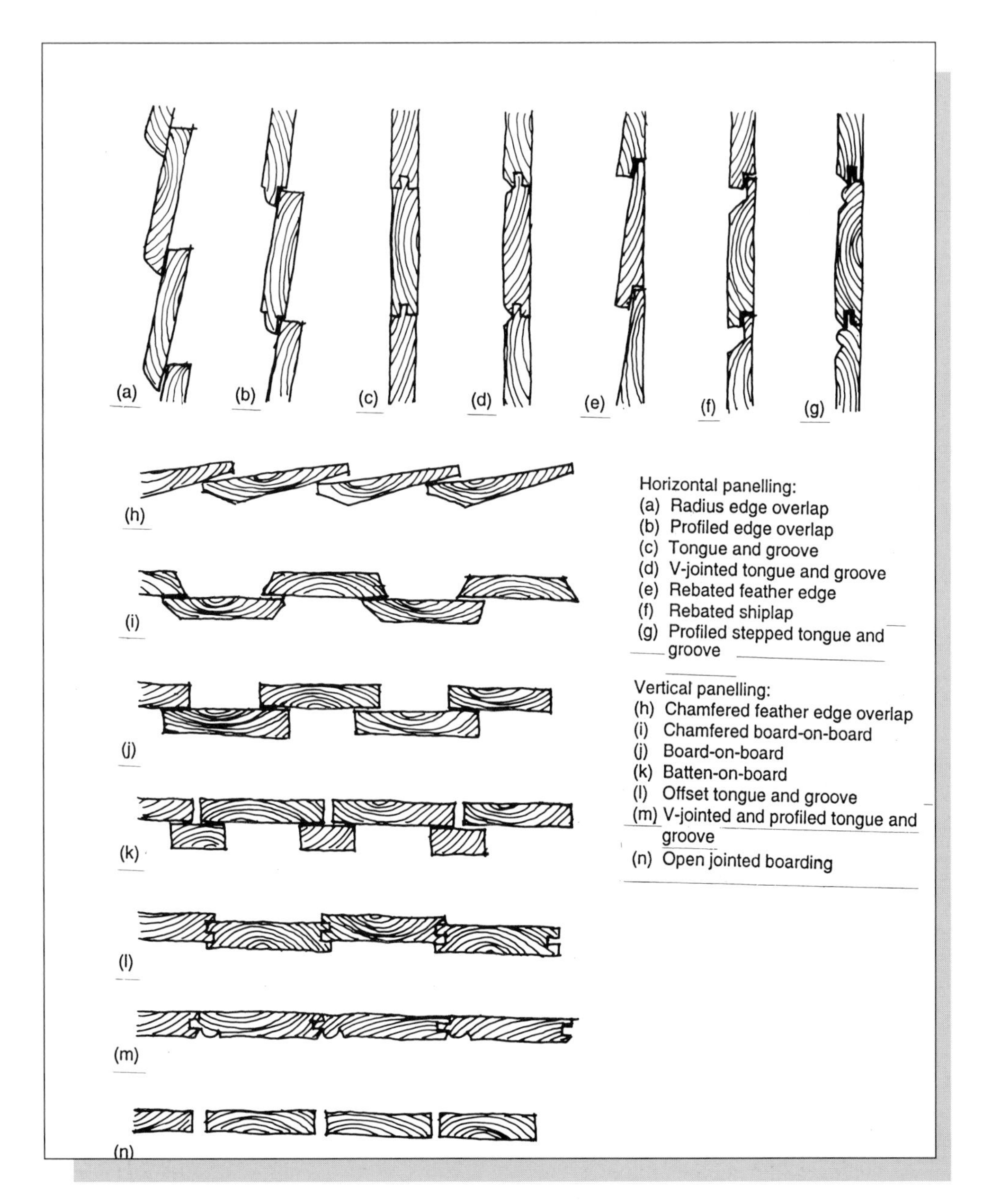

Figure 15.22: Timber panelling: (a–g) horizontal panelling; (h–w) vertical panelling.

Wattle-walling

This has been used since prehistoric times. The dimensions in this sort of construction can vary a great deal. The key to working is elasticity. If the branches are flexible enough, they can be plaited on poles for several metres. There are two types of wattling: rough and light wattlework.

Rough wattlework

Rough wattlework has been done in birch, ash, pussy willow and rowan. The bark is removed and the ends burned until they are black, achieving a sort of impregnation. The usual length of branches to be plaited is about 3–4 m. Poles are fixed between the top and bottom plates at a distance of about 50–60 cm, then the branches are woven in between so that the top ends and root ends alternate. The layers are pushed down to make them compact. Weaving can also be vertical on poles fixed between vertical studwork.

In Denmark and further south in Europe this wattlework is used as an underlay for clay finishing between the posts in timber-framed buildings. In its pure form this technique can be used for visual barriers or windbreaks on terraces and balconies, or for walling in sheds, etc.

Figure 15.23: Detail of bracken cladding. Source: Dag Roalkvam

Lighter wattlework

Lighter wattlework consists of twigs, usually juniper with leaves, but birch and heather can also be used. The juniper is cut around midsummer, as that is when the twigs are toughest and the needles most firmly attached to the tree. The same can be said for birch, which can also be used with the leaves attached.

Branches of about 50 cm in length and 1–1.5 cm thickness are cut and woven on horizontal poles at 20 cm intervals so that each branch lies inside one pole and outside two. The tops hang wide apart enough so that the cladding forms three layers, two layers outside and one layer inside each pole. The wattlework is pushed together with a hammer to make it tight. An extra branch of juniper put straight across, over the poles on the outside, increases the strength of the wall. Finally the wall is cut, and battens placed against the roof and on the corners so that the wind cannot lift it. At first the cladding is green; in time it becomes brown and dark grey, and after 30 years so much wild moss grows that it becomes green again. The main use of light wattlework is as cladding for outhouses built of staves, but juniper clad wood stores and even log houses also exist, and the cladding acts as a very good protection against all weathers.

With the introduction of building paper and wind-proof boarding, wattlework can be seen as a viable alternative cladding. The wind-proofing qualities are then not so important, but the visual qualities and durability of this sort of cladding brings advantages.

Examples show that wattle-cladding is as effective and durable as timber cladding. Juniper cladding is particularly good, and has had a functional life span of between 50 and 60 years, and even up to 100 years in the western fjord landscape of Norway. During a period of this length in this particular area, it is usual to change timber panelling at least twice. A juniper wall also has the advantage of being maintenance free, but one major disadvantage is that the wall is relatively flammable, and sparks from a bonfire or chimney can ignite it.

Wooden floors

Wooden floors give good warmth and sound insulation. They are relatively soft, warm, physically comfortable and do not become electrostatically charged if not treated with varnish. In addition, they are hard-wearing and relatively resistant to chemicals, but they need to be kept dry. Maintenance requirements are moderate.

It is difficult to specify the period in which the wooden floor first appeared. In the country, rammed earth or clay floors were common as late as the Middle Ages, but in the towns, stronger, drier floors were needed. As well as stone or tiled floors, wooden floors were quick to spread during this period. In buildings with several storeys there was no alternative. Boards, planks and cleft tree trunks were used next to each other, usually on a system of joists or directly onto the earth.

Wooden floors are usually made of high quality spruce, pine, oak, beech, ash, elm, maple or birch. Aspen is less hard wearing, but is well suited for bedrooms, for example. Aspen floors are soft and warm and have also been used in cowsheds and stables where they tolerate damp better than spruce and pine and do not splinter.

A floor has to be treated after laying. This can be done with green soap, varnish, lye or different oils (see recipes for surface treatments in the chapter on 'Paint, varnish, stain and wax'). Wooden floors are hard-wearing and durable, but should be thick enough to allow sanding several times. Timber to be used for floors is artificially dried, unlike other solid timber products, involving an increased use of primary energy which is initially relatively small. With the batten floor system, the timber can be laid after being dried outside to about 16–17 per cent. This can also be done for ordinary floorboarding by letting the boards lie together unfixed for half a year, when they are fitted together again and fixed permanently.

Floors that are treated with lye, soap or linseed oil are warm and anti-static and good moisture-regulators. Varnished floors are cold and vapour-proof, but their shiny surface makes them easier to maintain. This is, however, only a short term solution as the layer of varnish will slowly but surely split, especially where there is heavy traffic, then the floor needs re-sanding and varnishing. Oiled floors are renewed by just repeating the treatment on the worn parts of the floor.

Nailed and screwed wooden floors can, in theory, be re-used. In practice it depends upon how the boards have been fixed. Pure timber floors which have been treated with soap, lye or linseed oil can be composted or energy recycled in ordinary furnaces. Laminated timber, glued and varnished floors can be energy recycled using a special filter system for emissions, or they can be dumped at special tips.

Different types of wooden floor

Solid timber floor

The floorboards usually are tongued and grooved and can be bought in thicknesses of 15–28 mm. They are preferably laid with the hard-wearing pith side upwards. There are two main principles for laying floors: the floating and the nailed floor.

On a floating floor the floorboards are glued together along the tongues and grooves. The floor lies free from the walls, possibly on an underlay, and is held down by strong skirting boards. This method reduces the chance of recycling as it is difficult to remove the floor without damaging or breaking it.

In the nailed floor the floorboards are fixed to the joists with nails and no glue. To make it possible to re-use the floorboards, it is important that the nails go through the boards from the top and straight down. This is, however, seldom done.

Batten flooring is a mixture of the first two methods (see Figure 15.24). The floorboards are locked into position by battens of hardwood. Re-use possibilities are very high. This floor can be laid without being dried in a chamber drier, because it is easy to put them closer together by loosening the battens. Unlike other timber floors, in battern flooring individual floorboards can easily be changed.

Floor base

Floor base provides a surface for different floor finishes. It usually consists of rough spruce or pine boarding; timber from deciduous trees can also be used. The boards are nailed to the joists. This type of floor should be allowed to settle for a year before laying the floor covering. It provides a good working surface for other carpentry work, even if it cannot carry heavy loads due to the lower quality of the timber. Low quality spruce is usually used.

Parquet

The material normally used for parquet flooring is hardwood such as oak and beech. Birch and ash can also be used. These are sawn into long boards of 50–130 cm, or short boards of 15–50 cm, and are tongued and grooved. The short board is 14–16 mm thick; the long board is 20 mm thick. The breadth varies from 4–8 cm. A number of laminated parquet floors have a top layer of hardwood 4–6 mm thick glued onto a softwood base of chipboard. Urea glue is usually used for this. Parquet flooring is nailed or glued directly to the floor structure or onto a floor base. It can also be laid with a bitumen-based glue onto a concrete floor or onto battens in a sand base.

Small timber cubes

These are placed on an underlay with the grain facing upwards. Spruce, pine or oak can be used. This type of floor is comfortable to walk on and it effectively dampens

Figure 15.24: Batten flooring under construction.

the sound of steps. It is hard-wearing, and very suitable for workshops. It is easy to repair and tolerates alkalis and oils, but expands in response to damp and water and should not be washed down. The cubes are usually 4–10 cm high. The proportion of length to breadth should not exceed 3:1. Offcuts from a building site can be used. The cubes are laid in sand, and the joints are filled with cork or sand and then saturated in linseed oil. On industrial premises it is usual to dip them in warm asphalt before setting them.

Timber boarding

There are, in principle, three types of boarding made from ground timber: fibreboard, chipboard and cork sheeting. Plywood boards are usually made of larger wood sheets glued together. Fibreboard and chipboard are almost exclusively used as underlay on either floors or walls. On floors, they can provide the base for a 'floating' wooden floor or soft floor coverings; on walls and ceilings they can provide a base for wallpapering, hessian or paint. Certain products are delivered from factories with these finishes already mounted. Cork sheeting is usually placed on this sort of boarding and is often coated with a protective layer of polyvinyl chloride. Veneer products are often exposed when used in false ceilings, etc.

Fibreboard for covering is produced in porous, semi-hard or hard variations from wood fibre. The porous products are glued by their own glue which is developed through heating. The same principle is usually also applied for the semi-hard and hard boards. Some products have up to 1 per cent phenol glue added. Cork sheeting is made from broken up bark from the cork oak. This, too, could utilize its own glue, but phenol or urea formaldehyde glue is often used. Chipboard is produced from ground timber waste with 10 per cent by weight of urea formaldehyde glue added. The veneer is made of thin veneer sheets which are glued onto each other. The usual glue is urea formaldehyde at 2 per cent by weight.

Low quality raw material is used for chipboard in particular. Even timber from demolition sites can be used. The timber for fibreboard has to be relatively fresh so that the natural glues are available. The quality of timber for veneers needs to be medium to good. The phenol and urea formaldehyde glues that are used are based on coal-tar. Fibreboard manufacture has a very high consumption of primary energy; other products use much less.

In the completed building fibreboards are not a problem, and because they are porous they have good moisture-regulating properties. The glued products, however, emit gases, e.g. from formaldehyde. This has caused a great number of problems in the indoor climate. Much work has been done recently in the chipboard industry to reduce these emissions, for example with so-called 'E1' boards, which do not damage the indoor climate as much as the earlier boards. Urea formaldehyde glue is only partly resistant to damp, so if it gets damp during transport, on site or while being painted with a water-based paint, even the E1 board will give off much higher emissions than a factory dry board. Phenol glued cork sheeting has also been known to cause problematic emissions. Other types of surface treatment and glued finishes can also cause problems and need to be evaluated individually. Cork coated with polyvinyl chloride can become quite heavily electrostatically charged.

There is little chance of these products being re-used, with the exception of those made of hard fibreboard and plywood. In theory, old chipboard can be ground for new production but the centralization of manufacturing plants makes it less practicable. Pure fibreboard can be burnt for energy in normal furnaces, while other products need special filter systems for the fumes. With the exception of products containing phenol, all others can be composted. Formaldehyde glue is quickly broken down by natural processes. Unused building and demolition waste must be deposited at certified waste tips, as these products can increase the nutrients in the water seeping from the tip. Products containing phenol have to be deposited at special dumps.

Production of fibreboard

The raw material used is relatively fresh waste timber from sawmills and the building industry. The most common timbers are pine, spruce and birch. Low quality timber that still has its bark is ideal. The machines at sawmills that strip the bark have caused this particular resource to become quite rare. Leftovers from sawing planks and boarding are not often used in fibreboard production, but can be used if they are cleaned of any cement and all the nails are removed. Waste paper is used for the surface layer for both porous and pressed sheeting, but can also be used in the main pulp used for the porous boards.

Porous boards are usually made in thicknesses of 12–20 mm, though thicknesses up to 40 mm are common. The thicker board needs more time to dry and is most commonly used for insulation. As a raw material spruce is best, but pine or a hardwood can be mixed in, up to a maximum of 10–15 per cent.

Semi-hard boards do not need such a high standard of raw materials, and can contain a larger proportion of pine. They are usually produced in thicknesses of 6–12 mm, and the hardboard in thicknesses of 3–6 mm.

The manufacturing process consists of the following stages:

1. The raw material is collected and shredded by a shredding machine.
2. The shredded wood is washed of any polluting substances.

3. It is heated and ground between two coarse steel rollers.

4. The mass of fibres is mixed with water to a thin pulp and made into sheets on a moving band.

5. The sheets are put through a press heated to 200°C at a certain pressure, depending upon the degree of hardness required. The natural glue liquor is extruded and binds the board.

6. The sheets are cut to standard sizes.

7. The boards are hardened by warm air, at about 165°C, for two to seven hours.

8. They are then conditioned in warm humid air to give them a moisture content of 5–8 per cent.

Production of chipboard

Chipboard can be made from many types of timber. There is no need for the timber to have its own active glue, as the process includes gluing. Urea formaldehyde glue is used for both economic and technical reasons, but melamine, phenol formaldehyde/resorcinol and

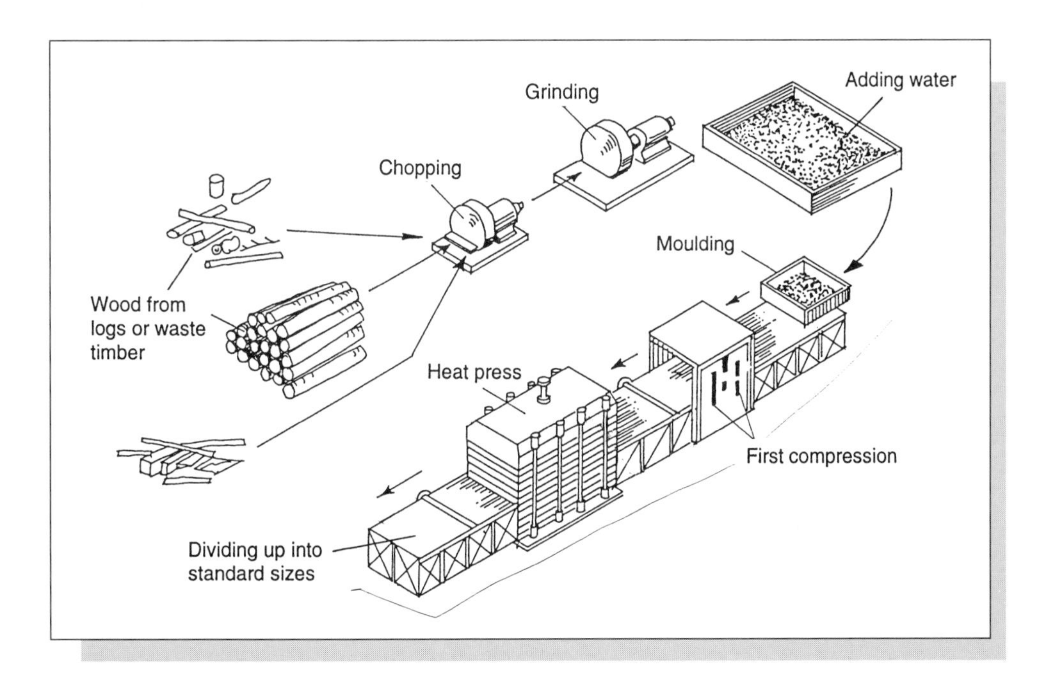

Figure 15.25: Production of hard and semi-hard wood fibreboards.

polyurethane glue may be used. A German manufacturer has recently introduced a chipboard glued with timber-based lignin glue. Waterglass glue can also be used, but is not available commercially at the moment.

The manufacturing process of chipboard is as follows:

1. The timber is shredded.
2. The shredded timber is ground to shavings.
3. The shavings are dried to a moisture content of about 2 per cent.
4. Glue is added. The amount of glue by weight is approximately 7–12 per cent.
5. The pulp is made into a sheet on a moving band.
6. The sheet is then pressed at 180–200°C.
7. The boards are dried and conditioned to the desired moisture content.

Production of plywood

Plywood is produced in different forms and from many different types of timber, including tropical species, through sawing, cutting by knife or peeling. Sawn plywood is mainly used in the production of furniture and is produced by sawing the log along its length in thicknesses of 1.5 mm or more. The other two types of cutting are used on logs that have been boiled or steamed until they are soft and pliable. Cutting by knife is done along the length of the log as with sawing. By peeling the veneer is peeled off the rotating log like paper pulled from a toilet roll. A plywood board is made by gluing the veneers together. This can be done in two ways, to make blockboard sheeting or plywood sheeting. Blockboard consists of wooden core strips glued together, usually of pine, which are covered both sides with one or two veneers. Plywood consists purely of different veneers glued together. There is always an uneven number of veneers so that the resultant sheet has an odd number of layers. The adhesive used nowadays is usually urea or phenol glue in a proportion of about 2 per cent by weight. Animal, casein and soya glue give good results as well.

Straw and grass sheet materials

Throughout European history many plants have been used as roof and wall cladding, mainly the different types of straw such as wheat, rye, flax, oats, barley, marram grass, reeds, ribbon grass, greater pond sedge and eelgrass; even the bregne species of grass. Plants can be used as they are, possibly cleaned of seeds and leaves, and some can even be used to make sheeting. In addition to the ordinary conditions a surface material has to fulfil, plant materials often give a good level of thermal insulation and good moisture-regulating properties. It has to be accepted that thatching is flammable. Eelgrass is less susceptible to burning because it contains salt and a large amount of lime and silica. Sheeting material made of eelgrass is considered more fire-resistant than the equivalent timber fibreboards.

In excavations made in Lauenburg, Germany, there are indications that buildings were thatched with straw as long ago as 750–400 BC. In Denmark this sort of roof is believed to have been in use for at least 2000 years, also, particularly on the islands of the Kattegatt, eelgrass has been traditionally used for roof covering and wall cladding.

The use of thatched roofs has decreased considerably since the turn of the century. This is partly due to insurance companies demanding higher premiums due to the higher fire risk, and partly because of the mechanization of agriculture. Straw that has gone through a combine harvester is unusable. In Germany and the Netherlands, reeds have almost become non-existent through land drainage. In Europe today the raw material is imported from Poland, Bulgaria and Romania. Even Denmark has difficulty supplying its local needs.

In England, Germany and the Netherlands thatching is still a living craft. Further south, roofs built from plants still dominate many cultures. In India, for example, 40 million houses are covered with palm leaves and straw.

Ecologically speaking these materials are very attractive. They are constant resources which are otherwise never used. The production processes do not require much energy and produce little pollution. In buildings the products usually have no problems. Sheeting products often have adhesives added, such as polyurethane glue at 3–6 per cent by weight. This reduces the environmental quality somewhat. As waste, the pure products can be composted or energy recycled. For the products containing adhesives filters are required for the fumes that come from their incineration, and waste has to be deposited at certified tips.

Roof and wall cladding with grass

Many different types of grass can be used for roofs and walls. Harvesting and laying methods for all coverings are labour intensive, although parts of the harvesting process for reeds could be mechanized relatively easily. The harvesting of eelgrass could also be made more efficient. In Denmark, a mobile harvesting machine for straw roof coverings is already in use. Here, the grain is removed without destroying the straw. During the three month long summer season this harvesting machine can produce straw for 200 roofs covering 180 m^2 each, but it is generally difficult to see any way of making the actual thatching process more efficient. Thatched roofs are and will always be labour-intensive.

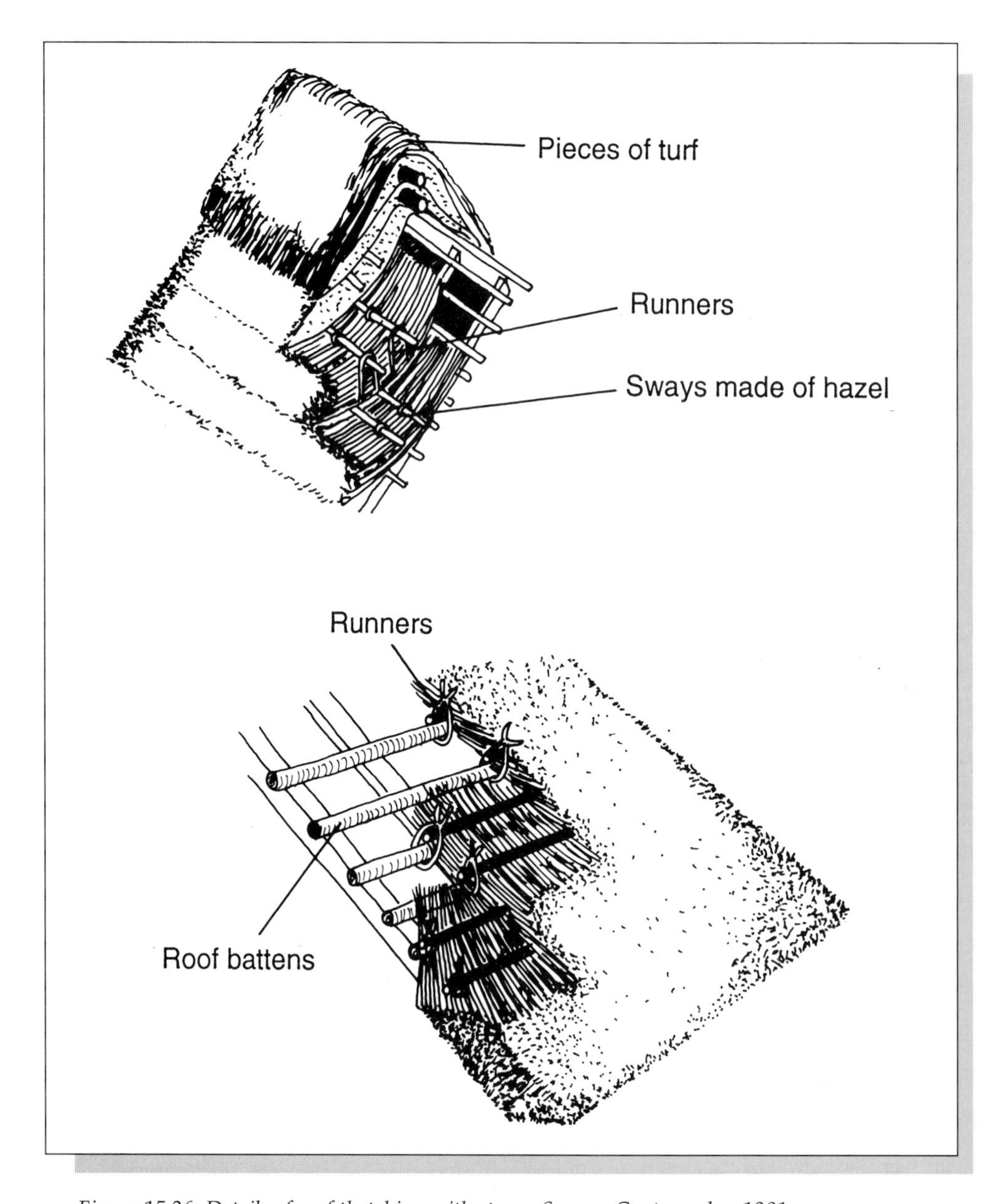

Figure 15.26: Details of roof thatching with straw. Source: Grutzmacher 1981

The durability of thatch depends upon where and how the plant was cultivated, especially in relation to heating and freezing cycles. Straw and reeds which are used on the continent today are nearly all artificially fertilized, which produces enlarged and spongy cell growth resulting in a far shorter life span than usual.

The durability of different roof coverings

Plant	*Artificially fertilized (in years)*	*Natural (in years)*
Reeds	30	50–100
Straw	10–12	20–35
Eelgrass	–	200–300
Bracken	–	8–10

(Hall, 1981; Stanek, 1980)

The long life span of eelgrass is due to its high content of salt, lime and silicic acid. It is therefore not so readily attacked by insects – a particular problem in normal thatched roofs. The most stable of the different cultivated grains is rye.

Strong sun generally causes splits and breaks down thatched roofs. They survive longer in northern Europe than further south. At the same time there can also be a different life span between the north and south facing parts of the roof. All organic material can return to earth as compost.

Straw

Thatching

When thatching with straw a series of battens (sways) are erected on the rafters at 30 cm intervals. Bundles of straw are laid edge-to-edge on these battens, one layer on each sway. Every layer is bound down by runners which are bound to the sways, preferably with coconut twine. The completed roof is evened out using special knives to a thickness of approximately 35 cm. The ridge is usually made with turf cut into 1–2 m-long pieces. On the inside of the rafters it has been the custom more recently to place fire-resistant insulation boards of woodwool cement. Good ventilation from the underside of the roof is important. As with timber roofs, the rule of the steeper the roof, the longer it lasts, applies. The usual slope in normal climatic conditions is 45°, while along coasts it should be up to 50°.

Wall cladding

This method of cladding has never been widespread. Traditionally the most usual material was rye, which was bundled together and threshed without destroying the stalks. Weeds and loose straws were combed out with a special comb. Then eight or nine hoops were bound together into a yealm and trimmed with a knife.

When cladding a house with straw, it is usual to start at the bottom. Every layer should be 30 cm high, and fastened by nailing the upper part to a batten. The bundle hangs down to cover the first batten. Every layer is cut at the bottom to make it straight and even. As long as the straw cladding is intact, it will give useful extra insulation, as it holds small pockets of air.

Eelgrass

Thatching

A layer of twigs (preferably pine or juniper) is placed on battens at 30 cm intervals. The eelgrass is worked and shaken to get rid of any lumps and to make the straws lie in the same direction. Sections of eelgrass are then wrung hard to form 3 m-long scallops, in the same way one wrings water out of a floor cloth. The scallops continue out into a long, thin neck which acts as a fastening loop to the battens. The scallops are fastened close into each other on the four to five lower battens, and the rest of the roof is built up with loose eelgrass laid in layers and pulled well together. By mounting a buffer along the roof's edge similar to the turf mound on a turf roof, it is possible to manage without scallops. The roof needs to settle for a few months before a second layer is added. The total thickness is usually 60–80 cm, but there are examples of 3-m-thick roofs, which must be one of history's warmest roofings. After the final layer the thatching is cut level with a special knife. The ridge is often covered with a long strip of turf. This could be replaced with a layer of eelgrass kneaded in clay. After a few years the roof will settle down and become a solid mass with the consistency of flaked tobacco. The time is then ripe for a new layer. Rain only gets through the outside layer and then trickles slowly down to the edge of the roof. At the same time the roof is open to vapour coming from the inside of the house.

Figure 15.27: Thatching with eelgrass, Denmark.

Wall cladding

Eelgrass was often used for wall cladding on gables, using 10 cm-thick layers of combed-out seaweed of about 60–70 cm in length. These bundles were stuffed between vertical battens at 30 cm intervals. Every layer was fixed by a horizontal branch woven between the battens. Finally the gable was cropped with a long knife so that it had a smooth even surface. Like eelgrass roofing, the eelgrass gable has a very high durability, but with time will settle, and cracks must be refilled.

Plant fibre and grass boarding

The raw material for boards made from plant fibre is usually straw, but also residue from corn winnowing and even certain types of leaf can be used. Many of the different types of straw contain the same type of natural glue which binds timber fibreboards. In some products, however, it is usual to add glue.

The most common raw materials for boards or sheeting are wheat, hemp, rye, oats, barley, reeds, rape, flax and maize. It is mainly their straw that is used. Decomposed plant fibres in the form of peat can also be compressed into boards. Hardboards are mainly used internally as a base cladding, but also in some cases as external cladding. More porous boards can also be produced for use as thermal insulation. (See 'Peatboards', p. 295, and 'Strawboards', p. 291.)

Boards are not particularly resistant to vermin, and when used externally they often have to be impregnated with fungicides. If they are rendered, the problem is considerably reduced. The alkaline properties of the render prevent the growth of mould. In Sweden there are examples of this external cladding lasting 40 years. The raw materials used in these boards is environmentally very attractive, as it is based mostly on waste from agriculture. There are exceptions, in which glues and impregnation liquids have been used.

In manufacture and use these products are environmentally sound. Within the building they are good moisture-regulators. Small amounts of non-reacted isocyanates can be emitted from products that contain polyurethane glue. Pure products can be composted or energy recycled. Impregnated products or those glued with polyurethane glue can be energy-recycled in incinerators with special filters for the fumes. These products cannot be composted, but should be dumped on a special tip.

Production of strawboards

Strawboards are best produced locally in small businesses. It does not matter what state the straw is in as long as it is not beginning to rot. The moisture content before the process starts should be 6–10 per cent. The procedure is as follows:

1. The straw is cleaned in a ventilation unit.

2. The fibres are straightened and put in the same direction. If extra adhesive is required, it is added at this stage, usually in the form of a polyurethane glue in a proportion of about 3–6 per cent by weight. It may be possible to find less damaging glues. Wheat, hemp and barley do not need any added glue, even if it would give greater solidity. Flax boards seldom contain glue. Flax straws have to be boiled under pressure for a few hours before they can be used.

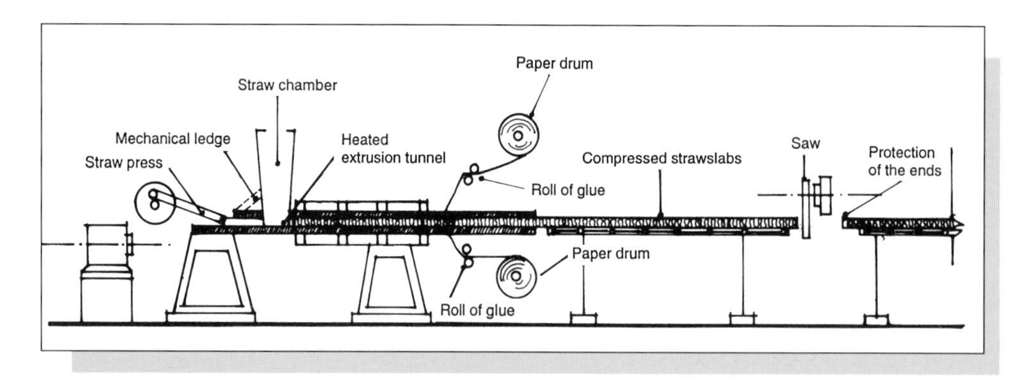

Figure 15.28: Production of strawboards. Source: Stramit

3. The boards are put under pressure in a closed chamber at a temperature of 200°C.

4. They are cooled.

5. They are cut to size.

6. The porous boards are coated with an adhesive and are then covered with a stiff paper, preferably recycled, which gives them rigidity.

Boarding from domestic waste

Boards made of waste are still at an experimental stage. There have been experiments with products to be used internally, for boarding under different finishes, on walls and on floors. The use of this raw material is very interesting environmentally speaking. It may contain contaminants such as plastic, which can affect the indoor climate negatively. There is also a risk of emissions from binding agents that may be used. There may even be a need to add fungicide to the boards. But if these boards are not treated in any way, then they can probably be recycled into the same sort of product again.

Manufacture of rubbish boards

The manufacture of these boards begins in the local authority rubbish tip and proceeds as follows:

1. The rubbish is crushed and ground. Iron is separated by electromagnets and heavy articles are sieved away.

2. The homogenized mass is dried at 148°C to a moisture content of 3–5 per cent.

3. Through centrifuging, the heavier rubbish is separated from the lighter. The light rubbish is mainly paper, plastic and food leftovers. The heavy part is returned.

4. The light rubbish is mixed with about 50 per cent wood shavings pulp.

5. The mixture is glued under pressure. Urea glue is used, but more harmless glues are being developed for this sort of use.

Soft floor coverings

Soft floor coverings are usually materials such as linoleum, plastic, rubber matting and cork. The latter is introduced on p. 351. All are dependent on having a solid, smooth floor base of concrete, timber, magnesite, rammed earth, boarding or the like.

Soft floor coverings are easily cleaned and comfortable to walk on. They are glued to the floor base, so cannot be re-used in any way. Changing these floors when they are worn out is a very labour-intensive and expensive process, almost as expensive as laying a new floor. Most of these coverings can, however, be taped to the underlay, which immediately improves their environmental profile at the waste stage as they do not become completely stuck to other materials. All soft floor coverings are delivered in rolls or as tiles.

Linoleum

Linoleum was first produced in England in 1864 and comes in thicknesses of 1.6–7 mm. A normal manufacturing procedure is to first boil linseed oil (23 per cent by weight) with a drying agent, usually zinc (about 1 per cent), and let it oxidize. This is mixed with 8 per cent softwood resin and 5 per cent cork flour, 30 per cent wood flour, 18 per cent limestone powder and 4 per cent colour pigments, primarily titanium oxide. The mixture is granulated and rolled while heated on a jute cloth (11 per cent) which is hung for oxidizing at 50–80°C. All manufacturers cover this with a layer of acrylate to make it easier to roll and stay clean. In certain cases polyvinyl chloride is used. Linoleum with no surface coating should be waxed before use.

It is normal to glue linoleum to the floor, but this should not be done before the base onto which it is glued is properly dry. A timber floor takes a year to dry, while concrete needs even longer! If a floor finish is glued too early, fungus can form in the glue, spread to the floor construction and walls and even eat the linoleum away. The adhesive usually used is Ethylene vinyl acetate (EVA) dispersion glue. A glue which contains natural latex in a solution of alcohol can also be used, or linoleum can be taped or fastened with small staples. The surface can be waxed, but too frequent waxing of linoleum can increase its static charge. Cleaning is simple and this is done with a damp cloth or with a weak solution of green soap.

Linoleum does not tolerate continuous exposure to water and is therefore not suitable for bathrooms etc.

The raw material situation for the production of the main constituents of linoleum is good; they are mainly renewable resources. The primary energy consumption is much lower than for the alternatives, plastic and synthetic rubber.

From the finished product there is a possibility that linseed oil can release oxidation products, such as aldehydes. There has also been evidence of emissions from added solvents, glue and the plastic-based surface-coating. The differences in these emissions are very large between the different manufacturers. With careful production techniques it should be possible to reduce the problems to a minimum.

Linoleum cannot be recycled, but can probably be energy-recycled or composted. Waste can lead to an increased amount of nutrients in groundwater and it should therefore be dumped at a special tip. The same applies if poisonous colour pigments have been used.

Natural rubber (latex)

The source of natural rubber products is the rubber tree. Rubber coverings contains 30 per cent by weight of sulphur powder, colour pigments and fillers of chalk and kaolin. It also contains vulcanizing agents, stabilizers, fire retardants (usually zinc oxide) and lubricants in the form of stearin, to about 2.5 per cent by weight.

Natural rubber is a renewable resource in southern climates. The primary energy consumption for these floor coverings is about half of the equivalent synthetic rubber products. Inside a building rubber flooring causes no problems. The material can be recycled if it can be removed and cleaned from the floor base.

Plastic and synthetic rubber

This flooring is delivered in three main types: polyvinyl chloride (PVC), polyolephine and synthetic rubber flooring.

Vinyl covering

This is produced with PVC mixed with fillers such as sand, chalk, kaolin, wood flour, zinc oxide, lime or powdered stone. Vinyl tiles with asbestos mixed in are still being produced in Eastern Europe. Colour pigment, softeners and stabilizers, which can often contain lead and cadmium, are also added. The usual softeners are di-oktylphtalate (DOP) and di(2–ethylhexyl)phtalate (DEHP). PVC coverings are hard and usually lie on a soft underlay of jute matting, polyester fibres, cork, foamed PVC or fibreglass.

Polyolephine covering
This is produced from ethylene and propylene. No softener is used, but stabilizers, fire retardants and colour pigments are added. It also has acrylate on its surface coating.

Synthetic rubber covering
This is based on styrene-butadiene-rubber (SBR) and has many additives: stabilizers, fire-retardants, vulcanizing agents and softeners.

These products are all based on oil which is a very limited resource. The primary energy consumption for all of the products is very high. In all phases, from production through use to waste, these products present pollution risks. In the indoor climate there is a high chance of the mucous membranes being irritated. The polyolephine flooring causes the fewest problems. From a newly laid PVC floor, up to 62 different substances are emitted, including solvents and phthalates. Phthalates are emitted for as long as the building stands, and there is clear evidence of a relationship between the occurrence of DEHP and asthma in children (Øye, 1998). Extremely high emissions have been measured from vinyl flooring on concrete because the alkali increases the breakdown of substances, including phenol emissions in some cases (Gustafsson, 1990). SBR flooring has been known to emit styrene and butadiene.

Floor coverings of PVC and SBR will shrink somewhat as the softeners evaporate, and damage can occur in the joints which makes them dirt traps and an attractive breeding ground for fungus. On all plastic surfaces, which are not moisture absorbers, the production of bacteria is generally 30 times greater than the equivalent damp absorbing surface, such as timber. Plastic flooring can normally also become highly electrostatically charged.

PVC and polyolephine floorings can be recycled, theoretically, but it is highly unlikely that this will occur in practice because of the difficulty of removing the material. SBR flooring cannot be recycled. Polyolephine flooring can probably be energy recycled at plants with particular filter systems for the fumes. All waste must be specially disposed of.

Carpets and textiles

Carpet as a floor covering has a particular function, providing a more comfortable surface to walk on. It is soft, has little thermal conductivity and a good noise absorption capacity. Carpeting can be woven, knitted, tufted or needle-punched in many different natural fibres: cotton, wool, bristles, sisal, coconut, jute and hemp, and in synthetic fibres such as nylon, acryl, polypropylene, polyacrylnitrile, polyester and rayon.

In the East, carpeting has been used for centuries. In Europe, hides have traditionally been used as floor mats. Here the first true carpets originated about 200 years ago. Until then, people managed with natural materials strewn on the floor: juniper or bracken, sawdust or sand, which absorbed dust and damp. This kept the floor clean, as it was regularly changed. Juniper also had a particularly fresh smell. In the 1960s, wall-to-wall carpeting was introduced, transforming the carpet from loose floor covering to an independent floor covering often laid directly onto concrete.

The spread of this type of covering was very rapid, in housing and in buildings such as schools, offices, public buildings etc. In the beginning natural fibres were used, but synthetic fibres soon took over, making up half of the market by 1967 and the majority of the market today.

Local raw materials for the production of carpeting are wool, flax, hemp and nettles. Timber is the raw material for rayon. Sisal comes from Mexico, while coconut is found on the coast of the tropics, where it is often an extra resource. Synthetic fibres, e.g. polyamide, polypropylene and polyacryl, are based on oil.

The first part of the manufacturing process is to clean the fibres. The procedure then varies according to the technique and material used. Weaving and knitting require spun thread. Needle-punched carpets are made of unspun wool. For needle-punched and tufted carpets, a binder is required to attach the top surface onto a woven underlay of fibreglass, or something similar. A natural rubber glue can be used for this, but a synthetic rubber glue is normally used.

All carpets, both natural and synthetic, can contain anti-static agents, and substances to protect them against moths and fungus – often ammonium compounds. Woollen products are often impregnated with pyrethrin to protect against moths. Jute can be sprayed during its cultivation or at the time of transport, in some cases with DDT. Loose carpets are laid directly onto the existing floor; fitted carpets are usually laid an underlay of PVC of foamed synthetic rubber, but even natural rubber, cork or woollen felt are possible alternatives. The carpets are pressed against the floor with skirting boards, or glued. Different sorts of adhesive can be used. Joints are sewn or glued. While natural fibre products have their origin in renewable resources, oil – the origin for plastic products – is a very limited resource. The primary energy consumption of plastic based products is also very much higher. In buildings, carpets can generally cause four particular problems:

- static electricity

- gathering of dust and the development of mould and mites
- emissions from plastic materials, diverse adhesives, impregnation substances and other additives
- loosening fibres from synthetic floor coverings.

Static electricity

The static charging is dependent on the type of fibre and to a certain extent on the material of the floor, and even the shoes of the inhabitants. There is a clear tendency for synthetic materials to produce a higher static charge than natural materials. Many methods have been tried to reduce this, but they have often been uneconomical or short term, e.g. anti-static agents. Attempts to reduce the problem by raising the relative humidity achieved negative results by increasing the production of mould and other micro-organisms.

Dust and the development of mould and mites

A connection has been made between wall-to-wall carpeting and allergies. The number of bacteria in a fitted carpet is 100-times greater than on a floor with a smooth surface. Synthetic carpets are the worst, with very few moisture-regulating properties; natural fibre carpets are a little better.

It is also difficult to clean a fitted carpet. About 35 per cent of the dirt remains in the carpet after it has been vacuum cleaned. A loose carpet that can be beaten has a great advantage over a fitted carpet.

Emissions

Up to 30 different substances have been registered in emissions from a needle-punched carpet, including formaldehyde. Levels of 4–phenyl cyclohexene and styrene measured in a needle-punched carpet on an underlay of styrene-butadien-rubber (SBR) have been so high that it has had to be removed (Gustafsson, 1990). Many coconut mats and other types of carpet have a PVC base which in turn adds to pollution of the internal air. Natural carpets can also have added poisons to combat mould and moths, which can be volatile. A commonly used adhesive such as EVA glue can release up to 34 different gases under normal circumstances.

Loosening fibres

Little is known about the effects of this phenomenon which is dependent upon the size of the fibres, their form and movement. It is assumed that it may cause risks.

The durability of carpets is relatively low, so they need to be changed regularly. If they are glued, this can cause problems.

Wall-to-wall carpeting has little chance of being re-used and has probably hardly any chance of being recycled because of the many different materials it contains. A few types can probably be energy recycled in incinerators with special filters for the fumes. Waste from plastic products and natural products with plastic-based glue, poisons against mould etc., have to be deposited at special tips. Carpets of pure natural fibres can be composted.

Wallpapers

Wallpapers have primarily a decorative purpose within a building, in the same way as painting, but can also have a role as a moisture-regulator or vapour-hindering membrane. This depends upon the type of material used. Wallpapering a room with a heavy pattern or an illustrated theme will make its mark on its inhabitants. Most of us can remember the rabbit wallpaper in our childhood bedroom! Oscar Wilde declared on his death bed: 'The wallpaper or me. One of us has to go!'

William Morris, the great wallpaper designer of the Arts and Crafts movement, stated: 'No matter what you are going to use the room for, think about the walls, it is these that make a house into a home.' (Greysmith, 1976.) There are four main types of wallpaper:

- wallpapers based on natural textiles
- synthetic textile wallpapers
- paper wallpapers
- plastic wallpapers.

Paper and textile wallpapers are best-suited to dry rooms, while plastic wallpapers are best used in bathrooms, washrooms, etc.

Wallpaper can be tacked or pasted onto different surfaces such as newspaper, plasterboard or smooth rendered concrete. It is important that the concrete has dried out properly so as to not cause damp patches or mould.

History

Textiles inside buildings have a long history. They were initially used for dividing rooms. The Assyrians and Babylonians were probably the first to paste them onto existing walls. In England, textile wallpapers were produced during the fourteenth century. In the beginning they were woven and embroidered like a tapestry, so they were in a price class that only kings could afford. During the fifteenth century the Dutch began painting simple figures and ornamentation onto untreated linen. The price of wallpaper dropped a little, and rich merchants, statesmen and higher church officials could afford it.

About 100 years later waxcloth wallpaper arrived, which consisted of a simple sacking of hemp, jute or flax covered with a mixture of beeswax and turpentine. A pattern could be printed on the surface. Waxed wallpaper was much cheaper than the earlier types of wallpaper, but it was only when it began to be made from paper that prices fell so that everyone had a chance of buying it. It was first available in 1510, initially as small square pieces of paper in different colours, pasted-up as a chequered pattern. During the eighteenth century the first rolls of wallpaper came on the market with hand-printed patterns, and around 1850 the first machine-printed wallpapers arrived.

An analysis of the many wallpaper patterns throughout history gives a good indication of cultural developments. William Morris's organic, flowery wallpapers tell of the great need to keep in touch with nature during industrialism's first epoch. Something of the same longing can be seen today, even if in a somewhat superficial way, on the panoramic photographic views of South Sea islands, sunsets, etc., which appear on some wallpapers.

Figure 15.29: A typical wallpaper pattern from the 'Golden Age' of wallpaper at the end of the 19th century. Source: Greysmith, 1976

Types of wallpapers

Wallpapers of natural textiles are usually woven with jute, but other plant fibres such as wool, flax, hemp and cotton can be used. The textile fibres are woven together and glued onto an underlay of paper or plastic. A wallpaper is also made consisting of rye straw woven together with cotton threads.

Wallpapers from synthetic textiles are mainly woven with fibreglass. The fibreglass is often used in combination with polyester thread. This is usually given a coating of plastic to prevent it from losing fibres. It is also quite normal to add fibreglass to an otherwise pure natural textile in order to strengthen it.

Wallpaper made of paper consists of cellulose, preferably in the form of recycled paper. In certain cases formaldehyde products are added to increase resistance to water. The printed pattern on the wallpaper is often glue-based paint, or emulsion, oil or alkyd paint. Until 1960, paint based on animal or plant glue was the usual paint used for printing wallpapers. Paper wallpaper often has a thin plastic coating to improve its washability.

Plastic wallpapers are based on a structure of paper or a natural textile, and usually consist of softened PVC. It can be smooth or textured. In Sweden, about 3000 tons of vinyl wallpaper is used every year.

Wallpapers of natural textiles are based mainly on renewable raw materials. Fibreglass fabric is made from quartz sand, which is considered to have rich reserves. Plastic products are based on oil, which is a very limited resource. Plastic production has a negative effect on the environment (see 'Plastics in building', p. 147).

If the wallpaper contains volatile substances, these can also cause a problem in the indoor climate. Considerable emissions of styrene have been measured from fibreglass reinforced polyester wallpaper, increasing in damp circumstances (Gustafsson, 1990). PVC coatings have a high level of emissions which can irritate the mucous membranes. Fibres from glassfibre paper are probably too coarse to be carcinogenic. Both textile and paper wallpapers cause no problems so long as no hazardous glue or other volatile substances have been used. However, if the glue is exposed to continuous damp, mould can arise.

The 'shagginess' factor can also cause problems. Large amounts of dust can gather on rough surfaces, giving rise to the growth of micro-organisms. Electrostatic charge also plays a role: the large negative charges in PVC wallpapers attract dust of the opposite charge. PVC wallpapers in themselves are also potential growth-beds for micro-organisms. It has also been observed that PVC wallpapers shrink as the softener loses its strength, allowing gaps to appear which can harbour dirt and give rise to mould.

Softeners in plastic wallpapers create a sticky layer if they are warmed which catches dust and soot.

When renovating or demolishing, it is usual to remove old wallpaper from walls. This is quite easy with paper wallpapers. Steam or hot water can be used on the soluble pastes. It is more difficult with plastic wallpapers. Wallpaper for bathrooms which has a foamed PVC underlay is difficult to remove, and will often take a piece of the wall or plaster with it. Wallpapers have no recycling value. Paper and natural textiles can be composted, providing they have no polluting or potentially dangerous additives or adhesives. Fibreglass wallpapers which contain polyester and PVC wallpapers have to be deposited on special tips.

Table 15.7: Environmental profiles of roof coverings

Material	*Quantity of material used (kg/m²)*	*Effect on resources*			*Pollution effects*				*Ecological potential*		*Environmental profile*
		Materials	*Energy*	*Water*	*Extraction and production*	*Building site*	*In the building*	*As waste*	*Re-use and recycling*	*Local production*	
Galvanized steel, from ore	6	3	2	2	3	1	2	2	✓		3
Aluminium, 50% material recycling	4	2	3	3	3	1	2	2	✓		3
Copper from ore	6	3	3	3	3	1	3	3	✓		3
Concrete tiles	50	1	2	2	2	1	1	1(1)	✓	✓	2
Sheets made of cellulose-reinforced concrete	13	1	2	2	1	1	1	1	✓	✓	1
Slate	85	1	1	1	1	1	1	1	✓	✓	1
Fired clay tiles	35	1	2	2	2	1	1	1(1)	✓	✓	2
Polyester roofing felt with bitumen	2	3	2		3	2	1	2			3
PVC sheeting	1.5	3	2		3	1	3	3			3
Timber boarding, without impregnation	18	1	1	1	1	1	1	1		✓	1
Timber boarding, impregnated	16.5	2	1	2	2	2	3	3		✓	3
Turf roof on polyethylene sheeting	300	2	2		2	1	1	2	✓	✓	2(3)
Straw thatch	25	1	1	1	1	2(2)	1	1			1

Notes:

(1) Certain colour pigments with heavy metals make it necessary to give the material a lower evaluation as a waste product.

(2) Exposure to dust.

(3) Higher score when used in urban areas, due to very positive effect on air quality

Table 15.8: Environmental profiles of external cladding

Material	*Quantity of material used (kg/m²)*	*Effect on resources*			*Pollution effects*				*Ecological potential*		*Environmental profile*
		Materials	*Energy*	*Water*	*Extraction and production*	*Building site*	*In the building*	*As waste*	*Re-use and recycling*	*Local production*	
Stainless steel, from ore	3.8	3	2	2	3	1	2	2	✓		3
Galvanized steel, from ore	3.7	3	2	2	3	1	2	2	✓		3
Aluminium, 50% material recycling	1.6	2	3	3	3	1	2	2	✓		3
Cement-based boarding	20.5	1	2	2	2	1	1	1			2
Lime sandstone	96	1	2	2	2	1	1	1	✓		2
Calcium silicate boarding	11	1	1		1	1	1	1			1
Hydraulic lime render	85	1	2	2	2	2	1	1			2
Lime cement render	88	1	2	2	2	2	1	1			2
Gypsum based render	52	1	2	2	2	1	1	2			2
Stone on steel support system	81	1	1	1	1	1	2	1	✓	✓	1
Brick	108	1	3	3	2	1	1	1	✓	✓	2
Timber boarding, without impregnation	13.7	1	1	1	1	1	1	1		✓	1
Timber boarding, impregnated	13.7	2	1	2	2	3	3	3		✓	3

Table 15.9: Environmental profiles of internal cladding

Material	*Quantity of material used (kg/m²)*	*Effect on resources*			*Pollution effects*				*Ecological potential*		*Environmental profile*
		Materials	*Energy*	*Water*	*Extraction and production*	*Building site*	*In the building*	*As waste*	*Re-use and recycling*	*Local production*	
Stainless steel, from ore	3.7	3	2	3	3	1	2	2	✓		3
Cement-based boarding	20.5	1	2	3	2	1	1	1			2
Lime sandstone	96	1	2	3	2	1	1	1	✓		2
Calcium silicate boarding	11	1	1		1	1	1	1			1
Plasterboard	11.7	1	2	2	1	1	1	2			1
Hydraulic lime render	85	1	2	2	2	2	1	1			2
Lime cement render	88	1	2	2	2	2	1	1			2
Gypsum based render	52	1	2	2	2	1	1	2			2
Brick	108	1	3	3	2	1	1	1	✓	✓	2
Ceramic tiles	10	1	2	3	2	1	1	2[(2)]			2
Timber boarding	8.3	1	1	1	1	1	1[(1)]	1		✓	1
Hard woodfibre boarding	5.4	1	2	3	2	1	1	1			2
Porous woodfibre boarding	3.6	1	2	2	2	1	1	1			2
Chipboard[(3)]	7.8	2	1	3	2	2	2	2			3
Plywood sheeting	4	1	1		2	1	2	2			2
Woodwool slabs	11.5	1	2	3	2	1	1	1		✓	2

Notes:
Wallpaper is not included in this table.
(1) Pine can give off formaldehyde during a period after fixing. This is most likely because of the drying method that has been used.
(2) Certain colour pigments make it necessary to give the material a lower evaluation as a waste product.
(3) Chipboard is often covered with a plastic laminate based on phenol or melamine. This reduces the product's environmental profile even more.

Table 15.10: Environmental profiles of flooring

Material	*Quantity of material used (kg/m²)*	*Effect on resources*			*Pollution effects*				*Ecological potential*		*Environ-mental profile*
		Materials	*Energy*	*Water*	*Extraction and production*	*Building site*	*In the building*	*As waste*	*Re-use and recycling*	*Local production*	
Terrazzo concrete	25	1	2	2	2	2	1	1	✓[2]	✓	2
Stone	30	1	1	1	1	2	1	1	✓	✓	1
Brick	90	1	3	3	3	1	1	1	✓	✓	2
Ceramic tiles	14	1	2	2	2	2	1	1[1]			2
Polyvinyl chloride PVC	1.3	2	2		3	2	3	3			3
Polyolephine[3]	1.3	2	2		3	2	2	2			2
Styrene butadiene rubber	3.6	2	2		3	2	3	2			3
Timber	10	1	1	1	1	1	1	1		✓	1
Linoleum	2.3	1	1	1	2	2	2	1			2
Cork[4]	1.3	1	1		2	1	1	1			1
Laminated chipboard	15	2	2		2	1	2	3			2
Natural rubber	3.6	1	1		2	1	1	1			1

Notes:
Carpets are not included in this table.
(1) Certain colour pigments make it necessary to give the material a lower evaluation as a waste product.
(2) Does not apply for terrazzo cast *in situ*.
(3) From polyethylene and propylene.
(4) Untreated.

Environmental profiles

Tables 15.7 to 15.10 are organized in the same way as the environmental profiles in Table 13.5.

References

BUGGE A, *Husbygningslære*, Kristiania 1918
DOERNACH R *et al*, *Biohaus*, Frankfurt 1981
GREYSMITH B, *Wallpaper*, London 1976
GRUTZMACHER B, *Reet- und strohdächer*, Callwey, München 1981
GUSTAFSSON H, *Kemisk emission från byggnadsmaterial*, Statens Provningsanstalt, Borås 1990
HALL N, *Has Thatch a Future?*, *Appr. Techn.* Vol. 8, no. 3, 1981
MINKE G, *Alternatives Bauen*, Gesamthochschule, Kassel 1980
PARRY J P M, *Development and testing of roof cladding materials made from fibre-reinforced cement*, Appr. Techn. Vol. 8, no. 2, 1981
PARRY J P M, *Hurricane Tiles. New economical type of roofing combining the best features of sheet and tiles*, Cradley Heath 1984
STANEK H, *Biologie des Wohnens*, Stuttgart 1980
VRIEM H, *Taksponog spontekking* Fortidsminnesmerke foreningen 1941

16 Building components

The following components will be discussed in this chapter:

- windows
- doors
- stairs

Windows and doors

Windows bring in light and sensation, and acting as a protection from extremes of climate. Glazing bars were once made of lead, often strengthened by iron, within a main frame of timber. From the beginning of the eighteenth century wooden glazing bars were used, and glass was kept in place with putty. Today there are three main types of window frame: timber, aluminium and plastic. These are also used in different combinations.

The word 'door' comes from Sanskrit and means 'the covering of an opening'. The entrance door to a house was traditionally formed in a very special and careful way. The door was for receiving guests, as well as for greeting greater powers, both physical and supernatural, or for keeping them out. The material most often used is wood, but steel, aluminium and plastic doors are also made.

Both windows and doors can be seen as movable or fixed parts of the wall. They require the same qualities as the external or internal wall they sit in: thermal insulation, sound insulation, resistance to the elements, etc. Not least, both windows and doors must be able to withstand mechanical wear and tear and keep their form and strength through varying moisture conditions. It has proved difficult to satisfy all these conditions. The thermal insulation of a modern outside door is three to five times worse than the external wall, and a window's thermal insulation is five to ten times worse.

Glass and methods of installation

Float glass is normally used in windows, though machine glass is still in production in some European factories. Cast glass is used indoors, often as a decorative product which doesn't need to be transparent. There are various types of energy glass, security glass, sound-insulating glass and fire-proof glass. Energy glass is often coloured or covered with a metallic oxide. Security glass is specially hardened or laminated with a foil of polyvinyl butyral between the sheets of glass. Sound-insulating glass is also laminated in two or more layers. Fire-proof glass usually consists of several layers laminated with sodium silicate.

The temperature of glass has to be even across its whole surface when it is cut, otherwise tension can occur within the glass and lead to splitting. Depending upon the level of insulation required, there will be one, two or more layers of glass in windows. There are several ways of achieving this. The easiest is to hinge two timber windows together, which is a traditional way of constructing windows in Scandinavia. The sheets of glass are placed in the frame with putty based on acryl plastic or linseed oil. Internal glazing can be mounted with special beading of wood or aluminium. Before using linseed oil putty on a window frame, the timber must be treated with oil or paint, otherwise the linseed oil will be absorbed by the window frame and the putty will crack.

Sealed units have become the most common type of glazing in the building industry. These consist of two or three sheets of glass with a layer of air sealed between them. The air can be replaced with an inert gas, such as argon, which improves the thermal and sound insulation of the window because it circulates more slowly than air. The sheets of glass are connected by plastic or metal sections and sealed with elastic, plastic-based mastic. Until the late 1980s polychlorinated biphenyls, PCBs, were widely used, but today silicones are more common. Sheets of glass can also be welded together. The sealed units are usually fixed into a window frame with beads of wood or aluminium, together with rubber packing.

More recently, alternatives to glass have appeared on the market. These are mainly polymethylmetacrylate (plexiglass) and polycarbonate, which are mainly used in roof lighting, greenhouses and conservatories. The sheeting products are mounted in a similar way to the sealed units.

Normal glass is based on raw materials with rich reserves, while the production consumes large amounts of energy and produces pollution. Ingredients of plastic and metal oxides used also cause problems. Transparent plastic products are based on oil, and they generally consume high levels of primary energy and produce pollution.

Plastic and glass products probably present no problem in the indoor climate, even though there may be small emissions from plastic-based putty, mastics and sealants, depending upon the type of plastic and the mounting technique. Little is known about the durability of plastic roof-lights. Normal glass has almost unlimited durability. Coloured heat-absorbing glass can break if part of it is permanently in the shade and the rest is exposed to sun.

Under special circumstances even sealed units can have problems: at low temperatures; low pressure occurs inside them which bends the panes of glass inwards in the middle, giving a lower insulation value. If the building is not heated during the winter, the tension within the glass can be so great that the glass can break, especially if there is a wide space between the panes of glass.

The weak link in these units are otherwise the seals or sealants. Breaking down of the seals occurs either through vapour getting in or through physical deterioration of the packing. In a penetrating durability test carried out in Norwegian building research in 1986, one third of metal-sealed windows were defunct after 20 to 32 years. For some of the plastic sealed types, nearly all were failing after four to five years. Glass sealed panes were without exception useless after 10 years because of wind pressure, vibrations and thermal tensions (Gjelsvik, 1986).

Another important aspect of sealed units is that if only one of the panes of glass splits, the whole window must be changed, whether it is double or triple glazing.

In terms of maintenance, there is little doubt that the Scandinavian model of coupled timber windows gives best results, preferably with a window divided into smaller panes on the outside, where the chance of breakage is highest. Maintenance costs are small and durability and recycling possibilities are high, although coupled windows are best used in domestic buildings, as larger buildings would incur very high window cleaning bills.

Pure clear glass can be recycled. This is not the case for metal-coated glass or glass containing laminations of foil, reinforcement etc. Many of these products have to be dumped at special tips, including coloured and metal-coated glass.

Timber windows

Timber frames used to be made of high quality timber with no knots – often pine heartwood. When constructing the window, the highest quality was selected for the most exposed parts, such as the sill. The components were slotted together and fixed inside with wooden plugs. Windows are still mostly made of pine, but without the same demands on quality or the same preparation. The proportion of heartwood used is often very low.

The present methods of sawing timber do not guarantee that the heartwood is used in the most appropriate parts of the window. To compensate for this, it is quite common to use pressure-impregnated timber. Adhesive or screws are used as the binder between the components. The window furniture and the hinges are usually made of galvanized steel or brass. Between the frame and the casement

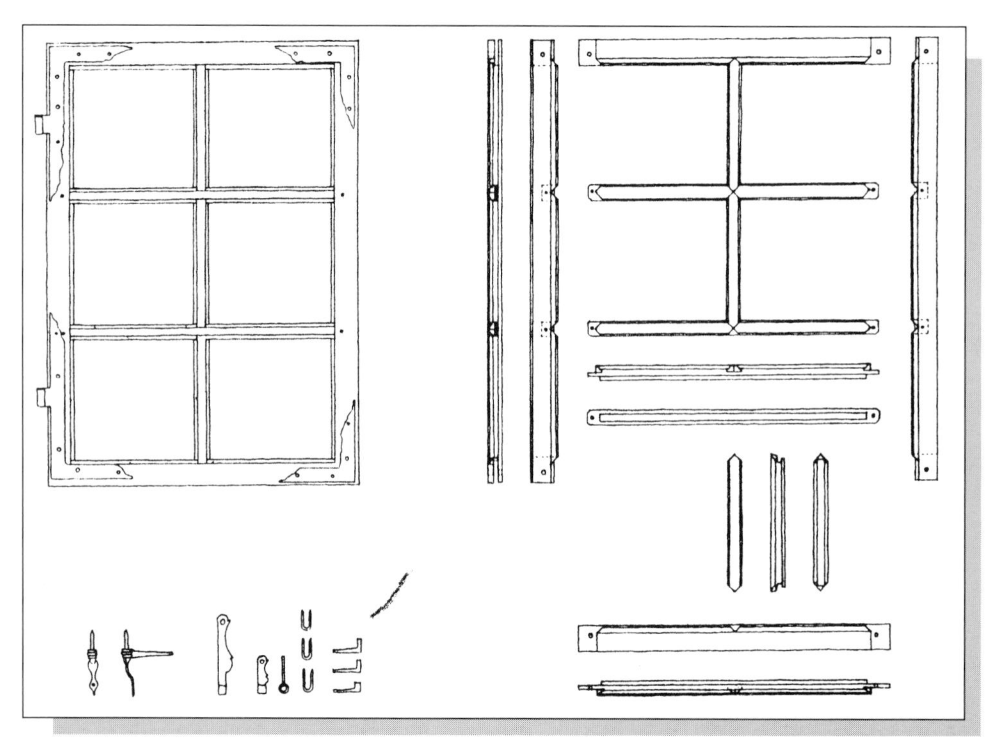

Figure 16.1: Traditional window construction for single-glazed windows. Source: Jessen 1975

in opening windows there is a bead, usually made of polyurethane or ethylene propylene rubber (EPDM), but it can also be made of silicone rubber, polyvinyl chloride, butyl rubber and chloroprene rubber. Woven wool and cotton beading is probably the most robust. These products can contain fungicide.

Timber windows are based mainly on renewable resources. The consumption of primary energy is low and production does not pollute the environment significantly. Pressure impregnation, plastic beading and metal furniture reduce this advantage. Timber windows are well suited for local production and create very few problems in a building, except for a certain level of emissions from impregnated timber and plastic.

Old quality timber frames have lasted for 250 years under favourable conditions. Until the middle of the twentieth century a timber window was considered to have a life span of 50 years. Since the 1960s, the rotting of timber windows has increased considerably. Serious damage has occurred as few as 10–15 years after installation. Sweden's State Testing Station has registered that linseed oil and alkyd oil paints give timber the best durability (Phil, 1990).

Timber windows of high quality are usually well suited for re-use. Copenhagen's local authority has calculated a loss of 70 million kroner over the last 10–15 years

Figure 16.2: Use of recycled windows in a house, between the living room and the garden. Source: Gaia Lista 1986

because they have not re-used windows in their building programme (Lauritzen, 1991). The calculation is based on the fact that cleaning up, repairing and repainting an old window represents only 80 per cent of new production costs. Older windows usually need a new sill; in some cases turning the window upside-down so that the previously exposed parts rest further up is sufficient. The recommended way to remove old paint is to use a blow-lamp. However, the vapour from a blow-lamp can cause acute allergies. Treating the paint and timber with acid or soda is also possible, but this is often quite aggressive to the wooden material method. Metal ironmongery and furniture can often be re-used or recycled. Pure timber waste can be energy recycled in normal incinerators. Impregnated timber and plastic materials have to be dumped at special tips.

The sustainable window

The modern sustainable window (Fig. 16.3) is manufactured as a three-layered, coupled window, where the middle and best-protected pane of glass is a low-energy glass with a coating of metallic oxide, preferably gold. By having this in the middle there is less chance of dust settling on the film, which would reduce the effective saving of energy.

The outer glass is held in place with linseed oil putty and the two inner panes are fixed with beading to make it easier for dismantling for re-use and recycling. The packing around the window is untreated wool.

The outside layer of the window is the part most directly exposed to an aggressive climate, e.g. burning sun or driving rain. The outside frame in a sustainable window is therefore designed so that it can be removed, and has smaller panes of glass in case of breakage. The sill is made from mature oak or pine heartwood. During the summer, the two inner window frames can be removed to improve the light inside. The window is fixed to the building structure with screws.

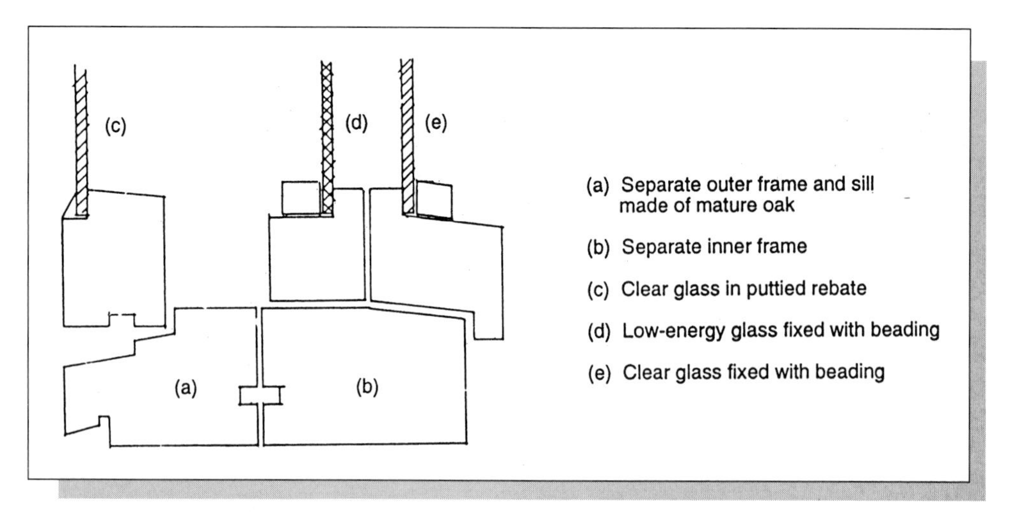

Figure 16.3: The principle section of a sustainable window construction for a cold climate. Source: Gaia Lista, 1995

Timber doors

Different types of timber can be used for doors: pine, spruce, oak, beech and birch, either as solid wood or as veneer. There are two main construction techniques for doors: framed and panelled doors and flush doors, both of which are built up with a solid timber frame. Both types usually have sealing strips as well as hinges, door handles, housing for the locks and other ironmongery.

Framed and panelled doors are built with a wide timber frame. This was traditionally fixed together with wooden plugs, but nowadays it is glued. In the spaces between the frame, solid timber panels are placed, or panels of chipboard, plywood, hard fibre-board or even glass. These are slotted into the groove on the inside of the frame. To stop the frame bending, it is usual to split it into two, turn half of it through 180° and glue it together again. This lamination is not necessary for internal doors between dry rooms.

This type of door has bad thermal insulation properties and is usually used internally. Two such doors with a porch in between, however, should give a good internal thermal climate in most conditions.

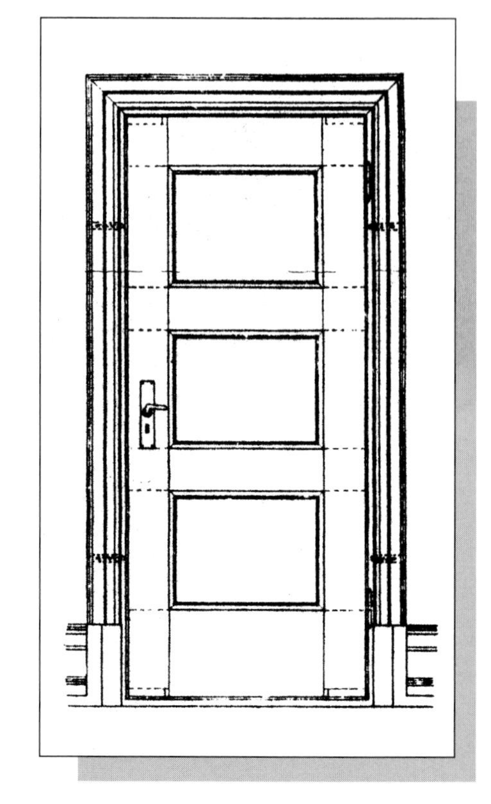

Figure 16.4: A framed and panelled door. Source: Bugge 1918

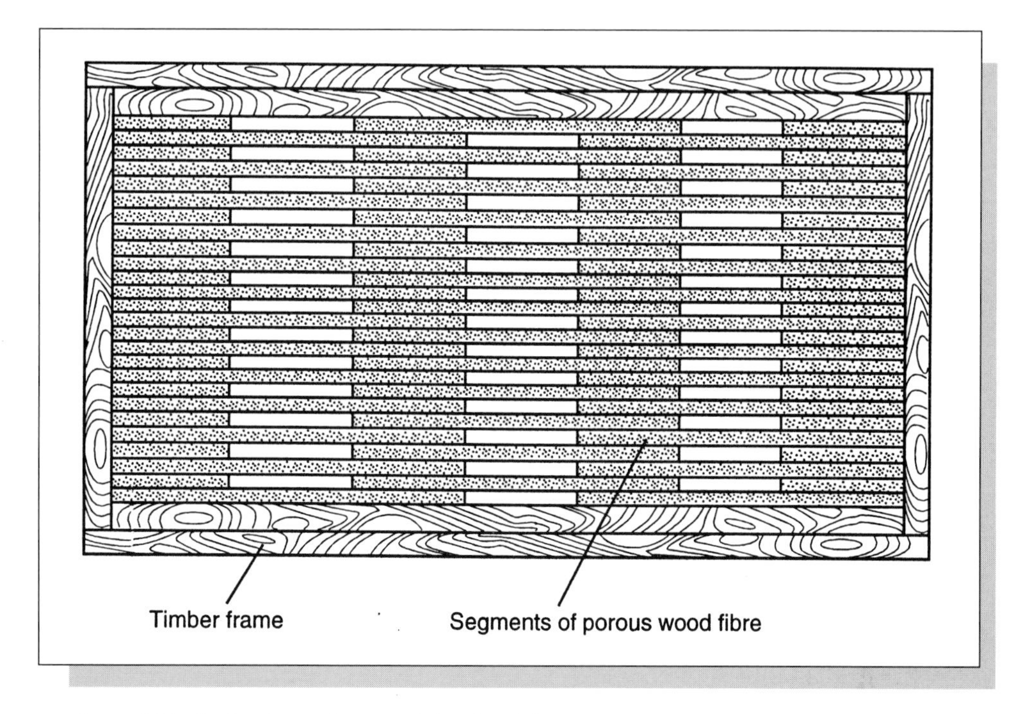

Figure 16.5: Section through a sound-insulated door.

Flush doors also have a frame, but not as large as the frame of a panelled door. A flush door is stiffened by thin layers of board, fibreboard or plywood, fixed with adhesive or pins on both sides. External doors must use a water-insoluble adhesive. Thermal insulation can be placed in the space between the layers of fibreboard, e.g. expanded polystyrene, mineral wool and porous fibreboard, woodwool slabs, wood shavings, etc. For light doors it is usual to add a sound-insulating layer of corrugated cardboard or layers of interlocking wood fibre bands, a waste product from the wood fibre industry. In fire doors non-flammable sheets of plasterboard or other heavy materials are inserted. A flush door can also have glazing, but glazing will need its own frame.

The normal adhesives used in door manufacture are resorcinol, phenol, polyvinyl acetate (PVAC) and polyurethane. Casein glue, animal glue and soya glue can also be used. Doors are often delivered ready to hang, so they have either a polyurethane varnish or an alkyd or linseed oil painted finish.

The environmental aspects of timber doors are good, but it is quite clear that the choice of insulating material, glued boards, sealing strips, surface treatment and ironmongery all play their part in production consequences and have effects on the internal environment.

Doors can often be re-used, especially robust, solid panel doors. It is also an advantage if the door frame can be dismantled with the door. The manufacture of a new door frame can be expensive, especially if its dimensions are not to the current standard. This assumes that the door was originally fixed for simple dismantling, preferably with screws.

Defective doors of solid timber can usually be energy recycled or composted, but laminated products have to be deposited at special tips, or energy recycled in incinerators that filter the fumes.

Plastic and aluminium windows and doors

Window frames of plastic and aluminium usually consist of profiles filled with foamed insulation of polyurethane or polystyrene. Some products use both aluminium and timber, where timber is the insulating material. Lower quality timber can be used, as the outer layer of aluminium protects it from the elements. Plastic windows are usually made of hard polyvinyl chloride (PVC) stabilized by cadmium, lead and tin compounds and added colour pigments. All these products have very limited reserves, and pollution during processing is considerable.

The manufacture of an aluminium window uses 30–100 times more primary energy input than a timber window; a PVC window uses about six times as much (Phil, 1990). The annual cost, taking into account the investment and maintenance, favours timber windows with an estimated life span of 30 years. There have been some problems with aluminium and plastic windows because condensation can easily occur within the frame, due to a profusion of cold-bridges. In a building the products are not a particular problem. The hard PVC has no softeners that could emit unpleasant gases.

Both PVC and aluminium windows can be re-used if they are initially installed for easy dismantling. Pure aluminium windows can be recycled. This is unlikely for the other products, as they have sealed, complex combinations of different materials. Waste has to be deposited at special tips if products can contain cadmium, lead and tin.

Stairs

Stairs are, in a way, part of the floor. The main materials used are timber, stone, brick, concrete and cast iron. The steps have structural properties, at the same time must provide a comfortable underlay for the foot. Common finishes include linoleum and ceramic tiles.

Wooden stairs

Stairs of non-impregnated timber are used mainly indoors, but they can also be placed outdoors if they are under shelter. Pine, oak, ash, beech and elm are hardwearing materials and can often be used without treatment. The timber should

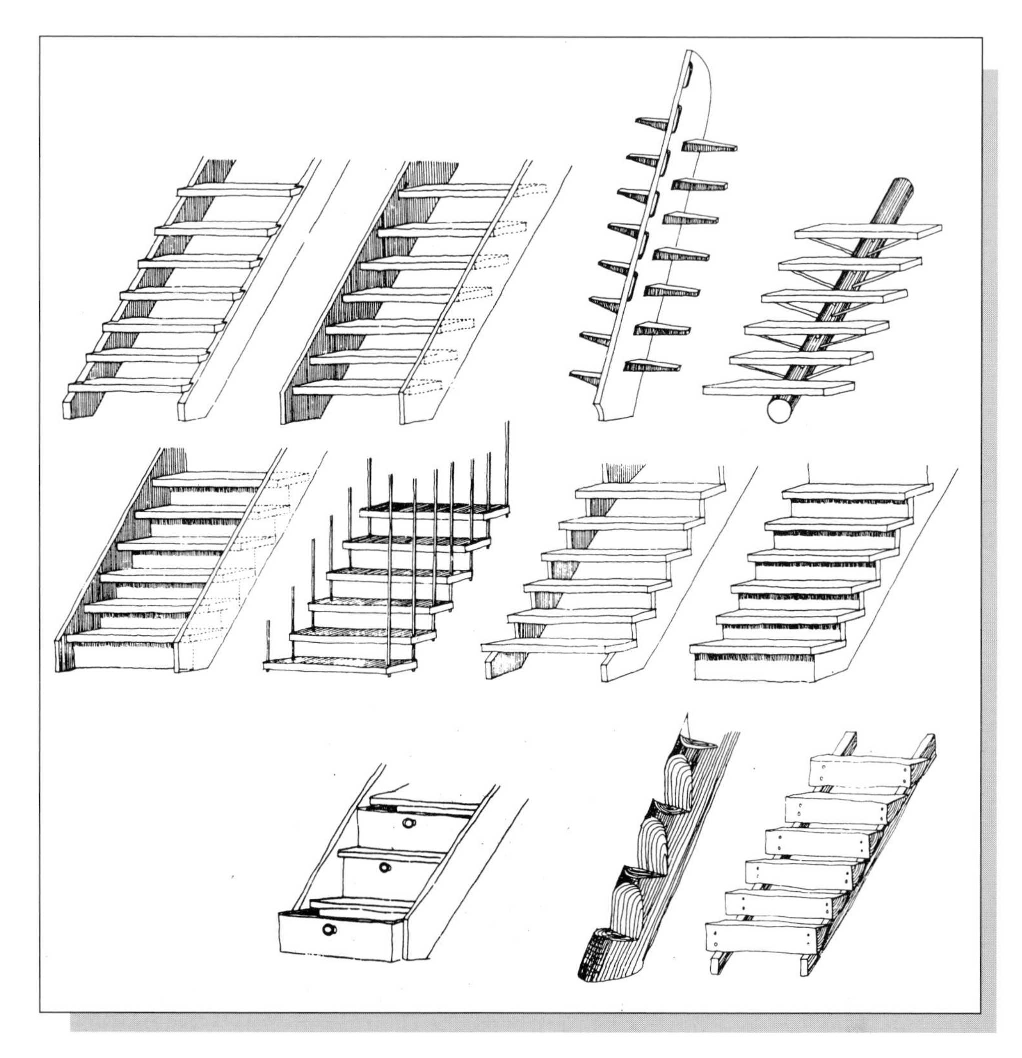

Figure 16.6: Different ways of constructing wooden stairs.

be of a high quality and should not have any knots. Handrails and banisters can be made of maple, which has a smooth surface well suited for this purpose.

It has become more common to use laminated timber in recent years. Resorcinol glue is widely used, but casein glue is also suitable. Outdoor wooden stairs are often pressure-impregnated.

Stone stairs

Stone stairs are particularly well suited for outdoor use. Stones can be used direct from the quarry, or cut. Granite is the most hard-wearing variety. It is also possible to use pieces of quartzite slate for the steps. It is usual to have a forged iron balustrade with natural stone stairs. This is set fast in pre-bored holes with floating sulphur. The sulphur solidifies in a few minutes and prevents any rust getting to the foot.

Brick stairs

These can be used inside and outside. They are usually short and built from ordinary bricks.

Concrete stairs

Concrete can be used inside and outside. Uncovered concrete stairs have a tendency to be dusty. It is normal to lay ceramic tiles on them, or terrazzo topping which is later sanded.

Cast iron stairs

Cast iron stairs came into use at the turn of the century and are often used for fire escapes. They are usually galvanized or painted.

Wooden and stone stairs use the most favourable raw materials, environmentally speaking. They also have low levels of pollution and primary energy consumption. Surface treatment and impregnation of timber stairs will reduce the environmental profile somewhat.

Within a building these products are relatively harmless. The only exceptions are impregnated or painted timber staircases. Steel stairs and reinforced brick and concrete stairs can increase the electromagnetic field in a house.

All types of stairs have a re-use potential, e.g. wooden stairs mounted in modular parts for simple dismantling, dry stone stairs, brick stairs laid in a weak mortar, standardized steel stairs, etc. Certain prefabricated concrete steps are also suitable for re-use. Products cast *in situ* can be recycled as fill or aggregate for low quality concrete work. Steel products can be easily recycled through smelting.

Stone, brick and concrete are inert and relatively problem-free as waste. Impregnated timber must be deposited at special dumps.

References

BUGGE A, Husbygningslare, Kristiania 1918

GJELSVIK T *et al*, *Four papers on durability of building materials and components*, Byggforsk, Oslo 1986

JESSEN C, *Landhurst*, Kobenhavn 1975

LAURITZEN E *et al*, *De lander på genbrug*, Copenhagen 1991

PHIL Å, *Byggnadsmaterial utifrån en helhetssyn*, KTH, Stockholm 1990

17 Fixings and connections

All materials and components in a building have to be fixed in some way, using either mechanical or chemical means. Mechanical fixings include nails, pins or staples, screws, bolts and wood or iron plugs. Chemical fixings bond materials together when set. They can be divided into glues and mortars.

Mechanical fixings

Even though forged iron has been known in Northern Europe since AD 1000, neither iron nor steel was used as a building material until the industrial revolution. Houses were built in earth, stone, brick and timber. The three first materials fastened together with mortar, whereas timber components which were to be lengthened, strengthened or connected were joined together with locking joints.

A common quality of locking joints is that they reduce the strength of the timber as little as possible. Certain joints are used to preserve the timber's tensile and bending strength, others to preserve the compressive strength. Wooden plugs were an integral part of locking joints, often integrated with the locks, but their most important role was as fixings for both structure and claddings. Today, nails and screws in steel are the sole components used for the majority of mechanical fixings in timber building. Steel bolts are used in buildings with large structural elements. Fixing products are also made of aluminium, copper, bronze and stainless steel.

A normal sized timber house will contain about 100–150 kg of nails, screws and bolts. Steel structures are joined mainly through welding, but bolts can also

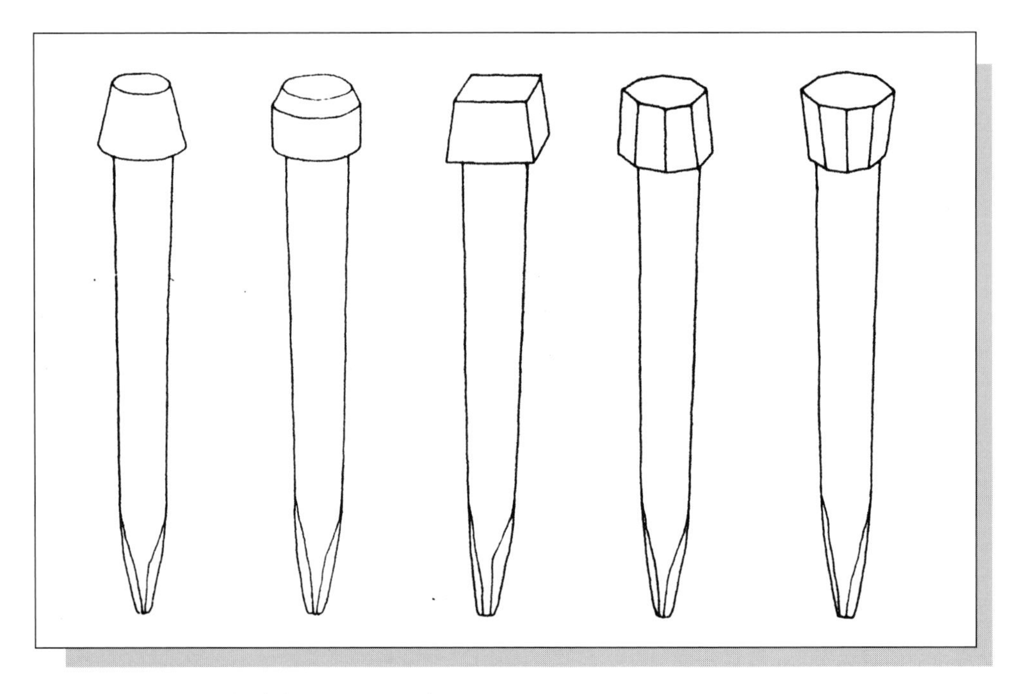

Figure 17.1: Timber bolts. Source: Myhre 1982

be used, as was the case during the nineteenth century, before mobile welding was common place.

Timber

Joints

Timber joint technology is particularly well-developed in Japan, with a choice of some 600 types. In Scandinavia there is a tradition of log construction with 10 to 20 different jointing techniques. Some structures such as stave buildings and vertical load-bearing panelling often use grooves to fix external panelling (see 'Types of structural walls', pp. 231–233). Nails are not necessary in this form of construction, and where the fastening is part and parcel of the whole structural system it is known as a 'macro-joint'.

Pins and bolts

The use of timber pins and bolts is particularly widespread in areas with early forms of timber frame and stave building tradition. Pins of juniper, oak and maple are considered the best, although other types of wood can be used. The pins in stave buildings are often 25–30 cm long with heavy heads, while pins for fixing cladding are smaller.

There is a lack of steel in India, and wooden bolts are often used as a fixing component in timber structures. The Forest Research Institute in Dehra Dun has researched the strength of wooden bolts and found that they were consistently about 68 per cent as strong as steel of the same size. The timber bolts used in the research had nuts 12 mm in diameter and 100 mm in length. The timber was taken from various trees of normal strengths (Masami, 1972). However, timber is simply not as homogeneous as steel and its strength properties are less standardized and difficult to record. Timber plugs disappeared from the market in Europe during the middle of the nineteenth century as a result of new standards specifying strength properties.

To a certain extent, the use of timber plugs is on its way back into building, for example in military radar stations where metal components would disturb radio signals. There are guidelines for their production and dimensions. Industrial production of timber bolts and pins is not necessarily less efficient or expensive than for the equivalent steel products (Kessel 1994).

Fixings made of timber are based purely on renewable resources. Primary energy consumption and pollution from production are low. The quality of timber used for jointing, pins and bolts is normally so good that no impregnation is needed.

Durability of the products is also very good. While connections in steel in certain situations can lead to condensation and decay of the adjacent timber, timber fixing components are neutral and stable.

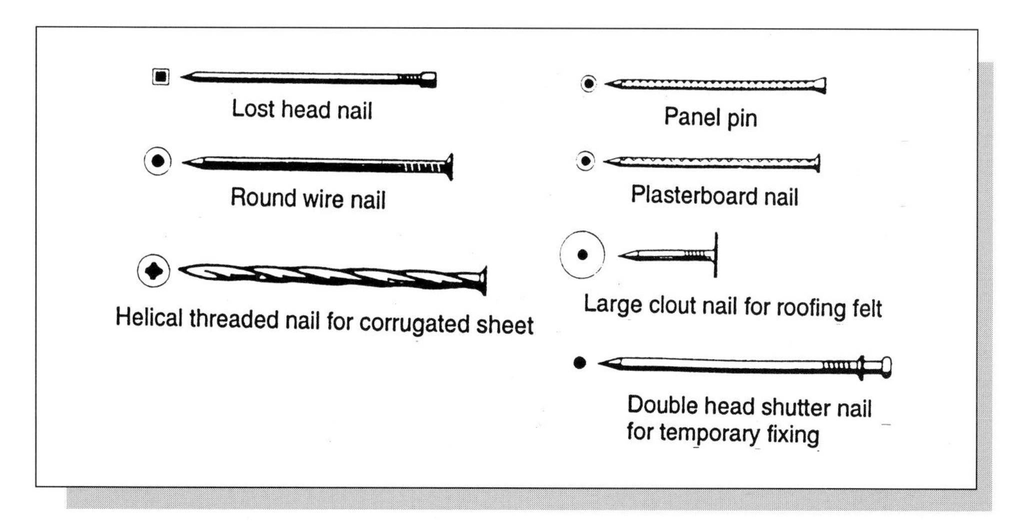

Figure 17.2: Standard nails.

The products are not a problem for the indoor climate. The joints can usually be easily dismantled so that the materials they join together can be re-used. Wooden pins are usually glued or swollen into the components they bind. They can be sawn off or drilled out for re-use of structures. Pure timber waste can be energy recycled or composted.

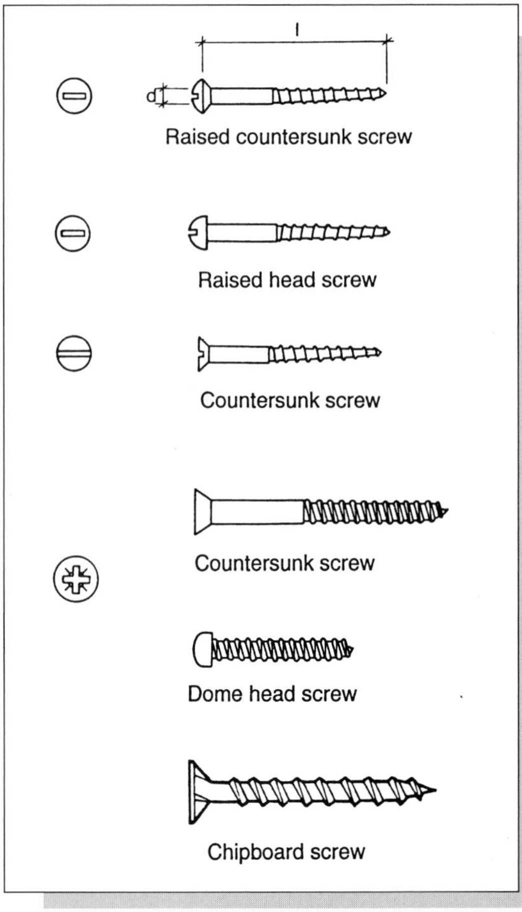

Metal

Nails

There are two main groups of nails: cut nails and wire nails. Cut nails are the oldest and original type and usually have a slight wedge form. They were used in all situations until the end of the nineteenth century, when the manufacture of wire nails began. Wire nails are ubiquitous nowadays. In the UK they are round or oblong; in Scandinavia they usually have a square cross-section with a pyramidal tip. Galvanized nails are used on external surfaces to cope with recurring dampness. They are also used internally, galvanizing is usually unnecessary.

Figure 17.3: Standard wood screws.

Gangnailplates

These are made for fixing larger components together, such as the timbers within a roof truss. The gangnailplate is a galvanized steel sheet punched to form many nails, which makes a good fastening and prevents the timber from splitting.

Screws

Screws draw themselves into the timber as they are turned, and are used in finer joinery work, ironmongery and internal detailing. The work is more demanding than nailing, but screws damage the timber far less.

Bolts

Metal bolts are used in connections where strong forces are to be transferred. Toothplate timber connectors are often laid between structural parts to increase the capacity of a bolt to transfer loads. These connectors have spikes which are pressed into the timber so that the forces are transferred to the surface of friction between the two parts. The bolt's task is thereby reduced to simply holding the two structural parts together.

Generally speaking, metals have limited reserves. In certain cases scrap metal is used. The use of primary energy and pollution during production is high. There is an over-investment of quality in the use of galvanized steel products in dry, indoor environments. Untreated steel products have a far better environmental profile. Metal products do not cause environmental problems in buildings. In a fire, however, they will quickly become red hot and burn through adjacent timber.

Their durability of metal products is generally good. If a metal component is exposed to great variations in temperature, condensation can form on it. This has a deteriorating effect on the adjacent timber through electrolytic activity. If timber is damp when a metal component is added, the same effect could occur. Timber impregnated with salt can also corrode metal.

Nails and nailplates have no re-use value, and will probably not be saved for material recycling. Exceptions can occur when demolition material is burnt and metals are cleaned from the ashes. Screw and bolts can be retained and re-used or recycled. Use of screw and bolt connections also means that materials they join together can be easily dismantled and re-used.

Metal that cannot be recycled should be deposited at special tips.

Chemical binders

Mortars, adhesives and fillers are important binders in the building industry. Mortar and adhesives are used to bind together different or similar components; fillers are a sub-group used to fill cracks and stick to the surfaces that surround them.

Mortars

A mortar is usually a mixture of lime or cement with sand and water, sometimes with additives, used as a binder for different types of mineral building-stones, slabs, tiles, bricks, blocks and in certain circumstances roof tiles. (See also 'Hydraulic binders' and 'Non-hydraulic binders', pp. 94–97.) Fine or coarse sand is used, according to the smoothness of finish required. In lime mortar, fine sand is usually chosen, preferably beach sand. Small amounts of fibre can be added to increase its strength. Mineral fibres or organic alternatives such as

Table 17.1: Mortars used for masonry

Mortar	*Materials and proportions*	*Properties*	*Areas of use*
Lime	Lime: 1 Sand: 2/3 Water	Elastic, medium strength, not very resistant to water and frost, quick-drying, balances relative humidity well	Internal laying of bricks, stone, expanded clay blocks, brick floors, render
Hydraulic lime	Hydraulic lime: 1 Sand: 2/4 Water	Hydraulic, medium strength, elastic, frost-resistant, balances relative humidity, quick-drying	All types of internal and external masonry, render
Portland cement	Cement: 1 Sand: 3/4 Water	Hydraulic, strong not so elastic, frost-resistant, low moisture absorption, slow-drying	Internal and external laying of tiles, render
Lime cement	Lime: 1½ Cement: 2/1/1 Sand: 10/7/11 Water	Hydraulic, medium strength to strong, elastic, frost-resistant, medium moisture absorption, medium-slow-drying	All types of internal and external masonry, render
Anhydrite and gypsum	Gypsum: 1 Sand: 1/3 Water	Elastic, weak, not very resistant to water and frost, balances relative humidity well, quick-drying	Smaller internal walls, internal render/plaster, and external render
Clay	Clay: 5 Sand: 1 Water	Elastic, weak, not very resistant to water and frost, balances relative humidity well, quick-drying	Laying of earth blocks and low-fired brick
Sulphur	Sulphur	Elastic, medium strength, medium resistance to frost, watertight	Laying of sulphur blocks and bricks

hemp, sisal, jute or animal hair can be used, with a fine aggregate of granulated and foamed recycled glass, perlite, vermiculite or similar materials, added to increase the insulation value. In certain modern mortar mixtures extra additives provide elasticity, watertightness, etc. Lime cement mortar is often made using additives that bring air into the mix, giving it a waterproofing quality. (See also 'Additives in cement', p. 97.)

Aggregates must not react chemically with any other materials in the mortar, nor take an active part in the solidifying or curing of the mortar. Water used in lime and cement mortars should be fresh and must not contain salt, sulphur or other substances that can break down the mixture.

Blocks or bricks are usually laid with mortar between them. The Southwest Research Institute in Texas has developed a fibre-reinforced sulphur mortar which can be sprayed onto both sides of a wall built completely dry.

Mortars have different elasticity coefficients and strengths. This is critical for the different tasks they perform, but also important for any later dismantling of components. Pure cement mortar is, for example, twice as strong as pure lime mortar; hydraulic lime mortar is somewhere between these. The use of lime mortars, hydraulic lime mortars and lime cement mortars rich in lime makes it possible to dismantle walls of bricks, concrete blocks and lightweight concrete blocks, etc for re-use. Lime cement mortars must contain a minimum of 35 per cent cement, partly because a smaller percentage does not strengthen the mortar and partly because the cement slows down the curing of the lime.

Mortar products are based mainly on materials with rich reserves. Their consumption of primary energy lies somewhere between that of timber and steel. Pollution during the production of binders is mainly in the form of dust and the emission of a considerable amount of nitrogen oxides, sulphur dioxide and carbon dioxide. Binders containing pozzolana create the least pollution.

Mortars were once entirely mixed on site with local aggregates; it is more normal today to use ready-mixed mortars. Centralized production means an increased use of transport energy, since even the aggregate has to be transported greater distances. However, the aggregates used are light and give better thermal insulation in the finished structures. Mortars cause no problem once in place, as long as no volatile organic compounds have been added.

Sulphur mortars can be recycled. This is true for pure lime mortars, in theory, because they can be reburned, but it is difficult to achieve in practice. Cement mortars can be ground into aggregate for low quality concrete structures.

As waste, mortars are normally inert and can be used as fill. Ground lime mortars can be used for soil improvement. Sulphur pollution can develop from gypsum waste because of microbial decomposition. Sulphur waste should be deposited at special dumps, preferably neutralized by adding lime.

Adhesives and fillers

Archaeological exploration indicates that animal glue adhesives were in use as far back as 3000–4000 BC. In China and Egypt casein glue was used in finer joinery. Somehow this knowledge disappeared, but was rediscovered in Europe around the sixteenth century. The first glue factory was built in the Netherlands in 1690. Around 1875 the manufacture of plywood started, and at the turn of the century laminated timber construction began. Synthetic resins

Table 17.2: The different types of adhesive

Type of adhesive	*Main constituents*	*Water-proof scale*[1]	*Areas of use*
Mineral adhesives:			
Waterglass glue	Waterglass, lime, stone dust, water	4	Ceramic tiles and paper, chipboard and fillers
Cement-based glue	Portland cement, stone dust, possibly acryl, water	5	Ceramic tiles, aerated concrete
Synthetic resins:			
Urea adhesive	Urea, formaldehyde, water	3	Chipboard, carpets on underlay
Phenol adhesive	Phenol, formaldehyde, organic solvents	5	Mineral wool, plywood, cork tiles
Resorcinol adhesive	Resorcinol, formaldehyde, possibly phenol, water	5	Laminated timber, finger-jointing of timber lengths
Polyvinyl acetate adhesive (PVAC)	Acetylene, acetic acid, polyvinyl alcohol. Possibly chrome compounds, water, organic solvents, possibly fungicides	3 5[2]	Furniture, joinery, fillers. Windows, doors, finger–jointing of timber lengths
Acrylate adhesive (two components)	Acrylate, water	5	Timber, plastics, ceramics, fibreglass, fillers
Ethylene vinyl acetate adhesive (EVA)	Ethylene, vinyl acetate, water, possibly fungicides	5	Plastic sheeting and linoleum on floors and walls
Polyurethane (one or two components)	Isocyanate, polyols	4	Wood, metal, plastic, strawboards
Epoxide adhesive	Epichlorhydrin, phenol, alcohol	5	Concrete, stone, glass, metal, plastic, ceramic tiles, fillers
Isocyanate adhesive (EPI)	Isocyanate, styrene butadiene rubber or polyvinyl acetate	5	Plywood, doors, windows, furniture, metals
Chloroprene adhesive	Acetylene, chlorine, organic solvents	5	Plastic
Styrene butadiene (SBR)	Butadiene, styrene, organic solvents	5	Plasterboard and chipboard, wood and concrete, needle-punched carpets
Plant-based and cellulose adhesives:			
Soya adhesive	Soya protein, possibly sodium silicate or fungicides, water	3	Plywood
Potato flour paste	Potato starch, possibly hydrochloric acid or fungicide, water	2	Wallpapering
Rye flour paste	Rye flour starch, possibly fungicide, water	2	Wallpapering, putting up hessian, linoleum
Cellulose paste	Methyl cellulose, water	3	Wallpapering, putting up hessian, linoleum, fillers
Cellulose adhesive	Derivatives of cellulose, organic solvents	5	Linoleum

Table 17.2: *continued*

Type of adhesive	*Main constituents*	*Water-proof scale*[1]	*Areas of use*
Sulphite lye adhesive	Lye from waste, water	3	Fibreboard, building paper and linoleum
Natural rubber adhesive	Natural rubber or recycled rubber, organic solvents	4	Ceramic tiles, linoleum
Natural resin adhesive	Lignin or shellac or copal, possibly organic solvents or water	4	Linoleum, timber
Animal glues:			
Animal glue	Protein from tissue, possibly calcium chloride, water	3	Veneer, furniture
Casein glue	Milk protein, lime, possibly fungicide, water	4	Plywood, laminated timber
Blood albumin glue	Blood protein, ammonia, hydrated lime, possible fungicide, water	4	Veneer

Notes:
(1) Sensitivity to moisture is divided into a scale from 1 to 5:
5: For outdoor use
4: Outdoor use, but sheltered from rain
3: Indoor use, in relatively dry places
2: Indoor use, in permanently dry situations.
(2) When chrome compounds are added.

came into production around 1930 and today are used across the whole industry. There are between 100 and 300 different building adhesives available on the market. A normal Swedish home contains about 700 litres of adhesive, either as pure adhesive or as part of other products.

Filler came into use well into the twentieth century when smooth, even surfaces were required. Fillers differ from putty in that they harden and do not retain any elasticity. Adhesives and fillers used inside buildings in their soft state can cause considerable problems in the indoor climate during their hardening period, and sometimes even afterwards.

Glued components have very little relation to re-use strategy, as the possibilities of dismantling are few. Both adhesives and fillers pollute their products in such a way that possibilities for recycling and energy recycling are also greatly reduced.

Adhesives are usually divided into mineral adhesives, synthetic resins, animal adhesives and plant adhesives. Fillers are produced on the same basis as ordinary glues, with powdered stone, fossil meal, wood dust, chalk, perlite and similar substances.

Mineral adhesives

Mineral adhesives are used mainly for ceramic tiles, but have also become an adhesive for masonry. They are then used for precision components with accurate dimensions, such as lightweight concrete blocks. The adhesive used is usually a cement glue with a large proportion of acrylate mixed in. The joint is so strong that attempts at dismantling the wall may be difficult without destroying the blocks. Waterglass glue can also provide the base for a filler by mixing it with clay powder.

Mineral adhesives are based on resources with rich reserves. Both the consumption of primary energy and the pollution caused during production are relatively low. Inside a building products containing acrylate can cause problems for the indoor climate during their curing process. Pure waterglass products create no problems at all. As waste, waterglass glue is considered to be inert, while cement-based glue containing acryl has to be deposited at special tips.

Synthetic resins

Synthetic resins are usually divided into thermosetting and thermoplastic adhesives. The former must have a hardener added in order to complete the gluing process, and include urea, phenol, resorcinol, epoxide, polyurethane and acrylate adhesives. Thermoplastic adhesives are delivered from a factory, often emulsified in a solvent. Important adhesives of this type are PVAC adhesive, EVA adhesive, chloroprene and SBR adhesive. The latter two represent a sub-group of contact-adhesives and require large amounts of organic solvents which can include aromatics and esters. EVA and PVAC adhesives are partly soluble in water, partly soluble in organic solvents.

Thermosetting adhesives are very widespread in the building industry, but they are less popular as building adhesives on site, except when gluing external components, or if high strength is needed. Thermoplastic glues are the most common glues used on site. Fillers for indoor use are based on PVAC adhesive or acrylate adhesive.

The synthetic resins are based on fossil resources. Their production consumes a great deal of energy and creates pollution.

Within buildings, these products can create problems for the indoor climate through the emission of solvents and other volatile compounds during the curing phase, and sometimes for a longer period, in some cases as a result of ageing. Waste from hardened and non-hardened adhesives usually requires disposal at special tips, as do glued products, depending upon which adhesive is used and in what quantity. As a whole, PVAC-glue and EVA-glue are the least problematic.

Synthetic resins and some non-solvent based pollution

Formaldehyde adhesives
Urea, melamine, phenol, resorcinol and phenol-resorcinol formaldehyde adhesives all represent a risk in the working environment. Formaldehyde can be emitted, for example, from a plywood press. Pure phenol is poisonous and can seriously damage health after long periods of exposure. Phenol and formaldehyde are also poisonous if they come into contact with water. Formaldehyde is not as problematic as phenol, as it quickly oxidizes to formic acid and then carbonic acid. The Academy of Science in the Czech Republic blame dead forests on phenol working with metals.

In buildings with products glued together with formaldehyde adhesives there will be emissions of formaldehyde. The adhesive with the weakest binder is urea formaldehyde, which therefore has the highest emissions. Relatively small doses of formaldehyde can give acute symptoms of irritation in the eyes, itching in the nose, a dry throat and sleep problems. This can, in the long run, develop into serious problems in the inhalation routes. The substance is also registered as carcinogenic and a cause of allergies.

Epoxide adhesive
Fresh epoxide, which is the main constituent of epoxide adhesive, is one of the most effective allergens that exists. At places of work exposed to it, up to 80 per cent of the work force have developed epoxy eczema. Epichlorhydrin, which is part of the adhesive, is registered as carcinogenic and allergenic. Epoxide adhesive also contains alkylphenols and bisphenol A compounds, which are suspected environmental oestrogens. The material is also soluble in water and is poisonous and corrosive to organisms in water in low concentrations. Hardened epoxide adhesive is chemically stable.

Polyurethane and isocyanate adhesive
Isocyanates can easily cause skin allergies and asthma. They can also cause a degree of sensitivity and mucous membrane damage such that later exposure can induce asthmatic attacks, almost totally independent of the degree of exposure. The problems are greatest in industry and on building sites, but there can also be emissions from inside a building where the adhesive has not completed its reaction.

Synthetic contact adhesives
Chloroprene and styrene butadiene adhesive are the main synthetic contact adhesives in common use. Butadiene is registered as carcinogenic. Styrene is mainly a nerve poison, but is suspected of being carcinogenic and mutagenic. Chloroprene in chloroprene adhesive is considered responsible for reducing fertility and causing deformities and sperm cell damage. These effects are most likely to occur in the working environment. In a completed building there can, however, be emissions from adhesives that have not completed their reactions.

Acrylate adhesive
This can emit excess monomers in a completed building, which can increase the frequency of over-sensitivity and allergies.

EVA and PVAC adhesives

These have softeners added, most often dibutyl phthalate (DBP). Together with the excess monomers of vinyl acetate, these can be released from ready glued surfaces and result in irritation in the inhalation routes. Softeners, particularly DBP, are also suspected of causing more serious damage, such as nerves damage, hormonal disturbances and reproductive problems. PVAC adhesive can also contain sulphonamides which can damage the immune system.

Animal glues

Animal glues are based on substances rich in protein such as milk, blood and tissues, and are divided into three main types: animal glue, blood albumin glue and casein glue. These are soluble in water. They are all good glues for wood, and can be used on everything from veneers and furniture to large laminated timber structures.

Animal glues are mostly based on waste from slaughterhouses and fisheries. Casein glue comes from milk. In buildings, under dry conditions the products cause no problem. In combination with damp cement they can emit ammonia which irritates respiratory passages. In continuous damp there is a good chance of mould or other bacteria developing and the rotting products can cause bad odours, irritation and allergies. This can also lead to the deterioration of the building structure. Waste from the glues can lead to the growth of algae in water, but this risk is insignificant because the amount is usually small. Glues that have strong fungicide additives must be deposited on special tips.

Materials glued with animal glue can normally be energy recycled in ordinary incinerators, or can be dumped without any particular restrictions.

Animal glue

This glue is made from the tissues of animals containing collagen, a protein. Collagen is not soluble in water, but boiling it at a low temperature in an evacuated vessel turns it into glue. This is then dried into a granulated powder or into small bars. Gelatine is animal glue which has been cleaned of colour, smell and taste. There are three different types of animal glue: bone glue, hide glue and fish glue. The first two are often called glutin glue. Bone glue is made of bones and knuckles, hide glue is made from waste hides from places such as tanneries. Fish glue is made of fish bones and other fish waste and has a characteristic smell. All of these glues are strong, but hide glue is considered the best.

Animal glue bars or powder can be placed in cold water to soften up and then dissolved in water at 50–60°C using about two to three times as much water as the weight of softened glue. The powder can also be released directly into warm water. Temperatures above 60°C decrease the quality of the glue. Bone glue and hide glue have to be used warm and the pieces to be glued must be put under pressure before it stiffens. The glue cures quickly when cooling. Fish glue can be used cold, as can the other animal glues when calcium chloride is added.

To make a good animal glue filler, sawdust or wood flour can be mixed in. Colour pigments can also be added. The filler works well on timber surfaces and is not as visible on untreated surfaces as on treated ones. Adding gypsum makes the filler white.

Bone glue and hide glue have been used a great deal for gluing veneer. Right up to the Second World War, animal glue was dominant in furniture making, and there are still craftsmen who say that the quality then was much higher than that achieved today with adhesives such as urea formaldehyde (Brenna, 1989).

Animal glue can be used on all woods. The disadvantage with the animal glues is their lack of resistance to damp, which restricts their use to dry interiors.

Blood albumin glue

Blood albumin is soluble in water. It is prepared from fresh blood or from blood serum which is allowed to swell in water. The glue is made by adding ammonia and calcium hydrate solution in certain proportions. Ammonia is corrosive and can cause eczema. The objects must be warmed up during the actual gluing. At certain temperatures the protein coagulates, and the glue joint becomes totally watertight. The joint should be kept dry, if the glue has no added fungicides.

Casein glue

Casein glue was used by craftsmen in ancient China. It is made from skimmed milk. The milk is warmed up and rennet is added to separate out the casein. The casein is then dried and mixed with 2.5 g of lime per 100 g casein. The powder is mixed with three times as much water so that the lime is slaked. A glue is then produced which, after setting, tolerates damp better than sinew glue. In permanently damp surroundings and with timber at 18 per cent moisture content, the glue can be attacked by micro-organisms. This raises the question of the addition of fungicides such as sodium fluoride.

Casein glue can be used for internal load-bearing structures, stairs, plywood, laminated timber, etc., without fungicide. However, it is seldom used nowadays. Producers of laminated timber prefer adhesives that can be used in all situations, and therefore choose resorcinol formaldehyde, which has a high resistance to moisture. Strengthwise, casein glue is as good, and there is proof of its long lasting qualities in structures that are 50 to 60 years old which have kept their strength (Raknes, 1987). A very impressive example of its use can be seen in Stockholm Central Station, where enormous laminated timber arches have been put together with casein glue. During the Second World War, casein glue was used in the manufacture of fighter planes.

There is a need for a renaissance for environmentally-friendly casein glue. This does not necessarily conflict with economic considerations: it has been shown that casein glue can be produced for less than 25 per cent of the cost of resorcinol formaldehyde.

Casein glue is often classified as poisonous, due to the addition of lime which can burn bare skin. By adding fungicide the whole situation is altered and the glue loses many of its environmental advantages.

Plant glues

Glues from plants include soya glue, natural resin glue and cellulose glues as well as glues based on rye flour and potato flour.

Soya glue is a water-based protein glue taken from the waste products of cooking oil production. Natural resin glues are based on the sticking properties of resinous substances, such as lignin from coniferous trees, and have to be dissolved in organic solvents. Cellulose glue is available in both water- and solvent-based variations. The water-based cellulose glue is usually called paste, and is mainly used for putting up wallpaper. The paste can also be made of potato starch or rye flour.

Cellulose glue is not attacked by micro-organisms, even in damp conditions. Soya glue and flour paste should be restricted in their use to dry areas. The solvent used for natural resin products and cellulose glue is turpentine or pure alcohol, the latter up to as much as 70 per cent.

These glues originate from renewable plant sources. Products usually cause little pollution in their manufacture, the exception being cellulose glue, the main which is methyl cellulose. The production of methyl cellulose involves chlorinated hydrocarbons such as methyl chloride, methyl iodide and dimethyl sulphate. Possible alcohol solvents can be produced from the plants themselves.

During building use, these products do not cause problems. Waste from glue can cause the growth of algae in water systems, but this risk is insignificant as the amount of glue in question is usually small. Glues with strong fungicides added are an exception to this. Materials glued together with plant glue can usually be energy recycled in normal incinerators or deposited without special restrictions.

Starch glue

Starch glue or carbohydrate glue is based on vegetable starch. The paste is relatively weak and is used primarily for pasting paper and wallpaper, but it can also be used for lighter woodwork and is used in the USA for gluing plywood. Potato flour paste and flour paste are starch glues.

Potato starch is dissolved in warm water and mixed to a porridge. The porridge is allowed to stand for 10 minutes so that the water is absorbed by the grains of starch and thickens. Afterwards cold water is added to make it easy to stir. The mixture is then boiled and thickens more; water is added until a workable consistency is obtained. The glue must not be used until it is cold.

If the paste has to stand for a time a little alum is added to prevent it turning sour. If hydrochloric acid is added to the potato starch, dextrine is formed, which gives a glue of a far higher durability. Dextrine is also used in fillers containing gypsum.

Flour from wheat, maize or rye is used to make flour paste, which is stirred in warm water to a white sauce, adding water carefully so that the paste does not become lumpy. The mixture can also be sieved. This glue must also be used cold, and it is a definite advantage to add alum. Paste from wheat flour is mostly used to stick paper and wallpaper. In commercial products, fungicides are often added. Rye flour paste is a little stronger and is used for sticking paper on hessian, linoleum and wallpapers, and as a filler. Sago flour is used for the gluing of wood.

Rye flour filler

Emulsion filler based on rye flour is based on 9 dl boiled linseed oil, 9 dl water and about 0.5 kg chalk. This is gently mixed and allowed to stand for half an hour without being stirred. A pinch of rye flour is sprinkled on the mixture and thoroughly stirred in. More chalk, which acts as the filler, is added, until the mixture has the consistency of porridge. Pigments such as umbra and ochre can be used to colour it.

References

BRENNA J, *Lakkhistorie*, Oslo 1989

KESSEL M H *et al*, *Untersuchungen der tragfähigkeit von holzverbindungen mit holznägeln*, Baven mit Holz 6/1994

MASANI N J *et al*, *Comparative study of strength, deflection and efficiency of structural joints with steel bolts, timber bolts and bamboo pins in timber framed structures*, Indian Forest Leaflet no. 3, Dehra 1972

MYHRE B *et al*, *Vestnordisk byggeskikk gjennom to tusen år*, AmS No. 7, Stavanger 1982

RAKNES E, *Liming av tre*, Universitetsforlaget, Oslo 1987

18 *Paint, varnish, stain and wax*

Paint, varnish and stain are used to make a building more beautiful. Traditional painting of buildings has to a great extent revealed a wish to imitate other more noble building materials. The light yellow and grey render or timber façade has imitated light stone façades of marble, lime or sandstone; dark red façades have imitated brick. Colour has in this way had an outward-looking, representational function. But it can be used in the same way inside.

Theo Gimbel is a well-known colour therapist with his own school in England. He believes that colours start chemical processes within us, and that each cell is a sort of eye that takes in colours. Hence blind people can also be treated with colour. Red helps tiredness and bad moods, but should be avoided by those with heart problems. Yellow stimulates the brain. Green has a quieting effect, and violet strengthens creativity and spirituality.

Colour coatings are also thought to protect the material underneath. This is not always the case: there are many examples of damage caused by surface treatments, such as render and masonry that quickly began to decay after treatment with vapour-proof paint or timber which is often attacked by mould after painting. Research has shown that the decay of untreated timber when exposed to ultraviolet radiation, wind and rain is relatively small. In very exposed areas only about 1 mm is worn down in 10 years; in normal weather conditions 1 mm is eroded in 10–100 years. A much more significant protection than even the most careful painting is obtained by the structural protection of materials (see 'Structural protection of exposed components', p. 431).

The most relevant justification for painting a house is aesthetic. Exceptions are internal surfaces such as the floor, frames and certain details where treatment with oils and waxes will ease cleaning and reduce wear. Colour can also be used to lighten wood panelling which, with the exception of aspen, lime and the sapwood of ash, will darken with time. Special paints are used for protection against rust, as internal vapour barriers, to protect against radon emissions from radioactive building materials to seal of volatile formaldehyde in chipboard, etc.

Ordinary paint consists of binder, pigment and solvents. The binder makes the coat of paint retain its structure, and binds it to the surface to which it is applied. The pigment gives the paint a colour, but also plays a role in its consistency, ease of application, drying ability, durability and hardness. The solvent dissolves the paint to make it usable at normal room temperatures. In addition, it is possible to add fillers to paint to make it more economical. Modern paints based on synthetic resins often need a large proportion of different additives in order to achieve technical and aesthetic requirements.

A *dispersion paint* contains particles so small that they are kept suspended in water – this is known as a 'colloidal solution'. An *emulsion paint* is a dispersion paint consisting of a finely divided oil made soluble in water by adding an emulsifying agent, usually a protein.

Lazure is painting with less pigment, used when the structure of the material needs to remain visible. Lazure painting can be achieved by using a larger proportion of solvent in the paint. *Varnish* is a paint without pigment, while *stain*, in its classic sense, is a paint with no binder, where the pigment is drawn into the surface. Stain is often used as if it were lazure. The terms used here are the classical definitions.

Wax and *soap* are also included in this chapter. They have nothing to do with painting, but are widely used in the treatment of wood surfaces. They saturate the wood so that dirt and moisture cannot get into it.

The necessary qualities of paint, varnish, stain and wax are:

- they must bind well to the surface
- they must not crack or flake off
- they must be elastic so that they can tolerate movement in the building.

Special conditions are often required by the materials and components to be treated, and in relation to their position within or on the building. Especially important are factors such as diffusion through the paint, sensitivity to water, resistance to wear, sensitivity to light and the risk of emissions. There is a big difference between interior and exterior paints in this respect.

Many types of paint are mainly based on raw materials from plants, while others are based on fossil raw materials. Pigments for painting buildings are usually mineral-based.

The consumption of primary energy and pollution during production varies a great deal from paint to paint, but is to a great extent dependent upon the choice of pigment and solvent. Organic solvents have been estimated to be responsible for about 20 per cent of the hydrocarbon pollution in the atmosphere, second only to the car (Weissenfeld, 1985). Binders and other additives also affect the environmental profile in manufacture, and there is a tendency towards plant products coming out best. It is mainly the organic solvents that cause problems in the paint trade, but various additives in modern synthetic resin paints are also problematic.

Inside buildings, the materials covering the surfaces have a large impact because they extend over such large areas. Emissions often continue several months after the work is completed. A whole series of different volatile substances can be emitted from certain synthetic resin products, their source usually being unreacted monomers and additives. As a general rule, the thicker the layer of paint, the longer the time taken for the paint to complete its emissions. In many cases, there are gases which have a very strong irritant effect on the respiratory system. Certain surface treatments can also be quite heavily electrostatically charged, which can make cleaning more difficult and increase the electrostatic charge of the inhabitants (see Table 15.1).

Materials that have had surface treatments are not easily recycled. Exceptions include treatments such as vegetable waxes or oils. The same principle applies to the potential for energy recycling and the problem of waste. Painted materials often have to be deposited at special tips. As waste, the pigments have the greatest impact, as they can contain heavy metals.

Paints in history

Surface decoration has been popular throughout the ages. Stone Age cave painters used paint based on binders of fat, blood and beeswax, using chalk, soot and different earth colours as pigments. Natural pigments were also used for Egyptian fresco paintings about 5000 years ago. Old Hebrew writings describe how casein was stored in the form of curd until the annual visit of the painter during the autumn; at harvest festivals, everything should be newly painted. In Pompeii, paint mixtures of chalk, soap, wax, pigment and water have been found.

It is generally assumed that timber buildings remained untreated up to the late Middle Ages, but as wealthier citizens began to have panelling installed in their houses at the end of the seventeenth century, surface treatments became more usual. The first coloured tar paintings came into being at this time. The object of painting was to make timber buildings look like stone and brick. The pigments were expensive, with the exception of the earth pigments English Red and Ochre, which after a while dominated the houses of craftsmen, farmers and prosperous citizens.

Around 1700, linseed oil came into use. During the nineteenth century many old and new pigments could be produced chemically. Painting a house became cheaper, and colours other than red and yellow, such as zinc white, became available to everybody. At this time, everyone had untreated floors, apart from scouring them with sand. Floor painting began around 1820. From the middle of the twentieth century, very rapid developments led to latex paint, synthetic oil paints and alloyed paints, based on raw materials of fossil origin.

The paint trade has changed a great deal over the last 100 years. During the nineteenth century painters prepared the pigments themselves from the raw materials. Even as late as the 1960s most painters mixed paints themselves, although ready-mixed paints had been on the market since the end of the nineteenth century. During the last 30 years everything has been industrialized, including the application of paint, particularly for windows, doors and outside panelling.

Conditions for painting

Painting should be done during a dry period when the surface is dry, preferably in the summer. The temperature does not matter too much, as long as it is above freezing. This is particularly important for linseed oil paints. Painting carried out during the autumn often seems to last longer than painting done during the summer, probably because the paint has dried more slowly. In hot sunny weather paint can easily crumple, because of a tension between the different coats.

It is important to choose the right paint for the right surface. Wood, for example, is an organic material which is always moving, swelling in damp weather, drying out and shrinking in dry weather, and these qualities should be taken into account.

The main ingredients of paint

Binders

Binders must be able to dry out without losing their binding power. Many different binders have been used througout history, including materials such as blood, sour milk and urine. According to a representative of the Norwegian custodian of national monuments, Jon Braenne, many of these 'improbable' paints gave 'amazingly good results' (Drange, 1980). Linseed oil and protein glue have been amongst the most popular, with a long tradition, and have been in continuous use up to the end of the 1950s. At this time synthetic resins arrived on the scene, replacing the old faithfuls. Different types of binder vary a great deal in terms of opacity, lustre, spreading rates and durability.

Solvents

Solvents are used to thin out thick paint mixtures and vaporize from the surface after painting. For certain types of paint, the binder is enough to dissolve the paint into a satisfactory consistency, as in the case of cold pressed linseed oil,

Table 18.1: Different types of surface treatment

Type of paint/ binder	*Solvent*	*Other groups of potentially toxic additives*[1]	*Areas of use*	
			Outside	*Inside*
Lime paint	Water		x	x
Silicate paint	Water	Possibly acrylate	x	x
Cement paint	Water	Possibly acrylate	x	x
Epoxide paint/varnish	Xylene, butanol, ethyl glycol, methyl isobutyl ketone, glycol, toluene	Epichlorohydrin, possibly phenol	x	x
Acrylate paint/varnish	Xylene, water	Acrylate	x	x
Polyurethane paint/varnish	Ethyl acetate, butyl acetate, ethyl glycol acetate, toluene, xylene	Amines, isocyanates	x	x
Alkyd oil paint/varnish	Xylene, toluene	Possibly phenols	x	x
PVAC latex paint (polyvinyl acetate)	Water, xylene, toluene	Different fungicides, different softeners, etc.	x[2]	x
Acryl latex paint	Water, xylene, toluene	Acrylate, different fungicides, different pH-regulating substances	x	x
Animal glue paint	Water			x
Casein paint	Water	Possibly lime		x
Linseed oil paint	Possible xylene, toluene	Possibly fungicides, siccative	x	x
Natural emulsion paints (binders: egg, animal glue, linseed oil, lime paint, casein paint, flour paste)	Water	Possibly fungicides	x[2]	x
Natural resin varnish	Ethanol, xylene, toluene		x	x
Cellulose varnish	Ethanol, glycol, acetone, xylene, toluene		x	x
Wood tar	Xylene, toluene	Possibly fungicides	x	
Starch paint	Water			x
Beeswax	Limonene			x
Green soap	Water			x
Chemical stain	Water		x	x
Water-based stain	Water	Metallic salts	x	x

Notes:
(1) Excluding the pigment.
(2) With fungicides

warmed up wood tar, etc. A few paints can be dissolved in light oils, such as fish oil, while some paints dissolve in water. Many paints, especially newer types and binders of natural resins and wax, must have an organic solvent, usually turpentine. There are two types of turpentine:

- *Vegetable turpentine,* distilled from the sap of coniferous trees or pressed out from orange peel. Sulphate turpentine is produced from sulphate cellulose.
- *Mineral turpentine,* distilled from crude oil. It is marketed, amongst other things, as white spirit. The ingredients for the most common oil-based solvents are xylene, butanol, metylisobutylene, butyle acetate, methyl glycol ether, toluene, methanol and petroleum.

Before crude oil-based solvents came on the market in the beginning of the twentieth century, only vegetable turpentine was available. The turpentine obtained from orange peel is widely used for dissolving natural resins, usually in combination with a mineral turpentine. The proportion of orange peel turpentine is usually 2–10 per cent. It can also be used pure.

While mineral turpentine has crude oil as its source, vegetable turpentine is based on renewable plant resources. In terms of primary energy consumption and pollution during production, vegetable turpentine is a more positive environmental choice, even if water is obviously a preferable solvent.

On the building site, vaporizing of mineral turpentines represents a major problem and is associated with nerve damage and other serious health problems. Many painters refuse to use paints with these solvents. The mineral turpentines with less acute emissions are the isoaliphates, which are obtained by boiling crude oil at a specific temperature. The vapour from vegetable turpentines is normally more irritating to the mucous membranes than that of mineral turpentine. One constituent, pinene, can cause allergies. There is, however, no proof that long exposure to vegetable turpentine can have the same chronic damaging effect on the nervous system as mineral solvents.

In freshly-painted buildings the solvents release gas for shorter or longer periods depending upon the drying conditions of the building. Solvents vaporize completely, so there are no waste problems.

Pigments

Pigments have to satisfy certain conditions such as opacity, strength of colour and spreading rate, and they must not fade with exposure to light. Pigment should neither smelt nor dissolve in the binders or solvents used in the paint. Not all pigments can be used in all paints, for example pigments in a lime paint have to be compatible with lime. White pigment is the most popular and represents about 90 per cent of all pigments used. Pigments can be inorganic or organic. There are two types of inorganic pigments: earth pigment and mineral pigment.

Earth pigment occurs ready-to-use in certain types of earth. It is composed of the decaying products of particular types of stone, and has good durability. Extraction occurs during washing of the earth. After it has been collected in a tub, water is added and the mixture stirred. When all the earth has sunk, the water is poured off and the uppermost layer of fine earth is treated in the same way. This is done five or six times. The earth is then ground in a mortar, adding water. It is finally dried and the binder is added.

Mineral pigment is obtained by cleaning natural minerals. Synthetic mineral pigments are extracted by burning (zinc white), calcination (ultramarine) or precipitation in a solution (chrome yellow). Compared with the natural earth colours, the synthetic variations are purer, which makes it difficult to reconstruct colours in ancient buildings. All inorganic pigments are made synthetically nowadays, with the exception of umber.

Organic pigments have less durability and fewer lasting qualities than the inorganic pigments. Pigments used in modern painting are usually made synthetically. One natural organic pigment is coal black, which is made of charcoal, preferably from willow, beech and maple. Organic pigments are not normally used nowadays for painting buildings, with the exception of some blue and green variations.

Many mineral pigments are based on limited or very limited reserves. The production of pigments normally has high energy consumption and pollution rates. This is particularly the case for cadmium, chrome, manganese and lead products; pollution occurs in the factory environment and when the waste is deposited in the surroundings. The production of white pigments also causes a great deal of pollution, particularly in the case of titanium white. The production of zinc white is also a polluting process. White pigments of chalk and ground glass, however, do not cause problems.

Pigments and siccatives are relatively well bound within paints, and they are less chemically active. When paint is sprayed, it is finely spread in the air as small drops and the pigments can be inhaled. Welding of painted objects, scraping, sanding or removing the paint with hot air can all produce the same problem. Warmed zinc can create so-called 'zinc frost', a very painful fever, but soon passes. Pigments containing chrome are strongly oxidizing and thereby irritating and damaging to the respiratory system. Zinc chromate can also cause chrome allergy. Chrome, cadmium and lead compounds are, amongst other things, strongly carcinogenic. Ferric oxides can be considered relatively harmless.

In buildings, pigments are normally harmless if they are well bound with the paint and not too exposed to wear and tear. Children have, however, been poisoned by licking painted surfaces. Alkyl phenoltoxilates are often used in

Table 18.2: Pigments in house paint

Pigment	*Constituents*	*Comments*
White pigments:		
Zinc white	Zinc oxide	Fungicidal effect in larger amounts, can only be prepared synthetically, usually from recycled zinc
Glass white	Ground, recycled glass	At an experimental stage
Lead white	Basic lead carbonate	Highly poisonous
Titanium white	Titanium oxide	Can only be prepared synthetically
Chalk	Calcium carbonate	From natural resources, not very strong in oil paint
Yellow pigments:		
Ferric oxide yellow	Hydrated ferric oxide	Originally prepared as an earth colour
Yellow ochre	Hydrated ferric oxide	Highly tolerant externally, originally an earth pigment prepared from an iron felspar
Cadmium yellow	Cadmium sulphide	Highly poisonous
Chrome yellow	Lead chromate, lead sulphate	Highly poisonous
Zinc yellow	Zinc chromate, potassium chromate	Poisonous, can only be prepared synthetically
Naples yellow	Lead antimonate	Highly poisonous
Red pigments:		
Red ochre	Ferric oxide	Originally an earth pigment
Iron oxide red, English red	Ferric oxide	Originally an earth pigment, by-product of iron production
Chrome red	Basic lead chromate	Highly poisonous
Cadmium red	Cadmium sulphide-selenide	Highly poisonous
Red lead	Lead oxide	Highly poisonous, hinders rust
Blue pigments:		
Prussian blue	Ferric ferrocyanide	Poisonous, synthetically prepared from ferric chloride and 'Blutlaugensalz' (German), or potassium ferrocyanide
Ultramarine	Sodium aluminium silicate	Occurs naturally as the mineral lazurite. Prepared synthetically with a mixture of kaolin, soda, sodium sulphate, sulphur, resin, charcoal and quartz
Manganese blue	Barium manganate	Poisonous
Cobalt blue	Cobalt aluminate	Siccative, somewhat poisonous, occurs naturally as a mineral
Mineral blue	Ferric ferrocyanide heavy spar	Poisonous, prepared from Prussian blue and heavy spar
Green pigments:		
Green earth	Silicates containing iron	Originally an earth pigment
Chromium oxide green, viridian	Chromium oxide	Poisonous, can only be prepared synthetically

Table 18.2: *continued*

Pigment	*Constituents*	*Comments*
Chrome green	Lead chromate, lead sulphate, ferric ferrocyanide	Highly poisonous, prepared from a mixture of chrome yellow and Prussian blue
Zinc green	Zinc chromate, potassium chromate, ferric ferrocyanide	Poisonous, prepared from a mixture of zinc yellow and Prussian blue
Brown pigments:		
Umber	Clay containing iron and manganese	Some is prepared from earth pigment, but the majority is done synthetically from ferric oxide
Burnt sienna	Hydrated ferric oxide, silicic acid	Originally prepared as an earth pigment
Black and grey pigments:		
Slate grey	Slate flour	Seldom used, can be easily obtained through grinding and making a paste of the slate
Iron oxide black	Iron oxide	Can only be prepared synthetically
Ilmenite black	Iron titanium oxide	Can be prepared from ilmenite minerals
Lamp black	Carbon	Prepared from amorphous carbon which occurs from burning oil and tar products
Bone black organic pigment	Carbon, calcium phosphate	Prepared by charring different organic materials, animal bones and wood

pigment pastes as a dispersal agent. These are thought to be considerably harmful environmental oestrogens, i.e. chemicals that can affect the development of a foetus.

Energy recycling of painted products can lead to the emission of poisonous pigment vapours. Material painted with paints containing heavy metals represent a considerable pollution hazard and must be treated as special waste. The same is true of zinc white, whereas titanium white is not a problem as a waste product.

Other additives

Many other additives are used, depending upon the type of paint and where it is to be used.

Fillers

These are simple, colourless materials with the primary function of economizing and spreading the paint, and in some cases of improving the opacity. They also make the paint more matt. Important fillers are kaolin, dolomite, talcum, sand, fossil meal, diabase, heavy spar, barite and calcite. In the original earth colours, neutral clay silicates from the earth were used as natural fillers.

Fibre materials

Fibre can be added to paint to make it tougher and provide reinforcement on difficult surfaces.

Thickeners

Thickeners are added in water-based plastic paints to give the paint a slow flowing consistency. Water-soluble cellulose glue or derivations of polyurethane and polyacrylate are used for this.

PH-regulating agents

These can be added to water-based plastic paints to increase the pH value and reduce the chance of mould growth. Ammonia or triethylamine are usually used.

'Skin preventers'

These are added to stop a skin forming on top of the paint in the tin. The substances used are buthyraldoxime and methylethylketoxime, added in proportions of 0.1–0.4 per cent.

Rust preventing agents

These prevent rust being formed on the tin or when painting metal surfaces such as nails, etc. Traditionally they contain chrome and lead compounds. In water-based paints a mixture of sodium benzoate and sodium nitrite is used in proportions of 10:1, and makes up around 0.5 per cent of the paint.

Fungicides

Fungicides are often necessary to prevent the paint from attack by mould during storage and after application. The least toxic alternatives are lime, and metal sulphates such as alum and ferrous sulphate, which are used in many paints with organic or even mineral binders. Some pigments also have preservative capacities. Paints with 50 per cent zinc white are not attacked by mould. Water-based plastic paints often contain a fungicide of many different compounds, including chloric-organic substances. Up to the end of the 1970s polychlorinated biphenyls (PCBs) were used. In certain water-based plastic paints pentachlorophenol is probably still being used. Other common fungicides are sodium nitrite, formaldehyde, tributyltin and isothiazolone. Fungicides make up about 0.5–1.0

per cent of the paint mixture. All fungicides are volatile to a certain extent and can cause problems in the indoor climate. Many of them irritate the mucous membranes and in some cases cause allergies.

Foam reducer

Foam reducer is often added to water-based paints so that the paint does not froth.

Drying retardants

These are added to water-based latex paints. They help reduce the evaporation of water while painting is taking place, and usually consist of glycols and glycol ethers. For a long period after painting is complete, glycols can be emitted and irritate the respiratory system.

Drying agents/siccatives

These are added to various oil paints to shorten their drying time, particularly in linseed oil paints. Normal siccatives are found in zirconium, cobalt salts and manganese. Calcium can also be used as a siccative, preferably in combination with other substances. Lead salts were once often used. Cobalt and manganese form from 0.02–0.1 per cent of the dry content in the binder. Lead forms about 0.5–1.0 per cent of the dry content. The alternative is a drying oil such as wood oil, added in the proportion of 2–10 per cent (see 'Drying oils', p. 419).

Softeners and film-forming agents

These agents are used in water-based plastic paints and consist of microscopic plastic particles dispersed in water. When the paint dries, these particles fasten to each other and form a film. Softeners, usually of the type dibutylphtalate, are mostly used in PVAC paints without acrylic additives. Other types are dioctyl phtalate and tri-n-resyl-phosphate. Softeners can release gas within a building and can be both irritating for the mucous membranes and cause allergies. Phthalates are known to be environmental oestregens, capable of affecting the development of a foetus.

Perfume

Perfume is added to a few water-based paints, mostly to neutralize the unpleasant smells from chemicals such as amines.

Paints with mineral binders

Mineral paints are matt and are best suited for painting on mineral surfaces, although they can be used on unplaned timber surfaces. The most common types are based on binders of lime, cement and waterglass, all of which are soluble in water.

The products are based on rich reserves. The environmental consequences of the production techniques can be acceptable, e.g. when using water as a solvent. All the alternatives are strongly alkaline and when damp have a corrosive effect on bare skin. Compared with other working environments and indoor climates, mineral products produce favourable results.

In buildings the products are environmentally sound, partly because they are open to vapour transport and do not mask the moisture-regulating properties of the materials underneath. An exception is lime or cement paint not well bound to the surface, which can loosen and flake off into the room and cause respiratory irritation. Mineral paints cannot cause electrostatic charging.

As waste the products are inert, and as long as there are no poisonous pigments in the paint they can be used as fill. The paint will not lessen painted products potential for recycling, or for energy recycling.

Lime paint

In lime paint the binder is slaked lime which can be bought separately in tins. Curing is based on carbonizing slaked lime with carbonic acid in air, forming a united crystalline layer. The pure lime colours give matt, absorbent surfaces which are difficult to wash. The paint is porous to vapour and not elastic. It binds best to a lime render but can be used on pure cement or rough timber. Brick can best be painted with lime if it has a rough surface. Lime paint cannot bind to plastic. The best results are obtained by applying lime on fresh render. Old lime paint can be removed by brushing.

It is important that lime paint is applied in thin coats. It can be used both inside and out, but walls painted with lime paint cannot be painted over with any other type of paint – the lime paint must be completely removed. It is important that the pigments are compatible with lime. If the lime contains more than 5–10 per cent additives, it has a lower binding capacity.

The following pigments are considered compatible with lime: titanium white, yellow ochre, ferric oxide yellow, cadmium yellow, red ochre, ferric oxide red, chrome red, ultramarine, cobalt blue, earth green, chrome oxide green, umber, brown ochre, terra de sienna, ferric oxide black, ilmenite black, bone black.

Factory-manufactured lime paint has dolomite added to improve its durability, plus a little sinew glue or cellulose paste to improve ease of application and opacity. Water-soluble glue is eventually washed out.

Lime paint gets dirty easily in urban environments. It is very sensitive to acids, which break it down to gypsum. It is therefore debatable whether this paint should be used in an area with an acidic atmosphere. The surface underneath is, however, protected from acidic attack, and the lime acts as a sort of sacrificial layer.

Recipes for lime paint

The following recipes are well tested and recommended. Painting should be carried out in damp periods, and the painted surface protected from direct sunlight for at least 14 days after it is complete. The walls to be painted should be moistened beforehand with lime water – part of all lime paint recipes. Lime water is made as follows:

1. 'Wet' slaked lime is mixed with water in a proportion of 1:5.

2. The mixture is stirred well until all the lumps have disappeared.

3. After 24 hours all the lime has sunk to the bottom. The water above the lime is lime water. The layer of crystals that has formed on the surface must be removed. Lime water is strongly alkaline, with a pH of about 12.5.

Lime milk is also an important ingredient in the paint. It is quite simply a dispersion of solid slaked lime and lime water in the form of lime solution. A very fine-grained calcium hydroxide with particles of about 0.002 mm arises through slaking. Lime milk is prepared in the following way:

1. Fresh 'wet' slaked lime is mixed with lime-water in a proportion of 1:5.

2. The mixture is stirred well until all the lumps are removed. After about 10 minutes a good lime milk is created. It can stand several days before use.

Lime surfaces rub off, but this can be retarded by adding a little sinew glue, see p.259 (Animal glues) to the lime solution. This method is only for use inside a building.

The pigments best suited for lime paint are ferric oxide colours: yellow, brown, red, black and ultramarine, which tolerate lime. The pigments should be mixed with water and made into a thick gruel.

Lime paint can best be directly applied onto completely fresh render, and there is seldom the need for a second coat. Old, decayed render, or lime or cement paint, must be brushed clean of dust and dirt if the paint is to bind properly. Lime needs several days to become properly bound to the surface. It is important that the render and the layer of paint do not dry out during this period. In particularly dry weather, the wall should be watered when it feels dry, especially if the sun is shining on it.

Recipe 1: White lime

The surface is painted with lime water, followed by two or three coats of lime milk, then another coat of lime water.

Recipe 2a: Red lime

The earth pigment ferric oxide is soaked in two parts water overnight to become a pigment pasta. The soaked pigment is then mixed with lime water in a proportion of 1:9, to become a lime paint. The wall is first given a coat of lime water, then a coat of lime paint, and is finished off with another coat of lime water.

Recipe 2b: Ochre lime
The earth pigment ochre is soaked in two parts water overnight to become a pigment pasta. The soaked pigment is then mixed with lime water in a proportion of 1:9 to become a lime paint. The wall is first given a coat of lime water, followed by two coats of lime paint and finally another coat of lime water.

Recipe 2c: Lilac, brown or green lime
This is made with the pigments ultra-marine, umber and burnt umber. The production and application are the same as for ochre lime, above.

Recipe 3: Yellow lime with green vitriol
This paint has a certain antiseptic effect even in addition to the actual effect of the lime. A solution of green vitriol and water in a proportion of 1:5 is made, then a separate mixture of 'wet' slaked lime and water is made in the proportions 1:5. The two mixtures are then stirred together to become a thick porridge, and water are added. Before painting, the surface is treated with one or two coats of lime water.

Recipe 4: Lime casein paint
By adding casein to the lime, a casein glue is formed which, apart from having a better opacity, is also more elastic than ordinary lime paint. This is the type of paint that is used in fresco painting and for wooden surfaces. The paint is waterproof. One part 'wet' slaked lime is mixed with half to one part curd (containing about 12 per cent casein), and all the lumps are pressed out. For a purer casein paint, four parts curd are used. The mixture is added with 20–40 per cent stirred pigment of titanium oxide, ferric oxide, umber or green earth and thinned out with skimmed milk. The surface is given a coat of lime water before painting.

Recipe 5: Floor treatment with lime.
Lime treated floors are light and easy to maintain. First sand the floor and vacuum clean it. Slaked lime and water are mixed in a proportion of 1:10. The gruel is brushed evenly over the floor with a broom. When dry, the floor is sanded and vacuum clean again, then washed with a 5 per cent solution of green soap in lukewarm water. Cleaning of the floor is also done with a 5 per cent green soap solution, but soaps containing sulphates or phosphates must not be used.

Silicate paints

Silicate paints have their origin in the binder potassium waterglass and were patented in 1938 by A. W. Keim. They can be used on all mineral surfaces, but also gives good results on rough wood. They can be used as an opaque paint or a lazure paint. Waterglass paints react with lime on a painted surface and form calcium silicate, which acts as a binder. The paint film forms a crystalline layer which has a high resistance against acids. The best results are achieved on fresh render. This paint is much more durable than lime paint and has a strong resistance to pollution. Its vapour diffusion coefficient is about as high as that of lime products. Some silicate paints available on the market have a few acrylates added (to a maximum of 5 per cent), which harden and form a 'dispersion-silicate paint'. As long as the surface material contains lime, added acrylate is not actually necessary. It can also be assumed that the added acrylate shortens the effective life span of the paint. For pure waterglass paints, pigment has to be added on site; paint with acrylate additives can be mixed at the factory.

Cement paints

Cement paints were first used in the 1940s and usually consist of Portland cement and possibly some lime, which is mixed with a small amount of water and then added to the pigment and water. They give their best results on newly-struck concrete or fresh render, but can also be used on brick. In durability and quality they fall somewhere between lime and silicate products. Pure cement paint is mainly used nowadays for special treatment of pools and various concrete structures, with large quantities of added polymers. Cement paints containing lime only have dolomite and cellulose glue added.

Recipe: Original cement paint
5 litres skimmed milk is mixed with 0.5 to 1.5 kg Portland cement and pigments that are compatible with lime are added up to a maximum of 5 per cent by weight, to make a gruel. The mixture is suitable for rough wood panelling and masonry. It has to be stirred while being used. The pigments suited for this paint are chalk for white, English red, ochre and other earth colours. The paint is very durable. It was used a great deal by American farmers.

Paints with organic binders

Organic binders consist of synthetic resins, protein glue, drying oils, tar, natural resins, cellulose products, starch and emulsion.

Synthetic resins

As with glues, synthetic resins can be divided into thermoplastic and thermosetting products. Thermoplastic products must have a hardener added before painting, and include paints and varnishes based on epoxide, acrylate and urethane (better known as DD-varnish). Alkyde oil paint is the most important thermosetting product. The thermosetting products can be dissolved in organic solvent or in water. Synthetic resin products are based on fossil raw materials with the exception of some linseed oil in alkyde oil. Their production uses a lot of energy and causes a high degree of pollution.

Products containing solvents, epoxide and isocyanates create a very bad working environment for painters. In buildings, they cause problems for the indoor climate. In many solvent-based products, the solvents can be emitted up to six months after application. Water-based paints contain volatile additives which can be released over an even longer period, e.g. softeners. Many products emit excess monomers, quite independent of the solvent.

Plastic-based paints and varnishes are able to induce considerable electrostatic charging, especially if they are used on floors.

Waste paint should normally be deposited at special tips, even if the pigments are inert. Painted products have little re-use value and normally have no recycling value. Flammable material can be energy recycled in incinerators with filters.

Epoxide

Epoxide is one of the most infamous materials known for causing allergies. At workplaces where people are exposed to epoxide, up to 80 per cent of the work force developed epoxide eczema. Epoxide products also contain alkylphenols and bisphenol A compounds which are suspected environmental oestrogens. Epichlorohydrin, another constituent of the mixture, is registered as carcinogenic and allergenic. It is soluble in water and in low concentrations has a poisonous and corrosive effect on water organisms. Ready-cured epoxide paint is probably chemically stable, although a certain amount of solvent is emitted at first. Certain makes of epoxide can also emit phenols during application, which can quickly lead to bad skin irritation.

Polyurethane

Polyurethane products contain isocyanate which can easily cause skin allergies and asthma. High sensitivity causing damage to the mucous membranes can develop, and asthma attacks can occur independent of the level of exposure. The most exposed places are industrial and building sites, but unreacted residues can also be released within buildings.

Acrylate

Thermoplastic acrylate paint can emit excess monomers of butyl methacrylate in a completed building. This can cause frequent sensitivity reactions and allergies.

Alkyde oil

Alkyde oil is a chemical compound between linseed oil and a polymer, usually glycerole or phthalic acid. This type of paint came into widespread use during the 1950s and contains large quantities of solvent, usually aromates like toluene and xylene. Alkyde oil paint with less pigment can also be called a stain. It does not penetrate material as well as pure linseed oil paints, but it adheres well to wood even if the surface is not completely dry. Alkyde oil paint is also considered hardwearing, and is used on concrete, render and galvanized iron. With no pigment, the paint can be used as varnish.

Alkyde oil is very thick and because of its high percentage of solvent (between 50–70 per cent), alkyde oil paint is a big risk in the working environment. The emission of solvents in the building can continue from a few days to several months, depending upon the climate of the room, how the paint has been applied and the type of solvent. In certain alkyde oil products, mainly the varnishes, alkylphenols are present in the binder in a proportion of about 1 per cent by weight. Alkyl phenols are confirmed environmental oestrogens.

Water-based synthetic resins

These resins are based on the dispersion principle where the plastic constituents polyvinyl acetate (PVAC), vinyl acetate, polyacrylate or styrene acrylate move freely about in the water in the form of microscopic plastic pellets. To make the mixture work as a paint, additives such as pH-regulators, fungicides and softeners are needed. Paint with polyvinyl acetate as a binder was introduced early in the 1930s, and it is still used as a cheap interior paint. Styrene acrylate is a sampolymer of polystyrene and acrylate and is used both inside and outside. But the most common binding agent nowadays is polyacrylate. It is more expensive than PVAC products, but has a better resistance to alkalis, better weather-proofing and adheres better to smooth surfaces. Most of the water-based synthetic resin paints today contain a mixture of polymer binders. Binders form about 30–40 per cent, pigment 30–35 per cent, fillers 16–20 per cent, water 20–25 per cent and different additives about 5 per cent. A homopolymer PVAC paint has to have softener added to make it suitable as a paint. If a co-polymer of PVAC and acrylate is made the necessary softness will be achieved without having to use a softener. Most paints also contain a small number of organic solvents to increase the possibility of forming a film.

Water-based synthetic resins and emulsion products are matt. They are not suited to surfaces that are exposed to damp, and often get thicker with age.

Acrylate monomers in water-based paint can cause eczema through contact with the wet paint. Some paints emit ammonia when damp which is very irritating to the mucous membranes. Many of the water-based synthetic resins can emit volatile compounds for long periods after painting is complete, such as excess monomers of acrylates, styrene, softeners and volatile components of fungicides such as formaldehyde, and even excess solvents. All of these can stimulate over-sensitivity and to a certain extent, allergies. The emissions reduce with time, depending upon the temperature, the moisture situation and the thickness of the paint. After twelve months most of the emissions cease. Additives in PVAC products often include sulphonamides, which can damage the immune system, and nonylphenoletoxilates which are confirmed environmental oestrogens.

Water-based synthetic resins are today the most widespread form of paint used indoors. They have also taken over a large proportion of the market for external work on masonry and timber.

Protein glue paint

Good paints which allow the passage of vapour can be based on protein glue (see 'Animal glues', p. 396). These cannot be overpainted with other types of paint. Pure glue paints are well suited to indoor painting. There are two types: animal glue paint and casein paint. The protein molecules consist of both fat-soluble (hydrophobic) and water-soluble (hydrophilic) parts and can therefore be used in emulsion paints.

Protein glue paint is dissolved in water. Animal glue paint is based on waste from slaughter-houses; casein paint is based on milk.

In buildings under dry conditions the products are inert and do not lead to electrostatic charging. In combination with damp cement, protein glue paint can emit ammonia which can irritate the inhalation routes. Long-term damp can easily lead to attacks by fungus and other bacteria. The bacteria break down the protein, and the rotting products emit a bad smell and cause irritatation. This can partly lead to the decay of the structure, and partly to problematic pollution in the indoor climate. Decaying products that contain protein cause allergies.

Waste from the paint can cause the growth of algae in streams and rivers. Animal glue paint and casein paint can normally be washed off painted materials, so the materials can be easily prepared for re-use. Painted materials can normally be energy recycled in conventional incinerators or be dumped without any particular restrictions.

Animal glue paint

Animal glue paint is not waterproof, but is well suited to use in dry interiors on masonry, wood, hessian and paper. The surface has to be cleaned of any fats before painting; otherwise a small measure of cal-ammoniac can be added. The opacity is good, and many pigments can be used. Though the paint is not waterproof, experience shows that washing down an animal glue painted wall and applying a new coat is no more work than meticulously cleaning a wall painted with waterproof paint. Animal glue paints can also be used in emulsions, usually with linseed oil. This produces a waterproof paint.

Glue paint recipes

Glue paint should not be used in bathrooms or similar areas, or on surfaces exposed to a great deal of wear.

When painting on render it is usual to soap the surface with a thin solution of green soap, consisting of 2 dl green soap to 10 litres of water. This has to sink in and dry in order to give the glue paint a chance to penetrate evenly into the render. Painting can begin when the surface is dry.

Painting should be done wet-on-wet so as not to cause stains. The pigment should be mixed with a little water to a thick colour paste with no lumps.

Recipe: Glue paint based on bone glue and skin glue (10 litres)

The ingredients are 200 g bone and skin glue and a little water, with 5 litres of water, 10 kg of painter's chalk and pigment. The paint is prepared in the following way:

1. The chalk is first soaked, in a bucket for example, and left to stand overnight without stirring.

2. The glue is made. Bone and skin glue should be left to soak in a little vessel overnight, with water just covering the glue. The glue should then be warmed in a water bath until it floats.

3. The glue solution should be poured into the chalk and the mixture stirred well with a whisk.

4. The pigment should be stirred into lukewarm water then added to the glue and chalk mixture. Certain fatty pigments are not easy to dissolve, but this can be improved by adding a teaspoon of alcohol, which breaks down the surface tension. The stronger the colour required, the more chalk must be replaced by pigment.

Casein paint

This is used mainly in emulsion paints. Milk protein is used for binding, and will react to a certain extent with surfaces containing carbon to lime casein, which is waterproof. Pure casein products are not waterproof and must be used indoors. (See also 'Recipe 4: Lime casein paint', p. 414.)

Drying oils

A drying oil dries in the air, at the same time keeping its elasticity. The most common and the best is linseed oil, but even Chinese tree oil and hemp oil make good quality paint. To some extent soya oil, olive oil and fish oil can also be used, but these are not actually drying oils.

Linseed oil dries by oxidizing in air and is transformed to a strong and solid linoxine. This oil has been used in painting since the beginning of the seventeenth century and can be used on wood, concrete, render and to a certain extent, steel. Linseed oil is also used on stone façades to close the pores and protect it from aggressive air pollution. Render and concrete should not be painted in the first year, as moisture pressing out from the inside can push the paint off. Oil paint can be produced in matt, half-lustre and full lustre. The half-lustre and lustre types are very strong and easy to clean.

Linseed oil products are generally waterproof but allow the passage of vapour. The porosity increases with time, and is optimal after a couple of years. In some cases, the paint may not be porous enough initially for painting masonry.

Cold pressed oil is better than warm pressed oil. Cold pressing, however, only frees about 30 per cent of the oil in the seeds. In warm pressing, the seeds are finely ground, and pressed while warm. Both methods are usually combined.

Raw linseed oil is probably the most firm, especially when cold pressed, but it dries very slowly because of the large amount of protein substances it contains. It is therefore mostly used out of doors. Boiling linseed oil to 150°C removes the majority of the protein substances, making the product dry more quickly. The paint can be used both indoors and outdoors. Stand oil is linseed oil which is boiled without air to 280°C and thereby polymerized. It is considered to be more firm and elastic. It also dries more quickly than the other two types. But even so, drying time is a problem with linseed oil products. In factory-produced oils, drying agents (siccatives) are added to a proportion of about 0.5 per cent. This also applies to products for outside use, even if the drying time there is not too critical. For indoor use it is normal to add drying agents to all qualities of linseed oil, but drying oils can achieve the same end. One such oil was originally a mixture of linseed oil, Chinese tree oil and natural resins, but today this is partly replaced with synthetic resins. Another possibility for reducing the drying time is to use linseed oil in a water-soluble emulsion with casein paint.

Linseed oil paints often have fungicides added, but this is not necessary for interior painting. Organic solvents are added to increase penetration and spreading rate. This is usually unnecessary for easy-flowing oils such as cold pressed linseed oil. The amount of solvent varies from about 10–30 per cent, and is much lower than the equivalent in alkyde paints.

The raw materials for drying oils are renewable, and environmental problems relating to their production are minimal. Products containing a high percentage of drying agents or solvents such as mineral or vegetable turpentine are an exception to this. Products containing solvents present a risk for painters in the working environment.

In buildings, linseed oil products are not a problem, except for solvent emissions during the period directly after painting. During the curing period the linseed oil will emit oxidation products, mainly aldehydes, which irritate the inhalation routes. Linseed oil paints are relatively open to vapour transport and only slightly reduce the vapour-regulating potential of the substrate. The products do not cause electrostatic charging.

Materials treated with linseed oil are difficult to clean and therefore have less chance of being re-used. The same is true for recycling. Energy recycling is possible without filters for the fumes, as long as fungicides and problematic pigments have not been used. As waste, products with no pigments or fungicides can be ground and composted. Consideration must also be given to what drying agents have been used, before treating the waste.

Recipes for linseed oil paints

Linseed oil paints can be used both inside and outside. For interior use, linseed oil emulsion paint is the best choice. Linseed oil paints are particularly well suited to external walls of timber panelling. It swells in damp weather and creates an elastic film which is never completely hard. When linseed oil has set, it is porous to water vapour and allows moisture to evaporate. The choice of pigment is important if the paint is going to retain these properties. Zinc white should not be used as an outdoor pigment. It is easily washed down by acids, rain and dew, and when exposed to ultraviolet radiation the paint starts crazing, especially on a south-facing surface.

Recipe 1: Normal linseed oil paint for outdoor use

To start mixing linseed oil paints prepare a colour paste where the pigments are well mixed with a small amount of linseed oil to an even consistency. The amount of pigment depends upon how transparent and shiny the paint is to be; more pigment will give a more matt paint. The pigment paste is mixed with the oil.

The first coat usually contains about 15 per cent vegetable turpentine to help the oil penetrate the substrate. The final coat does not need solvents, especially if cold pressed oil is used. Adding solvents to the paint generally shortens its life span.

Recipe 2: Linseed oil treatment of timber floors

The floor should be sanded. The first coat usually consists of a mixture of mineral or vegetable turpentine with boiled linseed oil or stand oil in the proportions 1:1 and in the final coat in the proportions 1:2. After application, all the excess oil be dried up after about 20 to 30 minutes.

Paint with fish oil binder

Fish oil has been used a great deal in coastal regions up to the beginning of the twentieth century.

Recipe: Normal paint with fish oil

Fish liver is laid in a barrel and put in the sun with a sack pulled over the top. The liver melts quickly to liver oil, is mixed with English red or other pigments and can be dissolved in alcohol if necessary. The paint is very durable and has good resistance to salt water.

Tar

Wood tar gives a weak brown colour, due to the coal dust and pitch. Pigment such as English red or ochre is stirred in to give beautiful and durable colours, but it is impossible to paint other colours on top of a tarred wall. For tar stains the binder can be supplemented with linseed oil or alkyde oil and thinned out with organic solvents; fungicides can also be added.

Wood tar is extracted from coniferous and deciduous trees. Wood tar is usually rich in polycyclical aromatic hydrocarbons (PAHs). An exception is tar extracted from beech. PAH substances such as benzoapyrene are carcinogenic and mutagenic, so tar products should not be used indoors. When used outdoors, the PAH substances will filter into the soil.

Re-use and recycling of painted products can be a problem. As waste, the products should be deposited at special dumps.

Natural resins

Several different types of natural resins can be used for varnishing wood. To make the resin more fluid, organic solvents can be added. A varnish layer of natural resin is about as vapour-proof as synthetic products, but it is less durable and needs a longer drying time, and is also more expensive.

Cholophonium is extracted from the resin of pine trees after distilling vegetable turpentine oil, and consists mainly of abietic acid. This is seldom used as the only resin in varnish mixes. It can be dissolved in alcohol or vegetable turpentine. Copal, a fossil form of resin, is extracted in India, the Philippines, Australia and Africa. Alcohol or vegetable turpentine are used as solvents.

Shellac comes from the Bengal fig tree (*Ficus bengalensis*) when it is attacked by wood lice. Alcohol is used as a solvent. Dammar comes from special trees in East India and Malaysia (*Dipterocarpaceae*). Alcohol or vegetable turpentine can be used as solvents. Sanderac is drawn from the juniper gum tree (*Callitris quadrivalis*) in Morocco. It dissolves in alcohol, turpentine, ether and acetone.

Rubber mastic is extracted from the resin of the mastic tree (*Pistacia lentiscus*) found on Mediterranean islands. It dissolves in alcohol or ether. Elimi gum is resin extracted from amyris trees (*Burseraceae*) on the Philippines, Mauritius, Mexico and Brazil. It dissolves in alcohol, petroleum or vegetable turpentine. Acaroid resin is from the grass tree (*Xanthorrhoea australe*) in Australia. It dissolves in alcohol.

The products are mainly based on renewable resources, with the exception of some types of solvent. During application, substances can be emitted by the solvent vapour which can lead to irritation of the inhalation routes and allergies. These emissions can continue after the building is finished. As waste, these products are normally not a problem, depending partly upon the pigment used.

Cellulose paints

There are two different cellulose paints – one based on normal cellulose paste and the other on nitrocellulose. The latter is used mainly for varnish and must contain up to 75 per cent organic solvents and softeners. Paste paint has approximately the same properties as protein glue paint.

Cellulose paints are mainly based on renewable resources from plants. The products are made from methyl cellulose in a highly-polluting process using substances such as chlorinated hydrocarbons (CHCs). The production of nitrocellulose requires large amounts of solvents with heavy environmental consequences in production and in the painter's working environment.

Within buildings these products are not a problem. Painted material can probably be energy recycled in normal incinerators or dumped on domestic tips without any problem, with the exception of Nitro-varnish products with softeners.

Starch paint

Starch paint is based on starch glue (see p. 398) and is mainly used externally on unplaned timber, usually in the form of rye flour paste. The paste decays over time and only the pigment is left. This can rub off. To compensate, it is common to add about 4–8 per cent linseed oil. In damp environments 1–2 per cent green vitriol is added to prevent any mould attack.

Starch paint is based on renewable raw materials from plants and represents absolutely no environmental threat, either in production or use. Re-use and recycling of treated materials is acceptable, as is energy recycling. The materials can normally be composted. The favourable environmental profile can be reduced by the addition of environmentally damaging pigment.

Emulsion paint

Emulsion paints are waterproof. They can consist of sinew glue emulsified in linseed oil, or casein emulsified in linseed oil and dissolved in water. They usually produce a good matt surface with only a few strokes of the brush. This type of paint is very economical. When painting render and concrete flaking will occur, if not with the first treatment, then with a later one. This is because of tensions within the glue. Painting wet-on-wet avoids stains.

Emulsion paints are exclusively based on renewable raw materials and are soluble in water. Production causes no problems and application of the paint causes no health risks. Within buildings they do not create indoor climate problems. They are washable and hygienic and do not cause electrostatic charging.

The paints are relatively strong and difficult to remove to enable the re-use and recycling of painted components. Painted products can be energy recycled in normal incinerators and even composted. The addition of environmentally damaging pigments causes problems which could reduce the otherwise good environmental profile.

Recipes for emulsion paint

In all the paints described below, the pigment is mixed with linseed oil. The paints should be applied directly after preparation.

Recipe 1: Animal glue/linseed oil paint

This paint is fairly strong, and can be used outside and inside, but is best used indoors. Protein glue is mixed in the same way and portion as in glue paint recipe, p. 419, and 2.5 litres of boiled linseed oil are added.

Recipe 2: Flour paste-linseed oil/casein paint

For interior and exterior wood and masonry: 10 parts flour are mixed with 10 parts cold water, then 50 parts boiling water, to form the glue. Linseed oil in 10–12 parts and 10 parts skimmed milk are added, with colour pigment to a proportion of 15–40 per cent volume.

Recipe 3: Casein/linseed oil paint (casein oil tempera)

For interior and exterior wood: 10 parts sour milk is mixed with four parts linseed oil and about four parts pigment. The paint has been said to last from five to 10 years externally.

Recipe 4: Egg/linseed oil paint (egg oil tempera)

For internal use on wood. It gives a hard shiny and easily cleanable surface. One part linseed oil is mixed with one part fresh egg and one part water, pigments to a proportion of 15–40 per cent.

Stain

Stains are used on wood and do not contain added binders. There are two main types of stain: chemical stain and water-based stain.

Chemical stain is based on a colour reaction with substances in the timber and is used mainly on wood such as spruce or pine. Tannic acid can be used. Lye treatment is also in the same category. (See also 'Bor salts from borax and boraad' and 'Green vitriol', p. 440.)

Water based stain is made with pigments that are soluble in water. Modern exterior stains usually also contain metal salts such as cobalt chloride, copper chloride, potassium dichromate, manganese chloride and nickel chloride, in order to impregnate the wood. Several ordinary pigments can be used in the stain; but even bark or onion peelings are used as stain colours.

Stains are the least resource-demanding treatments. They are also relatively problem free in production and use. The exceptions are water-based stains with metal salts added. These are usually poisonous, and can seep into the soil. The same can be said for the waste from these stains with added metal salts; they should be deposited at special tips. As far as the other products are concerned, re-use, recycling, composting and dumping are all relatively problem free. It is only the addition of poisonous pigments that reduce the quality of an otherwise very positive environmental profile.

Recipes for chemical stains

Recipe 1: Normal stain

10 g tannic acid is dissolved in 1 litre warm water. The stain is applied cold. It is usual practice to then apply a second layer consisting of a solution of 10 g potash (K_2CO_3) in a litre of water. The colour is light grey–green.

Recipe 2: Lye stain

5 g of tannic acid is dissolved in 1 dl lukewarm water. 50 g potash is dissolved in 5 dl hot, almost boiling water, and 4 dl cold water and 1.25 dl lye solution (12 per cent lye in water) are added. The tannic acid solution is mixed with the potash solution. This must be prepared in a stone vessel. It gives a stronger grey–green tone than the first recipe. The final colour emerges after eight to 14 days. The stain should stand a few days before use. Lye stains are highly alkaline, and protective clothing must be used.

Recipes for water-based stains

Recipe 1: Onion peel stain

The onion peel is boiled in water for 15 minutes and to a weak pink colour. The solution is applied to wood, giving a faint yellow colour.

Recipe 2: Bark stain

The bark to be used has to be gathered during the summer. The colour is extracted by pouring a 5 per cent soda solution over the bark and letting it stand for four weeks. For 250–500 g bark, use 250 g soda and 5 litres boiled water. After four weeks the mixture has a very strong smell, but after an hour's simmering the smell disappears.

A brown colour comes from beech, apple and spruce bark and a yellow colour from poplar and cherry bark. The latter needs 10 per cent soda solution. The bark of ash gives a grey–green colour, and birch an apricot colour (using a 10 per cent soda solution).

Beeswax

Beeswax is particularly well suited to the treatment of floors and bathroom walls. It fills splits and pores in timber and prevents vermin from laying eggs. Wax is usually dissolved in mineral turpentine or orange peel turpentine and can be thinned out with linseed oil. It can be coloured with earth or mineral colours. The wax is easy to wash with soap, but does not have much resistance to water, so the wood should be saturated with oil first.

Beeswax is a renewable resource which creates no problems in production or use. If organic solvents are added, they can be a health risk for the working environment during their application, and can even cause problems for the indoor climate later on, although these are relatively small. Re-use of treated materials, recycling, energy recycling and composting or dumping create no problems.

Recipe for wax treatment: Beeswax on wood

Three parts wax are melted in a water bath of 70–80°C, then one part turpentine is mixed in. The mixture can be applied directly onto wooden walls. Floors have to be well sanded first, and the surface temperature should not be under 20°C. When the surface is dry after a couple of days, it is polished. It must be waxed once a month where it is most worn. It can be cleaned with a damp cloth and warm soap water.

Soap

Green soap is used for the treatment and saturation of wood, usually floors. It consists mainly of fats from linseed oil or timber oil which are boiled out and saponified with lye. Fats from maize, cotton seed and soya oil can also be used. Small amounts of waterglass and soda can be added: soda increases its washing ability somewhat, but the same time decreases the effective amount of fats. Green soap is relatively alkaline, and hinders the growth of bacteria and mould.

Green soap is based mainly on renewable resources from plants and is free of problems both in production and use. The same is true for the re-use of treated materials, recycling, energy recycling and composting or dumping.

Recipe for green soap treatment

The floor must be dry and preferably newly sanded. A mixture of 2 dl solid green soap per litre of hot water is poured over the floor. The gruel is worked into the timber in the direction of the floorboards. The floating soap water is dried up without completely drying the surface. The surface is allowed to stand overnight, and the treatment is repeated four or five times. Before the final treatment, the raised fibres can be sanded with a paper of grade 120–150 in the direction of the boards. A stronger treatment can be achieved by adding chalk (see chalk paint, Recipe 5, above).

References

DRANGE T *et al*, *Gamle trehus*, Universitetsforlaget, Oslo 1980

WEISSENFELD P, *Holzschutz ohne Gift?*, Ökobuch Verlag, Grebenstein 1983

19 Impregnating agents, and how to avoid them

Organic materials are easily attacked by insects and fungus in damp situations. In northern Europe, six types of insects can damage timber buildings (see Table 19.1). Fungus is a type of lower plant species which lacks chlorophyll. Many fungi attack buildings, especially the timber in buildings. They can be divided into two main groups, discolouring fungi and disintegrating fungi. Discolouring fungi give timber a superficial discoloration, without decreasing its strength. Disintegrating fungi attack the cell walls in timber and destroys the wood.

Spores from disintegrating fungi are ubiquitous. They drift around with the wind in the same way as pollen, and attach to everything. These fungi have very important functions. They belong to nature's renovating corps, their main operation being the breakdown of dead organic material, which regrettably includes many building materials. The optimum conditions for this phenomenon relate to

Table 19.1: Vermin

Type	*Comments*
House longhorn beetle (*Hylotrupes bajulus*)	Does not attack heartwood in pine
Carpenter ants (*Camponotus herculeanus*)	Does not live on wood, but uses it as its home and lays eggs, even in pressure-impregnated wood
Common furniture beetle (*Anobium punctatum*)	Prefers a temperature of 20–25°C and a relative humidity of 50%, only found in coastal areas
Woodworm (*Dendrobium pertinax*)	Attracted to wood that has already been attacked by fungus
Violet tanned bark beetle (*Callidium violaceum*)	Dependent on bark left-overs for its survival
Bark borer (*Ernobius mollis*)	Dependent on bark for its survival

dampness, temperature and acidity. Dampness in organic material need to be from 18–25 per cent. Dampness quotients above and below these figures are not attractive for these spores. The majority of fungi, however, survive long dry periods. A temperature between 20 and 35°C makes an attack possible – attacks cannot happen below 5°C. Disintegrating fungus does not strike in environments with a high alkaline content, i.e. with a pH over 6.0. One exception is the fungus *Merulius lacrymans*.

There are four principal ways to avoid attack from insects and fungus:

- Use of high quality material in exposed locations
- Structural protection of exposed materials
- Preventive treatment of materials and passive impregnation
- Use of impregnating substances: active impregnation

Impregnating substances are usually divided into insecticides and fungicides. Their main task is chemically to prevent or kill vermin and micro-organisms. If the guiding principle for creating the substances was simply to make surfaces uncomfortable for insects or fungus, then impregnation would not be a cause of environmental concern. But the whole concept behind them is the creation of biological poisons that kill, something that has caused unforeseen consequences for other animal species, not least mankind. The main task of this chapter is to show how impregnation can be avoided. The main subject will be timber, as this is the most widely-used organic material in the building industry, especially in northern Europe. Other organic products will also be discussed. Fungicides in paint are discussed in the previous chapter.

Choosing quality material

In old trees a large part of the trunk consists of heartwood, which has a strong resistance to fungus and insects. Not even the house longhorn beetle can penetrate the heartwood of pine. Heartwood was traditionally used in log construction and external panelling, and until the nineteenth century in windows and doors.

Initially pine was thought to be more durable than spruce, but this conclusion has been modified. The core of pine has almost no moisture absorption capacity, whereas the sapwood has a moisture absorption, lengthwise in the cells, 10 times greater than that of spruce. Pine cladding from the young core is therefore less protected than spruce. Birch cladding is even weaker, with a permeability about 1000 times greater than that of spruce. Generally speaking, the absorption of moisture increases in relation to the breadth of the growth rings.

Figure 19.1: Nettleton Cabin in central Scotland, constructed from locally-sourced, untreated timber. Source: Howard Liddell

Timber should be felled in winter, because wood felled in summer has a much higher sugar content, making it more attractive to insects and micro-organisms. By removing the bark from trees, attacks by bark-eating bugs are avoided. Sawn timber should be dried to 20 per cent moisture content before spring, and logs that are not going to be sawn should be stored in water. Timber for log construction should be felled in September and profiled on both sides during the spring. It should be dried during the summer and used as building material in the autumn.

Material from a building which has recently been attacked by the house longhorn beetle or the common furniture beetle should not be re-used.

As with the choice of straw as reinforcement in earth structures or as roof covering, it is important to choose the quality of material with care.

Structural protection of exposed components

If buildings are constructed with materials in a way that lets air circulate and keeps them dry, then fungus will not attack. This is true also for paint, as shown in the choice of paints for outdoor and indoor use. There is, however, a definite trend towards all-round products which have fungicides added to protect them in all possible situations.

All types of timber should be used in a way that allows movement to take place, otherwise splitting and gathering of moisture will occur. The heartwood side, which is usually the least moisture-absorbent, should be on the outside. Moisture is usually most quickly absorbed at the end of the timber. The end grain must therefore be protected. Exposed ends of beams and any other exposed timber can be cut at an angle or preferably covered.

Vertical exterior panelling can end at ground level, sawn at an angle so that a drop is formed on the outside face of the timber. The distance of the panelling from the ground should be at least 20–30 cm. Above concrete paving, asphalt, brick paving and other hard surfaces, and where the wall is protected by a roof overhang, the distance can be reduced.

Table 19.2: Minimum slope of roof to prevent water seeping in

Type of roof covering	*Normal situation*	*Exposed location*
Profiled roof tile	22°	35°
Interlocking tile	20°	30°
Concrete roof tile	15°	22°
Natural slate, single layer	22°	30°
Natural slate, double layer	20°	25°
Roofing felt, two layers, the first with fibreglass reinforcement	3°	3°
Metal sheeting	3°–14°	3°–14°
Plank roof	22°	27°

Panelling should be well ventilated. The more damp and exposed a wall is to driving rain, the wider the air gap behind the panelling should be. This is usually 5 cm in very exposed areas and about half that in normal inland situations. Horizontal battens fixed directly to panelling should have a sloping top side, or be mounted on a vertical batten system against the wall.

On timber roofing and in vertical panelling, timber root ends should be pointing downwards. Water may collect in the joint between the two layers of vertical timber panelling. Along the coast where there is plenty of driving rain, this often results in rot, as drying periods are usually very short. This form of rot is not found inland. On the coast panels must be horizontal. This gives the advantage of less exposed end grain. Rot usually occurs at the bottom of the wall, and with horizontal panelling it is quite easy to remove and replace a few planks; with vertical boarding all the planks would be affected. On horizontal panelling, rot can easily occur at the vertical junction of two boards in the middle of a wall.

The profile and form of panelling is also important. With normal rough panelling the lower edge of the boards can be pointed in section so that they form drop profiles (see Figure 15.22). When using tongued and grooved, chamfered panelling outside, it is obvious that the tongue should point upwards and reach some way down the board to give better drainage.

Combinations of metal, lime and cement-based mortars and concrete can cause problems. Condensation can occur around metal components, while in combinations of timber and cement and lime, alkaline reactions can arise which increase porosity and moisture absorption in the timber.

In particularly damp areas, the colour on the exterior can also play a part. A dark ochre colour can reach a temperature up to 40°C higher than a white surface in sunny weather. This can be significant for drying time. In damp places, where even during the summer there are only short periods of sun between showers, the drying time should be as short as possible. If the temperatures get too high, however, splitting or cracking can occur, which can also increase the intake of moisture.

Methods of impregnation

There are many methods of treating timber in order to increase its resistance to decay:

- Self-impregnation of logs
- Cleaning out the contents of the cells
- Burning the outside layer
- Oxidation
- Application of non-poisonous protection to the surface
- Application of pH-regulating substance on the surface or through impregnation
- Application of poisonous protective layer on the surface or through impregnation.

The last method is a strategy with 'active' impregnating substances; the other methods can be characterized as 'passive'.

Self-impregnation of logs

The most common method is to chop the top of a pine tree and remove a few stripes of bark from the bottom to the top. Three or four of the highest branches are left to 'lift' the resin. After a few years the whole trunk is filled. This dramatically increases its resistance to rot. Houses built of these logs will probably last for hundreds of years without any further treatment. A double guarantee is achieved if this is complemented with a cleaning out of the cells and burning the outer layer.

Cleaning out the contents of the cells

Certain insects, mould and fungus live on the nutritious contents of cells, while other fungi live on the cell walls. Cleaning out the cells' contents can at least solve part, if not all of this problem. The method was conceived after the discovery that logs stored under water lasted longer than those stored in air. It must be assumed that the absorption of salt from sea water also has a positive antiseptic effect. Where timber boarding is laid on a roof, it has been common practice in

Scandinavia to boil the timber planks first. This is a very effective way of washing out the content of the cells.

Burning the outer wood

A traditional way of increasing the durability of wooden piles was to burn the part that was going to be placed underground. The carbon coating formed lacks nutrients and is almost impenetrable to insects and fungus. The burning also enriches the resin and tars in the outermost part of the pile. The greatest impact occurs on pine, which is rich in resin. Burning spruce and deciduous trees is not so effective. During burning the timber can easily split, and it is easy for fungus to creep in through the splits, so burning must be carefully controlled, preferably by using a blow lamp. The depth of the burning should be 1–3 mm, after which the surface is brushed with a bronze brush. This process takes a long time.

Julius Caesar described the technique in *De Bello Gallico* in connection with setting up fortification in the Roman Empire. The method has also been used for centuries by Portuguese and British timber warships, as it not only increased resistance to rot but also made the surface waterproof.

Oxidizing and exposure to the sun

As late as the nineteenth century it was unusual to treat external walls at all. Timber developed a silver–grey colour based on an oxidation process caused by ultraviolet radiation from the sun. Any material applied to a wall will reduce or block this effect. The oxidation penetrates a few millimetres into the timber making an effective, protective layer, but, particularly in damp climates, fungus may develop. If fungus is discovered, the wall must be washed with liquid green soap or cleaned by spraying steam.

Untreated surfaces are also exposed to splitting through drying out too quickly. Rubbing in elasticizing agents, preferably linseed oil mixed with a little lime, increases resistance to fungus. The same technique, using olive oil, was applied around the Mediterranean for more than 2000 years.

On older log houses the sunny side of untreated walls often becomes a sun-brown colour. The warmth from the sun draws resins to the surface, which also forms a protective layer.

Non-poisonous surface coats

The application of a non-poisonous layer on the surface is mainly to protect the timber from mechanical wear and tear and direct solar radiation. Exposure to these can lead to large or small cracks which, in a damp area, can lead to fungus attack.

Many different paints give a timber wall this sort of protection. The best of these are probably the pure linseed oil paints, which penetrate well into the wood. The effect of this layer depends upon reapplication at regular intervals.

A positive parameter in this treatment is the fact that the paint is considered water-repellent. Research has shown that this treatment does not necessarily make the timber last longer, however, and in certain cases it can have a directly negative effect by retaining moisture in the material and not letting it escape.

pH-regulating surface-coat or impregnation

Substances which regulate pH are an effective way of preventing or removing mould attack. Mould will not grow if the pH level is higher than 6.0. The same can be said of insect attack, except for the longhorn house beetle. The pH-regulating materials to use are alkalines such as clay, cement, lime and waterglass. They are not poisonous in themselves, so they do not cause problems in the indoor climate of the building.

Waterglass as a pH-regulating coat

Waterglass is very alkaline and in addition forms a coat so hard that insects cannot penetrate it to lay their eggs. Waterglass is, however, not waterproof when used on timber, and can therefore only be used indoors or on protected parts of the building. Waterglass needs a rough surface – it does not bind well to a planed surface. It is dissolved in boiled water and applied to the wood with a brush. It can also be applied to straw, using a solution of one part waterglass to two parts water. Waterglass is very open to water vapour. It is very fire-resistant and was therefore often used in loft structures in old town houses.

Poisonous surface-coats or impregnation

Experience has shown that timber with a high content of tar and resins lasts longer than timber with a low content of the same. This is partly because the timber is harder and partly because these substances have ingredients which are poisonous to fungus and certain insects. These natural fungicides and insecticides consist of, or are similar to, different types of tannic acid. Traditional types of timber protection aim to increase the quantity of such materials by covering the timber with tar. Extract from bark has also been used to impregnate oak, birch and spruce, with good results. This method was once so popular that bark extract became a major Norwegian export. Over 2000 years ago the Chinese tried using salt water as an impregnating agent, because the salt's action on the wood was slightly antiseptic. More recently, metal salts have been used for impregnation, and wood tar has mostly been replaced by derivatives of oil.

Since forestry has been industrialized, the quality of timber has reduced considerably, and the need for impregnating substances has rocketed over the last few years. New fashions in architecture, which include exterior timber structures, have accelerated this trend.

The following functional qualities are expected of a good impregnating substance, independent of the organic material it is to protect:

- Enough poison to prevent attack from fungus and insects; wood ants (*Camponotus herculeanus*) are not usually deterred, whatever poison is used
- Not poisonous to people or animals
- Ability to penetrate into the material
- Resistant to being washed out or vaporized from the material
- Free from damaging technical side-effects such as miscolouring, corrosion of nails, etc.

An impregnating substance with all these qualities does not exist. Effective poisons such as metal salts have particularly damaging effects on humans. Less damaging substances such as bark extract and cooking salt are at the same time less effective.

Preventive impregnating agents must be differentiated from biological poisons, which are used after the material has been attacked. The same material can, however, often be used in both cases. In Table 19.3, both main groups are treated as one group. The poison categories 'medium' and 'high' represent strong biological poisons. There is generally a clear connection between a poison's strength and its effectiveness. Different impregnating poisons are used in larger or smaller proportions in different mixtures, often in reciprocal combinations. To make the mixtures fully effective, both fungicide and insecticide may be needed in the same mix. They are dissolved in water or solvents. The substances are applied to the timber by pressure impregnation or by brushing on. About 90 per cent of pressure impregnation uses water-soluble metal salts; the rest uses solvent-based creosote. For external application, solvent-based derivatives of oil are most commonly used.

Apart from creosote, permetrine is the most common oil derivative and has superseded such derivatives as pentachlorophenol, which were phased out during the 1980s and 1990s because of environmental and health risks. It is used mainly to protect timber, but also to protect against moths in woollen blankets.

The most important salts for pressure impregnation are arsenic, chrome and copper. There are different classes of impregnating substances; timber in contact with the ground requires strong substances in large doses, but in well-ventilated, outdoor cladding a much weaker mix will be effective enough. There is a clear tendency today to choose an undifferentiated all-round impregnating agent,

Table 19.3: Poisons used for impregnation

Type	*Fungicide*	*Insecticide*	*Level of poison*
Mineral:			
Zinc salts	x	x	Medium
Arsenic salts	x		High
Chromium salts	x	x	Medium
Fluorine salts	x		Medium
Copper salts	x		Medium
Potassium ferric sulphate	x		Low
Potassium aluminium sulphate	x		Low
Borax and boric acid	x	x	Low
Aluminium sulphate	x		Low
Ferrous sulphate	x	x	Low
Lye from soda or potash(1)	x		Low
Oil- and coal-based:			
Creosote	x		High
Carbolineum	x		High
Pentachlorophenol		x	High
Hexachlorobenzene		x	High
Pyrethrin		x	Medium
Xylidene	x		Medium
Endosulphane		x	Medium
Tributyltin	x		High
Parathion		x	High
Discofluamide	x		Medium
Tolufluamide	x		Medium
Plant-based:			
Wood tar:			
from softwood	x		Medium
from beech	x		Low
Extract from bark	x		Low
Wood vinegar (for treating already attacked wood)	x	x	Low

Note:
(1) Potash lye can be prepared from wood ashes.

preferably the strongest. This rationalizes production for manufacturers, but at the same time involves considerable 'over-impregnation'. A strong impregnating agent usually contains all three substances: arsenic, copper and chrome. For timber above ground level it is quite adequate just to use copper.

Both metal salts and oil products have very restricted resources.

Production of timber-impregnating substances and the work at manufacturing workshops can result in emissions of strong biological poisons into earth, air and water. For products based on metal salts, the chrome, copper and arsenic used are heavy metals with large biological amplification capacities. These substances can quite easily combine with earth particles, but do not combine so easily with sand, which delays the drainage and spread of the substances to some extent. Acid rain increases the rate of drainage. In the solvent-based impregnation industry, vaporized solvents can be released, such as aromates, phenols and different components containing chlorine. These substances, in the same way as heavy metals, have a capacity for biological amplification and bind much less with the soil.

In completed buildings creosote-impregnated timber emits, amongst other things, naphthalene. Considerable concentrations of naphthalene have been registered inside buildings even when the application has been outdoors (Gustafsson, 1990). Creosote combined with solar radiation can cause rapid and serious burning of the skin. A roof treated with creosote can heavily pollute the garden and groundwater. Pentachlorophenol will emit chlorinated hydrocarbons into the air and soil long after impregnation is finished. Permetrine is particularly poisonous for organisms in water, and can also cause considerable damage to the human's nervous system, including concentration problems and general illness. It takes a relatively long time for the emissions to fully break down.

Water-soluble metal salts are usually stable in buildings. They are, however, released from exterior surfaces exposed to rain. In Denmark it has been calculated that a couple of tons of arsenic are washed out in this way annually.

When impregnated timber burns, many of the poisonous substances are released, including about 80 per cent of the arsenic, so waste must be disposed of at special tips. Even here, slow draining of poisons into the soil will occur. In northern Europe there are, at the moment, several hundred thousand tons of copper, chrome and arsenic stored in impregnated timber.

The least dangerous impregnating substances

Tar

Wood tar is usually extracted from parts of pine that are rich in resin: the bole and the roots, which are burned to charcoal. It can also be extracted from other coniferous and deciduous trees. Tar from beech is widely used in mainland Europe. Modern extraction techniques give a very clear tar – previously, when burning took place in a charcoal stack, high levels of pitch and particles of carbon were included.

Extraction of wood tar in a charcoal stack

The stack is dug out in a sloping piece of ground. The bottom is shaped like a funnel and covered with birch bark. A pipe made out of a hollowed branch is placed in the bottom of the funnel. The timber is split into sticks about 18–20 cm long and 1 cm thick and they are stacked radially round a strong central log. The stack is then covered with earth and turf, and lit at the bottom. The stack is allowed to smoulder for up to 24 hours, depending upon its size. The tar gathers in the funnel and can be removed through the wooden pipe.

Wood tar can be used pure or mixed with boiled or raw linseed oil in a proportion of 1:1; pigment can also be added. Wood tar extracted from pine trees contains considerable amounts of polycyclical aromatic hydrocarbons (PAH) substances, for example benzo-a-pyrene, which is a well-known mutagen and carcinogen. Tar from beech is almost free from these substances.

Bark extract

Bark extract often has borax and soda salt added to increase its antiseptic effect. The extract is poisonous to insects and fungus, even though somewhat weak. It is not dangerous to humans. Bark extract is not waterproof, and is most useful on exposed materials indoors. Extract based on birch bark has the best impregnating properties. (See also 'Recipe 2: Bark stain' p. 425.)

Wood vinegar

Wood vinegar is corrosive and is not used as a preventative but for treating materials that have already been attacked by rot and insects. Wood vinegar is extracted by distillation from deciduous trees, although even coniferous trees contain wood vinegar, but in smaller quantities.

Soda and potash lye

These have been used for surface treatment in many Swiss villages for hundreds of years, and the buildings have kept very well. A drier climate is, of course, partly responsible for their success, but this treatment deserves discussion. Impregnation with lye brings the resins and tar to the surface of the wood in the same way as burning. The lye also has an antiseptic effect. The treatment has to be repeated every two to three years. Gloves and glasses should be worn during the treatment, as the material is very alkaline.

Recipe for lye made from soda and potash

The soda solution is made by boiling 5 litres of water with 250 g of soda powder. The liquid is applied when still warm. Potash solution is either based on pure potassium carbonate or on wood ashes, which contain about 96 per cent potassium carbonate. A potash solution is made up by boiling up 2.5 litres of pine ash with 5 litres of water and letting it simmer for 15 minutes. The solution is sieved and applied while still warm.

Bor salts from borax and boracid

These impregnating substances combine effectiveness against vermin with relative harmlessness to humans. The emission period from an impregnated surface is as short as 10 hours, so the interior of a building will be risk-free after a couple of days.

In Germany borax is the only one of the more effective poisons used indoors. It is also used to impregnate cellulose insulating materials where it also acts as a fire retardant. It is, however, quite easily washed out of materials. Borax is bought as powder, and usually used as a 5–10 per cent solution in warm water applied in two coats. Very dry timber is moistened first so that the borax will penetrate better.

Green vitriol

Green vitriol is a relatively harmless impregnating substance based on ferric sulphate. In liquid form it can irritate the skin and is slightly damaging to organisms living in water. A good impregnating solution consists of 10–13 g/litre of water, with a little alum added as a fix. Green vitriol is also a fire retardant and gives timber a shiny silver surface. It is often called acid treatment. Such a treatment can last up to 15 years but will in time be washed out of the timber.

References

GUSTAFSSON H, *Kemisk emission från byggnadsmaterial*, Statens Provningsanstalt, Borås 1990

Section 3: Further reading

ADDLESON L, *Materials for building*, London 1976

ASHURST J *et al*, *Stone in building. Its use and potential today*, London 1977

BECKLY A, *Handbook of painting and decorating products*, London 1983

BERG A, *Skifertekking og skiferkledning*, Forening til Norske Fortidsminnesmerkes Bevaring, Årsberetning, Oslo 1945

BILLGREN G *et al*, *Träfönsterets beständighet*, Byggforskningsrådet, Stockholm 1977

BOISITS R, *Dämmstoffe auf der ökologischen Prüfstand*, IBO, Wien 1991

BOKALDERS V, *Byggekologi* 1–4, Byggtjänst, Stockholm 1997

BRÄNNSTRÖM H, *Torv och spån som isolermaterial*, Byggforskningsrådet R 149:1985, Stockholm 1985

DANCY H K, *A manual on building construction*, Intermediate Technology Publications, London 1975

DAVEY N, *A history of Building Materials*, London 1961
DREJER C *et al*, *Färg och måleri*, Byggförlaget, Stockholm 1992
EISNER K *et al*, *Some experiences in research and manufacture of panels from agricultural waste and non-wood fibrous raw materials in Chzechoslovakia*, Wien 1970
ENGLUND A, *Zostera marina, isoleringsmatta och vattenrenare*, Ekoteknik, Östersund 1993
GRAUBNER W, *Encyclopedia of wood joints*, Tannton Books, Newtown 1992
GRÜTZMACHER B, *Reet- und Strohdächer. Alte Techniken Wiederbelebt*, Callwey Verlag, München 1981
GRÆE T, *Breathing Building Constructions*, Oklahoma 1974
HOUBEN H *et al*, *Earth Construction. A comprehensive guide*, Intermediate Technology Publications, London 1994
HUSE A, *Kartlegging av helse- og miljøskadelige stoffer i maling, lakk, lim m.v.*, SFT rapp. 92:09, Oslo 1992
KESSEL M H *et al*, *Untersuchungen der trägfähigkeit von Holzverbindungen mit Holznägeln*, Bauen mit Holz 6/1994
KOMAR A, *Building materials and components*, Moscow 1974
KÖNIG H L, *Unsichtbare Umwelt. Der Mensch im Spielfeld Elektromagnetischer Feldkräfte*, München 1986
LAURICIO J O *et al*, *Fabrication of hollow block from agri-forestry materials for low cost housing*, Appr. Techn. Vol. 5 no. 2, 1978
LEWIS G *et al*, *Natural Vegetable fibre as reinforcement in concrete sheets*, Magazine of concrete research 31/1979
LIDDLE H *et al*, *Pore-ventilation: Sports Halls*, The Scottish Sports Council, Research Report no. 43, Edinburgh 1995
LINDBERG C O *et al*, *Jordhusbygge*, Stockholm 1950
LUNT M G, *Stabilized Soil Blocks for Building*, Overseas Building Notes no. 184, Watford 1980
LÅG J, *Berggrunn, jord og jordsmonn*, Oslo 1979
MCDONALD S O, *A Straw Bale Primer*, private edition, Gila New Mexico 1991
MCINTOSH J D, *Concrete mixes for blocks*, Concrete Building and Concrete Products 1965
MINKE G, *Der Baustoff Lehm und seine Anwendung*, Ökobuch Verlag, Freiburg 1994
MOESSON T J, *Production of strawboards by the 'Stramit'-process*, Vienna 1970
MUIR D *et al*, *The energy economics and thermal performance of log houses*, Quebec 1983
NORGES BYGGFORSKNINGSINST, *Materialer til luft og damptetting*, Byggforskserien A573.121, Oslo 1986
NILSSON L, *Armering av betong med sisal och andra växtfibrer*, Byggforskning rapp. D14:1975, Stockholm 1975
PISTULKA W *et al*, *Baukonstruktionen und Baustoffe*, Wien 1982
PROCKTER N J, *Climbing and screening plants*, Rushden 1983
RISOM S, *Lerhuse, stampede og soltørrede*, Copenhagen 1952
ROAF S *et al*, *The ice-houses of Britain*, Routledge, London 1990
ROALKVAM D, *Naturlig ventilasjon*, NABU/Norsk Forskningsråd, Oslo 1997
RYBCZYNSKI W *et al*, *Sulphur concrete and very low cost housing*, Canadian sulphur symposium, Calgary 1974
STEEN S A, *The strawbale house*, New York 1994
STERLING P E R (Ed.), *Earth Sheltered Housing Design*, New York 1979
STOCKLUND B, *Læsøgården*, Nationalmuseet, Copenhagen 1962
STULZ R, *Appropriate Building Materials*, SKAT, St. Gallen 1983
VERMASS C H, *The manufacture of particle board based on unconventional raw materials*, Hannover 1981
VOLHARD F, *Leichtlehmbau*, C. F. Müller, Karlsruhe 1988
VREIM H, *Takspon og spontekking, stikker, flis og sjingel*, Foreningen til Norske Fortidsminnesmerkers Bevaring, Årbok, Oslo 1941
VREIM H, *Laftehus, tømring og torvtekking*, •• •••
WIESLANDER G, *Water Based Paints. Occupational Exposure and some Health Effects*, Acta Universitatis Upsaliensis, Uppsala 1995

Index